Design and Control of Electrical Motor Drives

Design and Control of Electrical Motor Drives

Editor

Tian-Hua Liu

MDPI • Basel • Beijing • Wuhan • Barcelona • Belgrade • Manchester • Tokyo • Cluj • Tianjin

Editor
Tian-Hua Liu
National Taiwan University of Science
and Technology
Taiwan

Editorial Office
MDPI
St. Alban-Anlage 66
4052 Basel, Switzerland

This is a reprint of articles from the Special Issue published online in the open access journal *Energies* (ISSN 1996-1073) (available at: https://www.mdpi.com/journal/energies/special_issues/electrical_motor_drives).

For citation purposes, cite each article independently as indicated on the article page online and as indicated below:

LastName, A.A.; LastName, B.B.; LastName, C.C. Article Title. *Journal Name* **Year**, *Volume Number*, Page Range.

ISBN 978-3-0365-9821-5 (Hbk)
ISBN 978-3-0365-9822-2 (PDF)

Contents

About the Editor

Tian-Hua Liu is currently a Distinguished Professor in the Department of Electrical Engineering, National Taiwan University of Science and Technology. He received his Bachelor's, Master's, and Ph. D. degrees from National Taiwan University of Science and Technology. He has been a Visiting Professor at the University of Wisconsin-Madison, Virginia Tech, and the University of Auckland. He has served as a department head, Dean, and Provost at National Taiwan University of Science and Technology.

energies

Editorial

Design and Control of Electrical Motor Drives

Tian-Hua Liu

Department of Electrical Engineering, National Taiwan University of Science and Technology, Taipei 106, Taiwan; Liu@mail.ntust.edu.tw

1. Introduction

This Special Issue contains the successful invited submissions [1–11] to a Special Issue of *Energies* on the subject of the "Design and Control of Electrical Motor Drives". Electrical motor drives are widely used in industry, automation, transportation, and home appliances. Indeed, rolling mills, machine tools, high-speed trains, subway systems, elevators, electric vehicles, and air conditioners all depend on electrical motor drives. However, the production of effective and practical motors and drives requires flexibility in the regulation of current, torque, flux, acceleration, position, and speed. Without proper modeling, drive, and control, these motor drive systems cannot function effectively.

To address these issues, we need to focus on the design, modeling, drive, and control of different types of motors, such as induction motors, permanent synchronous motors, brushless DC motors, DC motors, synchronous reluctance motors, switched reluctance motors, flux-switching motors, linear motors, and step motors. Therefore, relevant research topics in this field of study include modeling electrical motor drives, both in transient and in steady state, and designing control methods based on novel control strategies (e.g., PI controllers, fuzzy logic controllers, neural network controllers, predictive controllers, adaptive controllers, nonlinear controllers, etc.), with particular attention paid to transient responses, load disturbances, fault tolerance, and multi-motor drive techniques.

This Special Issue encourages and invites original contributions regarding the recent developments and ideas in motor design, motor drive, and motor control. Potential research topics include, but are not limited to, the following: motor design, field-oriented control, torque control, reliability improvement, advanced controllers for motor drive systems, DSP-based sensorless motor drive systems, high-performance motor drive systems, high-efficiency motor drive systems, and practical applications of motor drive systems.

To make it clear, the topics of interest for the call for papers included, but were not limited to, the following:

- Induction motor (IM) and drive;
- Permanent magnet synchronous motor (PMSM) and drive;
- Synchronous reluctance motor and drive;
- Switched reluctance motor and drive;
- Switching flux motor and drive;
- Linear motor and control;
- Step motor and control;
- Fault-tolerant drive;
- DSP-based motor drive;
- High-efficiency motor and drive;
- Sensorless drive system.

The published papers fell into five general areas—DC motor drives, PMSM motor drives, induction motor drives, flux-switching motor drives, and synchronous reluctance motor drives. The details are shown in Table 1.

Citation: Liu, T.-H. Design and Control of Electrical Motor Drives. *Energies* **2021**, *14*, 7717. https://doi.org/10.3390/en14227717

Received: 3 November 2021
Accepted: 12 November 2021
Published: 18 November 2021

Table 1. The broad spectrum of published papers.

Sub-Topics / Detailst	DC Motor Drives	Induction Motor Drives	Syncronous Reluctance Motor Drives	PMSM Motor Drives	Flux-Switching Motor Drives
The Main Contents	• Inverse Optimal Control • State Derivative Space • Nonlinear Control	• Fault-Tolerant Control • Three Level Inverter • Star-Delta Starting	• Rotor Position Estimating • Predictive Control	• Sensorless • High-Speed Control • Backstepping Control • Low Inductance PMSM Control • FPGA-Based Control • Harmonic Current Suppression	• Sensorless • Predictive Control

2. A Short Review of the Contributions in This Issue

Eleven papers were accepted for this Special Issue. Lee et al. [1] proposed that inverse optimal control in a state derivative space system in DC motor tracking control, without using a tachometer, but using the feedback of state derivatives, could reduce cost. In the paper [2] investigated by Kao et al., current harmonic control improved the three-phase current THD from 5.3% to 2.3%. Tsai et al. [3] implemented FPGA-based current control and SVPWM ASIC for AC motor drives, which could reduce power loss by 33% compared to a conventional method. Itajiba et al. investigated a Y-Δ starting of induction motors. By using a statistical method, the experimental results that used two forms of Δ connection were studied. Chao et al. [5] proposed an intelligent fault diagnosis drive system to improve the reliability of inverters. A real-time, smooth switching method was used for fault-tolerant control. Kasper et al. [6] investigated optimal torque feedforward and modal current feedback control for low-inductance PM motors. The method reduced torque ripples and motor losses significantly. Liu et al. [7] designed a high-speed PM motor, focused on rotor unbalanced radial forces, rotor power losses, and rotor mechanical strength. Lin et al. [8] used backstepping control to improve the chattering phenomenon of AC motor drives. In addition, the backstepping control could reduce nonlinear uncertaintyeffects. Mubarok et al. [9] implemented a wide-adjustable sensorless IPMSM speed drive system based on current deviation detection under space-vector modulation. By using the proposed method, the IPMSM motor could be operated from 0 r/min to 3000 r/min. Liu et al. [10] proposed three types of predictive controllers for sensorless flux-switching motor drive systems. An estimated rotor position method that had nearly ± 2 electrical degrees was developed. The adjustable speed range was from 4 r/min to 1500 r/min. Finally, Liu et al. investigated sensorless synchronous reluctance motor drive systems. A rotor position observer that was based on motor parameters was developed. The experimental results showed that the drive system could be adjusted from 30 r/min to 1800 r/min, with good dynamic responses.

We found the task of editing and selecting papers for this collection to be both interesting and rewarding. We would like to thank the authors, staff, and reviewers for their effort and time.

References

1. Lee, F.C.; Tseng, Y.W.; Wu, R.C.; Chen, W.C.; Chen, C.S. Inverse Optimal Control in State Derivative Space with Applications in Motor Control. *Energies* **2021**, *14*, 1775. [CrossRef]
2. Kao, W.T.; Hwang, J.C.; Liu, J.E. Development of Three-Phase Permanent-Magnet Synchronous Motor Drive with Strategy to Suppress Harmonic Current. *Energies* **2021**, *14*, 1583. [CrossRef]
3. Tsai, M.F.; Tseng, C.S.; Chen, P.J. Implementation of an FPGA-Based Current Control and SVPWM ASIC with Asymmetric Five Segment Switching Scheme for AC Motor Drives. *Energies* **2021**, *14*, 1462. [CrossRef]
4. Itajiba, J.A.; Varnier, C.A.C.; Cabral, S.H.L.; Stefenon, S.F.; Leithardt, V.R.Q.; Ovejero, R.G.; Neid, A.; Yow, K.C. Experimental Comparison of Preferential vs. Common Delta Connections for the Star-Delta Starting of Induction Motors. *Energies* **2021**, *14*, 1318. [CrossRef]

5. Chao, K.H.; Ke, C.H. Fault Diagnosis and Tolerant Control of Three-Level Neutral-Point Clamped Inverters in Motor Drives. *Energies* **2020**, *13*, 6302. [CrossRef]
6. Roland, K.; Golovakha, D. Combined Optimal Torque Feedforward and Modal Current Feedback Control for Low Inductance PM Motors. *Energies* **2020**, *13*, 6184.
7. Liu, N.W.; Hwang, K.Y.; Yang, S.C.; Lee, F.C.; Liu, C.J. Design of High-speed Permanent Magnet Motor Considering Rotor Radial Force and Motor Losses. *Energies* **2020**, *13*, 5872. [CrossRef]
8. Lin, C.H. Permanent-Magnet Synchronous Motor Drive System Using Backstepping Control with Three Adaptive Rules and Revised Recurring Sieved Pollaczek Polynomials Neural Network with Reformed Grey Wolf Optimization and Recoupled Controller. *Energies* **2020**, *13*, 5870. [CrossRef]
9. Mubarok, M.S.; Liu, T.H.; Tsai, C.Y.; Wei, Z.Y. A Wide Adjustable Sensorless IPMSM Speed Drive Based on Current Deviation Detection under Space-Vector Modulation. *Energies* **2020**, *13*, 4431. [CrossRef]
10. Liu, T.H.; Mubarok, M.S.; Xu, Y.H. Design and Implementation of Position Sensorless Field-Excited Flux-Switching Motor Drive Systems. *Energies* **2020**, *13*, 3672. [CrossRef]
11. Liu, T.H.; Ahmad, S.; Mubarok, M.S.; Chen, J.Y. Simulation and Implementation of Predictive Speed Controller and Position Observer for Sensorless Synchronous Reluctance Motors. *Energies* **2020**, *13*, 2712. [CrossRef]

energies

Article

Simulation and Implementation of Predictive Speed Controller and Position Observer for Sensorless Synchronous Reluctance Motors

Tian-Hua Liu *, Seerin Ahmad, Muhammad Syahril Mubarok and Jia-You Chen

Department of Electrical Engineering, National Taiwan University of Science and Technology, Taipei 106, Taiwan; seerinahmad@mail.ntust.edu.tw (S.A.); syahril.elmubarok@gmail.com (M.S.M.); johnny19770819@yahoo.com.tw (J.-Y.C.)
* Correspondence: Liu@mail.Ntust.edu.tw

Received: 18 April 2020; Accepted: 26 May 2020; Published: 28 May 2020

Abstract: A position observer and a predictive controller for sensorless synchronous-reluctance-motor (SynRM) drive systems are investigated in this paper. The rotor position observer, based on motor parameters, and stator currents and voltages, was designed and implemented to compute the rotor position. A pole-assignment technique was used to provide similar converging rates of the position observer, even when operated at different speeds. Furthermore, a predictive controller was designed to enhance performance. A digital-signal processor (DSP), TMS-320F-28335, was used as a computation tool. Several simulated results are provided and compared with the measured results. The measured results showed that the implemented predictive controller sensorless SynRM drive system could be adjusted from 30 to 1800 rpm with satisfactory performance, including quicker and better tracking responses, and a lower speed drop than that of a proportional-integral (PI) controller.

Keywords: rotor position observer; predictive speed controller; digital-signal processor; synchronous reluctance motor

1. Introduction

Synchronous reluctance motors (SynRMs) that have no magnetic materials mounted on their rotors have a rugged and simple structure, and are easy to control. These motors have been applied in vacuum cleaners, sewing machines, and electric vehicles. Several major sensorless techniques have been proposed for SynRM drive systems, including extended back electromotive force (back-EMF) modelling [1], correct-slope-estimating [2], adaptive-projection-vector [3], and search-coil-detection methods [4].

Previous research also found that a position observer is a very effective way to obtain rotor position because of its simple computation and hardware circuit. For example, Senjyu et al. investigated the flux-linkage observer for a SynRM. By suitably choosing a time constant of first-order lag compensation, a precise rotor-position observer was achieved [5]. Lim et al. proposed an online position observer of a SynRM by using a finite-element method [6]. Vagati investigated a flux-based observer to obtain an accurate magnetic model [7]. Tuovinen et al. implemented a back-EMF position observer method for a SynRM drive system that accounted for cross-saturation in the motor [8]. Tuovinen et al. analyzed a position observer with resistance adaption for SynRM drive systems [9,10]. Awon et al. investigated a discrete-time observer for sensorless SynRM drives [11]. Tuovinen et al. implemented a full-order observer for sensorless SynRM drive systems [12].

Moreover, some research studied predictive controllers for SynRM drives. Lin et al. used a predictive method to design a current controller for a SynRM drive system [13]. Hadla et al. proposed a predictive stator flux that did not employ any controller calibrations or weighting factors for a SynRM [14]. Carlet et al. designed a predictive current control that was based on predictive methods that did not require knowledge of the system model [15]. Antonello et al. investigated a hierarchical scaled-state direct predictive control of a SynRM drive system that kept the current ripple limited even though the motor was operated at a low switching frequency [16]. Lin et al. studied a dual-vector modeless predictive current control on the basis of predictive methods for SynRM drives without using any system models, parameters, or DC link capacitor voltages [17].

Predictive controllers were successfully applied for SynRM drives with an attached encoder or resolver [13–17]. However, to the authors' best knowledge, in previous research, predictive controllers were not employed or were rarely employed in sensorless-synchronous-reluctance-drive systems. In addition, previous predictive-control research only focused on current-loop controllers and not speed-loop controllers for synchronous-reluctance-drive systems [1–17]. To fill these gaps in research, this paper focuses on using a predictive speed controller and rotor-position observer to enhance the performance of a sensorless SynRM drive. A pole-assignment technique was employed to provide similar q-axis current responses. Due to smooth q-axis responses, the estimated position error was bounded in four electrical degrees. Figure 1 demonstrates the structure of the SynRM, which included an a-b-c phase stator, a rotor, and an air gap. Take the a-phase as an example. The "." means the a-phase-current flows out from this paper, and the "x" means the a-phase current flows into this paper. Compared to other AC motors, this synchronous reluctance motor uses its saliency characteristics to produce its output torque.

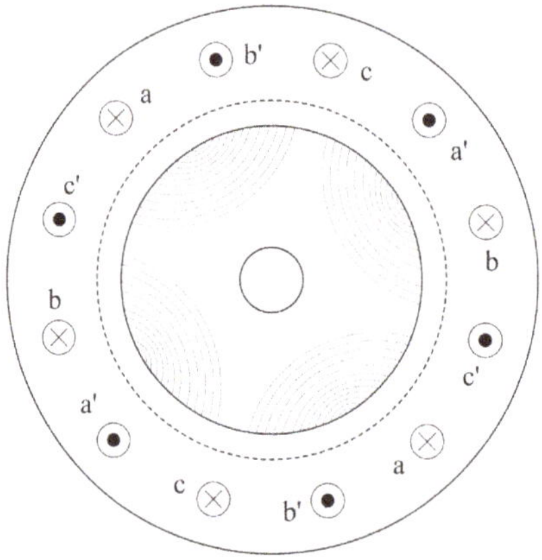

Figure 1. Synchronous-reluctance-motor (SynRM) structure.

2. Mathematical Model

In a synchronous frame, the differential equations of the d-q axis currents of a SynRM can be expressed as follows.

$$\frac{di_d}{dt} = \frac{1}{L_d}\left[v_d - r_s i_d + \omega_{re} L_q i_q\right] \tag{1}$$

$$\frac{di_q}{dt} = \frac{1}{L_q}\left[v_d - r_s i_q - \omega_{re} L_d i_d\right], \tag{2}$$

where i_d and i_q are d-q axis currents, $\frac{d}{dt}$ is the differential operator, L_d and L_q are d-q axis inductances, v_d and v_q are d-q axis voltages, r_s s the stator resistance, and ω_{re} is the electrical-rotor speed. When the d-axis current remained constant, the reluctance torque of the SynRM was proportional to i_q, which can be shown as follows:

$$\begin{aligned} T_e &= \tfrac{3}{2}\tfrac{P}{2}\left(L_{md} - L_{mq}\right) i_d i_q, \\ &= K_t i_q \end{aligned} \tag{3}$$

where T_e is the reluctance torque, and P is the pole number. The differential equation of the speed is

$$\frac{d\omega_{rm}}{dt} = \frac{1}{J_t}(T_e - T_L - B_t\omega_{rm}), \tag{4}$$

where ω_m is the mechanical-rotor speed, J_t is the inertia of the motor and load, T_L is the external load, and B_t is the viscous frictional coefficient of the motor and load. The differential equation of the rotor position is expressed as follows:

$$\frac{d\theta_{rm}}{dt} = \omega_{rm}, \tag{5}$$

where θ_{rm} is the mechanical-rotor position. The simplified block diagram of the SynRM is shown in Figure 2. By fixing the d-axis current command and controlling the q-axis current, torque T_e was easily controlled, and speed could then be easily adjusted.

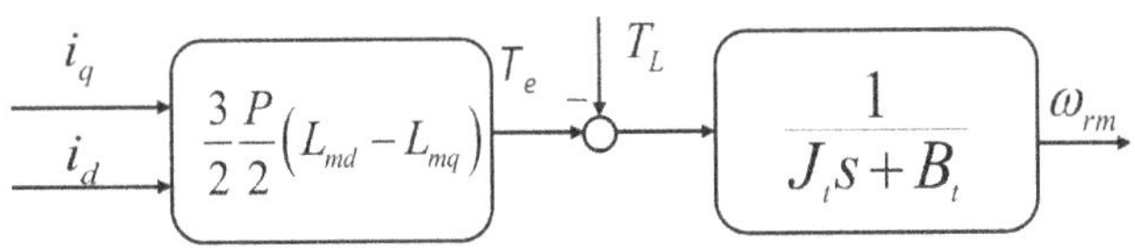

Figure 2. Block diagram of simplified SynRM.

3. Rotor-Position-Observer Design

In this paper, a closed-loop state estimator was derived on the basis of the d-q axis synchronous frame. From Equations (1) and (2), the following differential equation of the *d*-*q* axis currents could be derived as follows:

$$\frac{di_s}{dt} = Fi_s + Gv_s \tag{6}$$

In Equation (6), the related vectors and matrices are defined as

$$F = \begin{bmatrix} -\frac{r_s}{L_d} & \omega_{re}\frac{L_q}{L_d} \\ -\omega_{re}\frac{L_d}{L_q} & -\frac{r_s}{L_q} \end{bmatrix} \tag{7}$$

$$G = \begin{bmatrix} \frac{1}{L_d} & 0 \\ 0 & \frac{1}{L_q} \end{bmatrix} \tag{8}$$

and

$$v_s = \begin{bmatrix} v_d \\ v_q \end{bmatrix}, \tag{9}$$

where F and G are matrices, and i_s and v_s are current and voltage vectors. Figure 3 shows the proposed closed-loop observer that can be described as follows [11,12]:

$$\frac{d\hat{i}_s}{dt} = F\hat{i}_s + Gv_s - K(\hat{i}_s - i_s) = F\hat{i}_s + Gv_s - K\widetilde{i_s}. \tag{10}$$

From Equation (10), one can define the related vectors and matrix as follows:

$$K = \begin{bmatrix} k_1 & 0 \\ 0 & k_2 \end{bmatrix} \tag{11}$$

and

$$\hat{i}_s = \begin{bmatrix} \hat{i}_{ds} \\ \hat{i}_{qs} \end{bmatrix}, \tag{12}$$

where $\hat{i}_{ds}$ and $\hat{i}_{qs}$ are the estimated d-q axis currents, $\widetilde{i}_s$ is the estimated current vector, and K, including k_1 and k_2, is the gain matrix of the proposed observer. From Equation (10) and Figure 3, one can understand that a different matrix K can be designed to determine the converging rate of estimated current vector, $\hat{i}_s$ for different speeds. Detailed analysis is shown as follows.

First, from Equations (10) and (6), one can obtain the current vector error as follows:

$$\frac{d\widetilde{i_s}}{dt} = F\widetilde{i_s} - K\widetilde{i_s}. \tag{13}$$

From Equation (13), the equation can be rewritten as follows:

$$\frac{d\widetilde{i_s}}{dt} = M\widetilde{i_s} \tag{14}$$

and

$$M = F - K = \begin{bmatrix} -\frac{r_s}{L_{ds}} & \omega_{re}\frac{L_{qs}}{L_{ds}} \\ -\omega_{re}\frac{L_{ds}}{L_{qs}} & -\frac{r_s}{L_{qs}} \end{bmatrix} - \begin{bmatrix} k_1 & 0 \\ 0 & k_2 \end{bmatrix} = \begin{bmatrix} -\frac{r_s}{L_{ds}} - k_1 & \omega_{re}\frac{L_{qs}}{L_{ds}} \\ -\omega_{re}\frac{L_{ds}}{L_{qs}} & -\frac{r_s}{L_{qs}} - k_2 \end{bmatrix} \tag{15}$$

The eigenvalues of matrix M can be derived as follows:

$$\det(\lambda I - M) = \lambda^2 + (\tfrac{r_s}{L_{qs}} + \tfrac{r_s}{L_{ds}} + k_1 + k_2)\lambda + \tfrac{r_s{}^2}{L_{ds}L_{qs}} + \tfrac{r_s}{L_{ds}}k_2 + \tfrac{r_s}{L_{qs}}k_1 + k_1k_2 + \omega_{re}{}^2 = 0 \tag{16}$$

The two eigenvalues α and β of the $\det(\lambda I - M)$ are expressed as

$$\alpha,\beta = -\frac{(\frac{r_s}{L_{qs}} + \frac{r_s}{L_{ds}} + k_1 + k_2)}{2} \pm \frac{\sqrt{(\frac{r_s}{L_{qs}} + \frac{r_s}{L_{ds}} + k_1 + k_2)^2 - 4\left(\frac{r_s{}^2}{L_{ds}L_{qs}} + \frac{r_s}{L_{ds}}k_2 + \frac{r_s}{L_{qs}}k_1 + k_1k_2 + \omega_{re}{}^2\right)}}{2} \tag{17}$$

Assuming estimated current vector $\widetilde{i_s}$ has two poles, which are $-\alpha$ and $-\beta$, the following equation exists and can be described as

$$(\lambda + \alpha)(\lambda + \beta) = 0. \tag{18}$$

From Equation (18), one can obtain

$$\lambda^2 + (\alpha + \beta)\lambda + \alpha\beta = 0. \tag{19}$$

Comparing Equations (16) and (19), one can derive

$$(\alpha + \beta) = \frac{r_s}{L_{qs}} + \frac{r_s}{L_{ds}} + k_1 + k_2 \tag{20}$$

and

$$\alpha\beta = \frac{r_s{}^2}{L_{ds}L_{qs}} + \frac{r_s}{L_{ds}}k_2 + \frac{r_s}{L_{qs}}k_1 + k_1k_2 + \omega_{re}{}^2 \tag{21}$$

One can assign the two different negative poles, $-\alpha$ and $-\beta$, in Equations (20) and (21), at different operating speeds. As a result, the proposed SynRM drive system had similar dynamic responses for q-axis current errors at different operating speeds. Moreover, a Lyapunov function candidate V can be defined as

$$V = \widetilde{i_s}^T\widetilde{i_s}. \tag{22}$$

From Equation (22), V has a positive definite value. Because M^T and M are negative definite matrices, the differential value of the Lyapunov is expressed as follows [18–20]:

$$V = \widetilde{i_s}^T (M^T + M)\widetilde{i_s} < 0 \tag{23}$$

Equation (23) shows that current error vector $\widetilde{i_s}$ asymptotically converges and reaches zero at a steady state. The $\widetilde{\omega}_{re}$ is defined as

$$\widetilde{\omega}_{re} = \hat{\omega}_{re} - \omega_{re}, \tag{24}$$

where $\hat{\omega}_{re}$ is the estimated speed, and $\widetilde{\omega}_{re}$ is the estimated speed error. From Equation (24) and [7,8], one can use $\widetilde{i_q}$ to replace current vector error $\widetilde{i_s}$ and obtain the following equation:

$$\hat{\omega}_{re}(t) = K_P \widetilde{i_q} + K_I \int_{t_0}^{t} \widetilde{i_q} d\tau + \hat{\omega}_{re}(t_0) \tag{25}$$

where $\hat{\omega}_{re}(t_0)$ is the initial estimated speed, t_0 is the initial value of time, t is the value of the current time, and τ is the variable. Finally, estimated rotor position $\hat{\theta}_{re}(t)$ is

$$\hat{\theta}_{re}(t) = \int_{t_0}^{t} \hat{\omega}_{re} d\tau + \hat{\theta}_{re}(t_0) \tag{26}$$

After digitizing Equations (25) and (26), one can obtain estimated electrical-rotor speed $\hat{\omega}_{re}(m)$ and rotor position $\hat{\theta}_{re}(m)$ as follows:

$$\hat{\omega}_{re}(m) = K_P \widetilde{i_q}(m) + K_I \sum_{n=1}^{m} \widetilde{i_q}(n)\Delta t \tag{27}$$

and

$$\hat{\theta}_{re}(m) = \sum_{n=1}^{m} \hat{\omega}_{re}(n)\Delta t + \hat{\theta}_{re}(0), \tag{28}$$

where $\hat{\theta}_{re}(0)$ is the initial estimated position and n is the iterative number. The proposed position observer is illustrated in Figure 3. First, the current vector observer of the $\hat{i}_s$ was constructed. Next, the error of the current vector observer, $\widetilde{i_s}$, was computed. After that, motor-speed observer $\hat{\omega}_{re}$ was obtained via proportional-integral (PI) controller. Finally, motor-position observer $\hat{\theta}_{re}$ was computed via an integral controller.

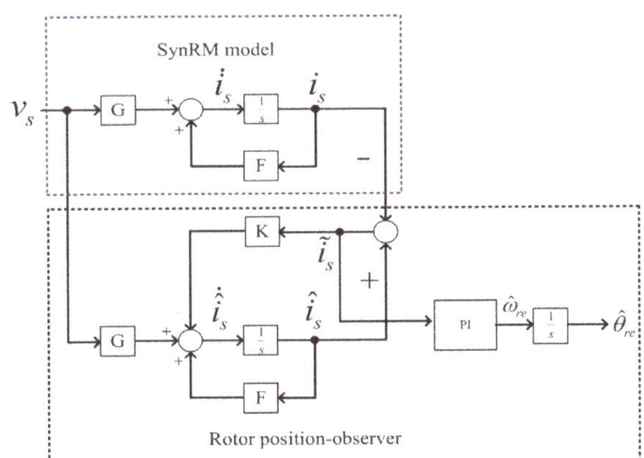

Figure 3. Proposed position observer.

4. Predictive-Controller Design

Predictive control is a new control technique that has had strong and widespread impact on industrial applications [20]. It has been used in process control, transportation control, and robotic control for more than three decades. Because of the powerful computation ability of digital-signal processors (DSPs), predictive control has been widely utilized for power converters and motor drives [21,22]. In our paper, predictive control was used for the first time for a sensorless SynRM drive.

The continuous-time dynamic equation of the speed of the SynRM without an external load is

$$\frac{d\omega_{rm}}{dt} = \frac{1}{J_t}(T_e - B_t\omega_{rm}) \tag{29}$$

Combining Equations (3) and (29), one can obtain:

$$\frac{\omega_{rm}(s)}{i_q(s)} = G_c(s) = \frac{K_t}{J_t s + B_t}, \tag{30}$$

where $\omega_{rm}(s)$ is mechanical-motor speed, $G_c(s)$ is the transfer function of the uncontrolled plant, and J_t and B_t are the inertia of the motor and the viscous frictional coefficient of the motor, respectively. After inserting a zero-order-hold device in front of transfer function $G_c(s)$, transfer function $G_d(s)$ can be rewritten as

$$G_d(s) = H(s)G_c(s) = \frac{1 - e^{-sT_z}}{s} \cdot \frac{K_t/J_t}{s + (B_t/J_t)}. \tag{31}$$

By converting the continuous into the discrete domain, discrete transfer function $G_d(z)$ is

$$\begin{aligned} G_d(z) &= \frac{\omega_r(z)}{i_q(z)} \\ &= \frac{K}{B_t} \cdot \frac{(1-e^{-\frac{B_t}{J_t}T_z})z^{-1}}{\left(1-e^{-\frac{B_t}{J_t}T_z}z^{-1}\right)} \end{aligned} \tag{32}$$

From Equation (32), one can obtain the discrete-time dynamic speed equation as

$$\omega_{rm}(m+1) = a\omega_{rm}(m) + bi_q(m), \tag{33}$$

where m is the sampling step of the digital control system. In Equation (33), parameters are shown as

$$a = e^{-\frac{B_t}{J_t}T_z} \tag{34}$$

and

$$b = \frac{K_t}{B_t}\left(1 - e^{-\frac{B_t}{J_t}T_z}\right). \tag{35}$$

To simplify the computations of the DSP in this paper, the predictive horizon and control horizon were selected as 1. Performance index $J_p(m)$ was defined in previous research as follows [21,22]:

$$J_p(m) = q[\omega_{rm}(m+1) - \omega^*_{rm}(m+1)]^2 + \left[\Delta i_q(m)\right]^2, \tag{36}$$

where q is a weighting factor. To prevent the positive speed error from canceling the negative speed error, or the positive control input from canceling the negative control input, we selected the square of the speed error and control input as the performance index. By taking the derivative of $J_p(m)$ to $\Delta i_q(m)$, one can obtain the following equation:

$$\frac{\partial J_p(m)}{\partial \Delta i_q(m)} = 2qb[\omega_{rm}(m+1) - \omega^*_{rm}(m+1)] + 2\left[\Delta i_q(m)\right] = 0 \quad (37)$$

From Equation (37), the q-axis difference is obtained as

$$\Delta i_q(m) = \frac{qb}{qb^2+1}\left[\omega^*_{rm}(m+1) - a\omega_{rm}(m) - bi_q(m-1)\right] \quad (38)$$

Ultimately, the optimal q-axis current is derived as follows:

$$i_q(m) = \Delta i_q(m) + i_q(m-1) \quad (39)$$

From Equations (38) and (39), one can construct a block diagram as demonstrated in Figure 4.

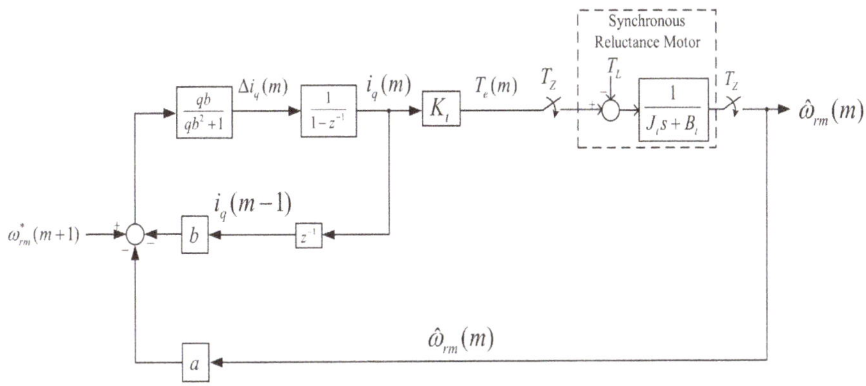

Figure 4. Block diagram of predictive speed control.

5. Implementation

To evaluate the theoretical analysis and simulated results, a SynRM sensorless drive system was implemented. The synchronous reluctance motor, type P56H5012, manufactured in the United States, is shown in Figure 5a. Figure 5b demonstrates the implemented hardware circuit board. The circuit was a three-phase inverter using six IGBTs, type PS 21265-P/AP, as its power devices. The power devices were used for providing voltages and currents to the motor. The inverter employed space-vector modulation with a 10 kHz switching frequency. In addition, the circuit consisted of voltage- and current-sensing circuits, 12-bit A/D converters, which converted analog signals to digital signals, an interfacing circuit, and a DSP that was a control center. Two Hall effect sensors, type LTS25-NP, manufactured in Switzerland, were used as current sensors. A protecting circuit was utilized to prevent an overload of the SynRM drive system. All hardware circuits were designed by us. The block diagram of the implemented drive system is demonstrated in Figure 5c. First, the voltages and currents were read by the DSP via A/D converters. Next, the DSP executed the algorithms of the position observer to obtain the estimated rotor position and speed. The DSP compared the speed command and the estimated speed to compute the q-axis command. After that, the DSP used the PI control to obtain the voltage commands. Then, the DSP used the d-q to an a-b-c axis for coordinate transformation and to compute the three-phase voltage commands. Finally, the DSP sent triggering signals T_1 to T_3' to the inverter.

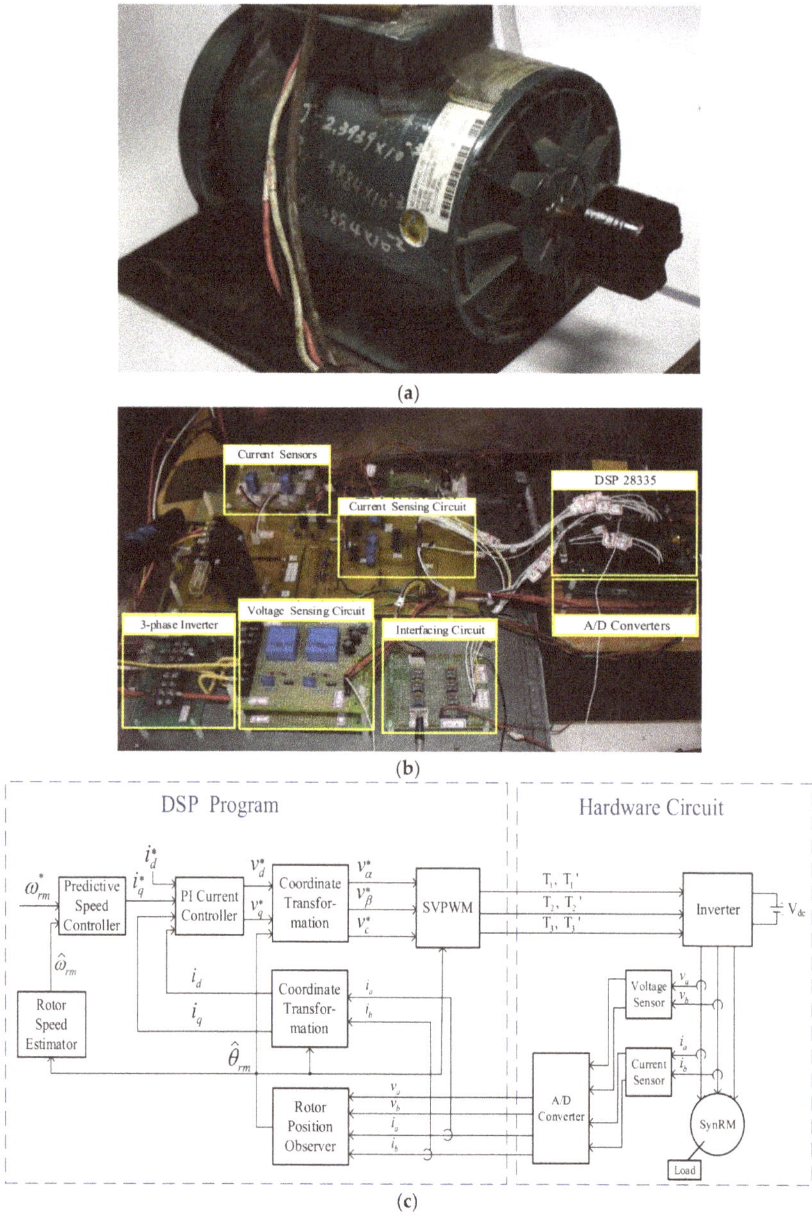

Figure 5. Implemented system. (**a**) Motor; (**b**) circuits; (**c**) block diagram.

A DSP, type TMS-320F-28335 [23], manufactured in the USA, was used as a computation tool to execute the position-estimating algorithm and predictive-control algorithm. The sampling time of the current-loop was 100 μs, and the speed loop was 1 ms. The d-axis current command was set at 0.5 A to provide the required flux linkage. A larger d-axis current command increased the output torque; however, it also created flux saturation. The q-axis was also adjusted to obtain a quick torque response. Then, the proposed position observer was used to compute the rotor position.

A dynamometer, type TM-5MT(DS), made by Neiwa Company, was used for adding an external load to the SynRM. A power supply was utilized to provide 320 V to the inverter. The motor used in this paper was a three-phase, four-pole, 220 V, 560 W, 3.4 A, 60 Hz, 1800 rpm, Y-connected SynRM. Its parameters were as follows:

Stator resistance r_s was 2 Ω; d-axis inductance L_d was 148 mH; q-axis inductance L_q was 67.2 mH; interia, J_t was 0.0024 kg.m^2; and friction coefficient B_t was 0.0015 N.m/(rad/sec).

6. Experiment Results

To evaluate the proposed methods, several results are illustrated in this paper. Input DC voltage was 320 V, the sampling time of the speed-loop control was 1 ms, and the current loop was 100 μs. The predictive speed control chose one step as the prediction and control horizons. PI speed controller parameters were determined by pole assignment, which were near $P_1 = -2 + j0.5$ and $P_2 = -2 - j0.5$. After that, one could obtain $K_P = 45$ and $K_I = 4$, and the d-axis current command was 0.5 A. The parameters of the PI controller of the current loop were determined by pole assignment, which uses a similar method as the PI controller of the speed loop.

Figure 6a demonstrates simulated line voltage v_{ab}. The line voltage was a pulse-width modulation (PWM) modulated voltage waveform with a 10 kHz switching frequency. Figure 6b shows measured line–line voltage v_{ab}. Both simulated and measured voltages were very close. In addition, the line–line voltage had positive and negative voltages sequentially because the space-vector pulse-width modulation (SVPWM) was applied. Figure 7a displays the simulated a-phase current, Figure 7b displays the measured a-phase current, and both currents were near-sinusoidal waveforms. Figure 7c displays the measured a-phase current harmonics. Figure 8a illustrates the comparison of the simulated results of the real and estimated rotor positions. Figure 8b illustrates the comparison of the measured results of the rotor positions. Figure 8c illustrates the measured estimated position error, which was between 0 and 4 electrical degrees. The estimated position error had a nonrepetitive waveform due to the influence of the PI controllers, which is shown in Figure 3. Figure 9a,b shows the q-axis current errors at operating speeds of 1100 and 1800 rpm, and the q-axis current errors had similar dynamic responses. The main reason was that similar poles were assigned for different speeds in this paper. In addition, the q-axis estimated current,$\widetilde{i}_q$, could reach zero at both operating speeds. Figure 10a,b illustrates the simulated and measured lowest operating speeds, which were 30 rpm. The SynRM drive system had obvious torque pulsations when it was operated below 30 rpm. Figure 11a,b illustrates the highest speeds at 1800 rpm, and they had similar transient responses. Figure 12a shows the simulated disturbance responses at 500 rpm and 2 N.m, and Figure 12b shows the measured load-disturbance responses in the same situation. The simulated and measured results were similar. The predictive speed controller performed better, including quicker recovery time and lower speed drop, than the PI controller. Figure 13a demonstrates the simulated responses, which reversed repeatedly between 200 and −200 rpm. Figure 13b demonstrates the measured responses under the same circumstances, and they had similar responses. However, the PI controller had an obvious time delay due to the influence of the integral parameter. Figure 14a demonstrates the simulated results of the triangular tracking responses, and Figure 14b demonstrates the measured results of the triangular tracking responses. The speed of the predictive controller could follow the speed command well. The PI controller, however, caused an obvious phase leg between its commands and speed responses. By choosing proper PI value coefficients, error in waveforms for Figures 12–14 could be minimized. However, the tuning of PI controllers relied on trial and error. In addition, each operation point required one parameter set of PI controller. Once the operation point changed, the PI controller could not work well. In this paper, the predictive controller provides a systemic design procedure. By suitably choosing weighting factor q, a systematic controller, which is illustrated in Equations (38) and (39), could be uniquely obtained, and the predictive controller parameters were related to the motor parameters, sampling time, and feedback speed.

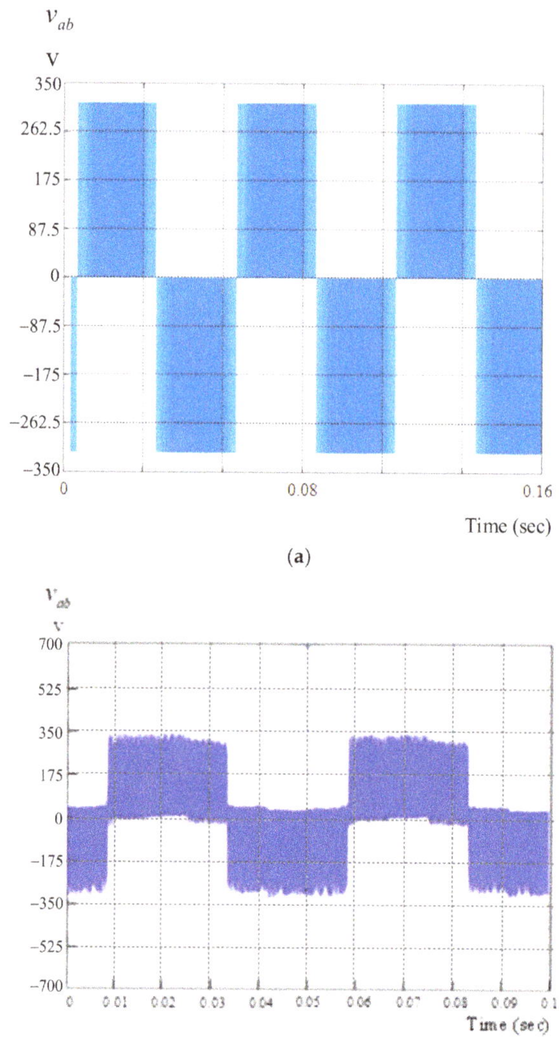

(b)

Figure 6. Line–line voltage of motor. (**a**) Simulated; (**b**) measured.

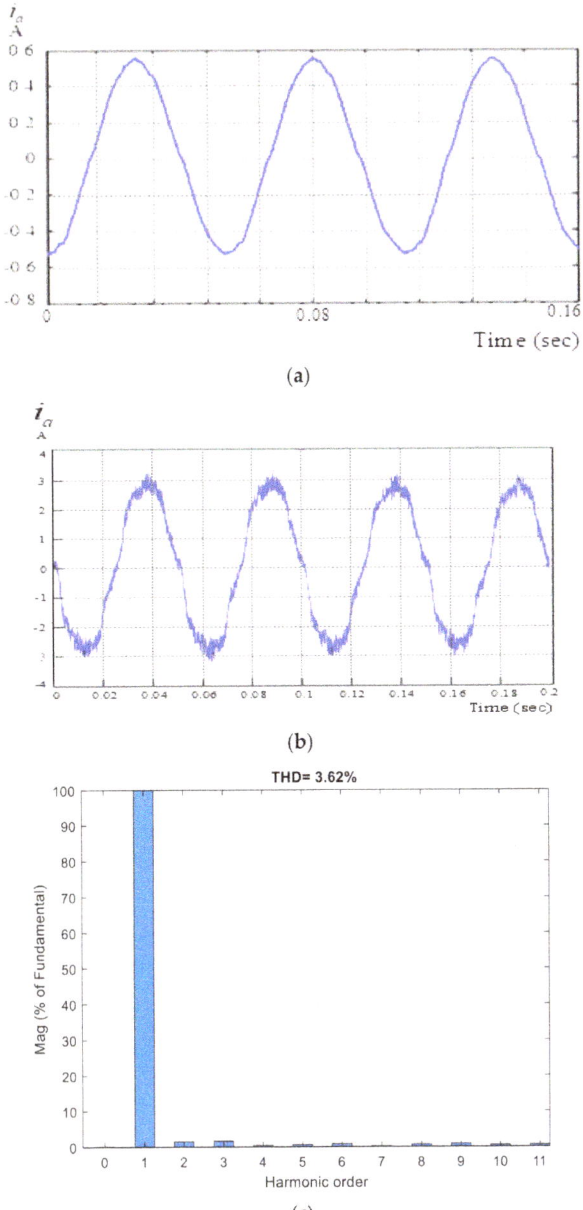

(a) (b) (c)

Figure 7. Phase currents. (**a**) Simulated; (**b**) measured; (**c**) harmonics.

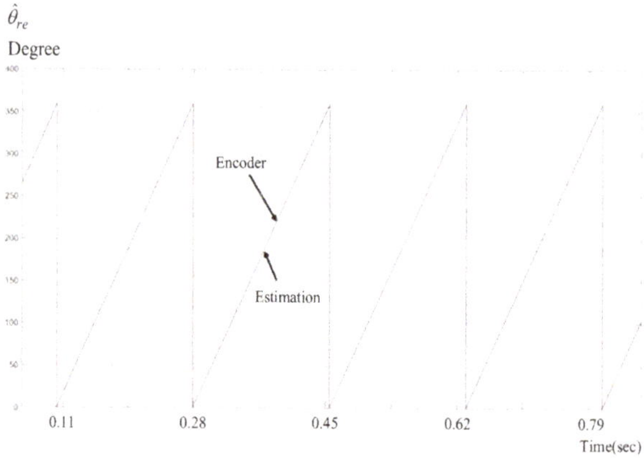

(**a**)

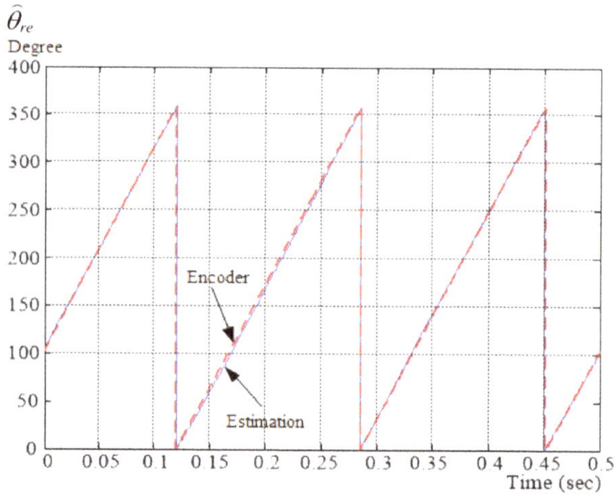

(**b**)

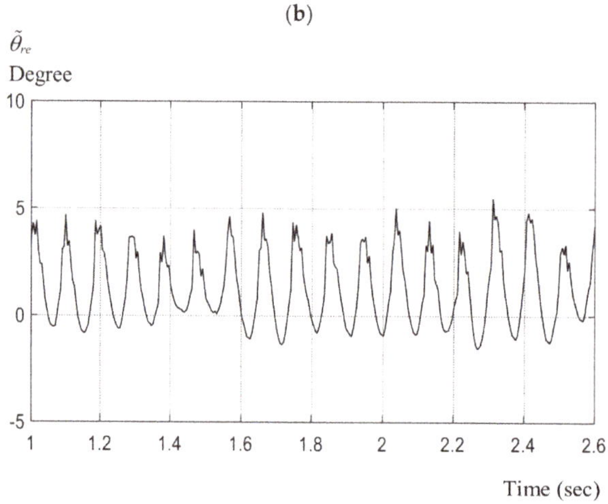

(**c**)

Figure 8. Responses at 500 rpm. (**a**) Simulated position; (**b**) measured position; (**c**) measured estimated error.

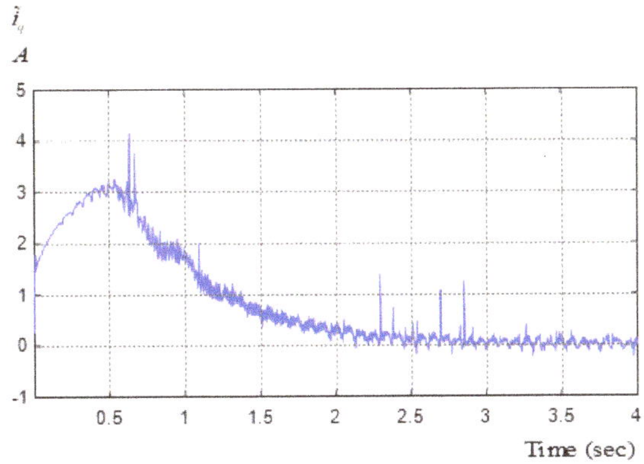

(a)

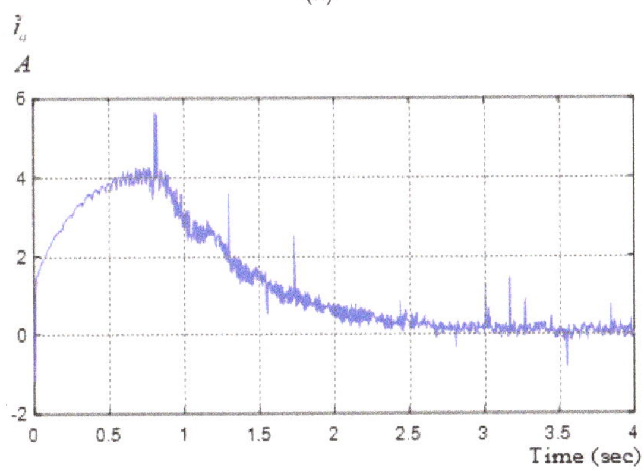

(b)

Figure 9. Measured dynamic q-axis current errors at (**a**) 1100 and (**b**) 1800 rpm.

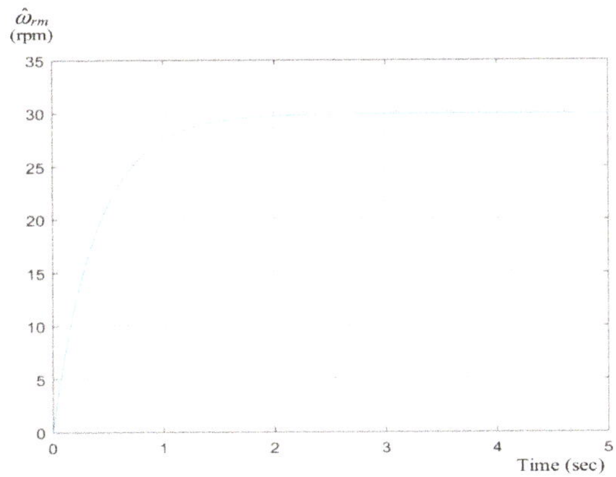

(a)

Figure 10. *Cont.*

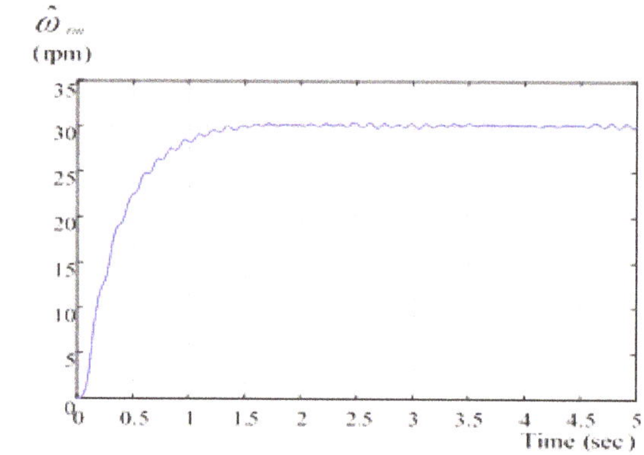

(**b**)

Figure 10. Lowest step-input speed at 30 rpm. (**a**) Simulated; (**b**) measured.

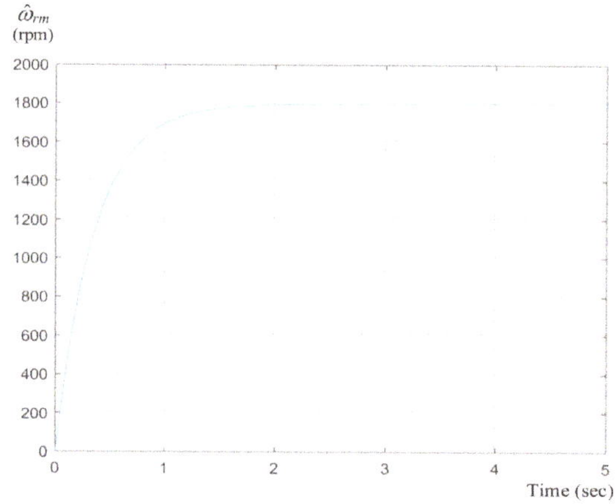

(**a**)

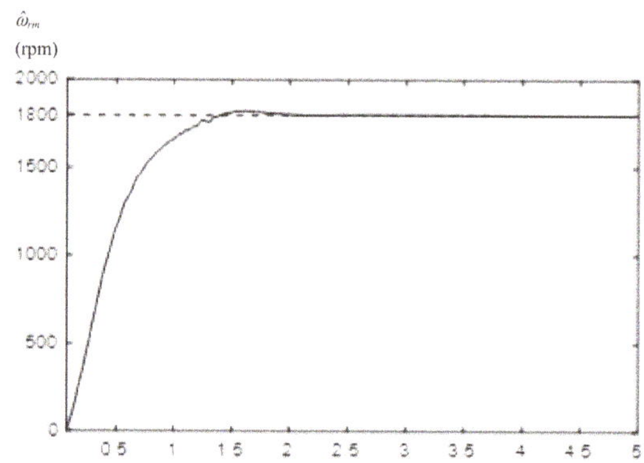

(**b**)

Figure 11. Highest step-input speed responses at 1800 rpm. (**a**) Simulated; (**b**) measured.

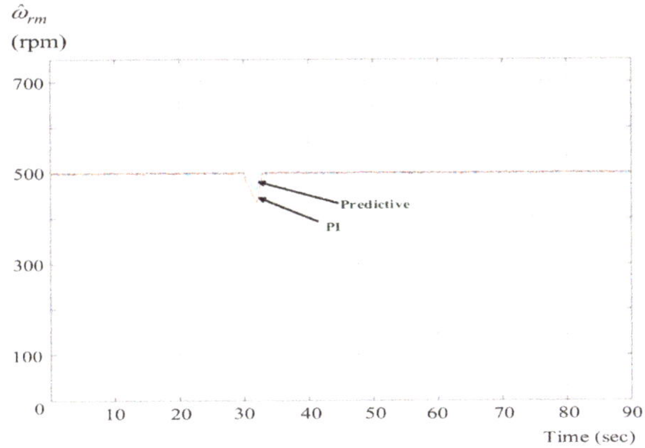

(**a**)

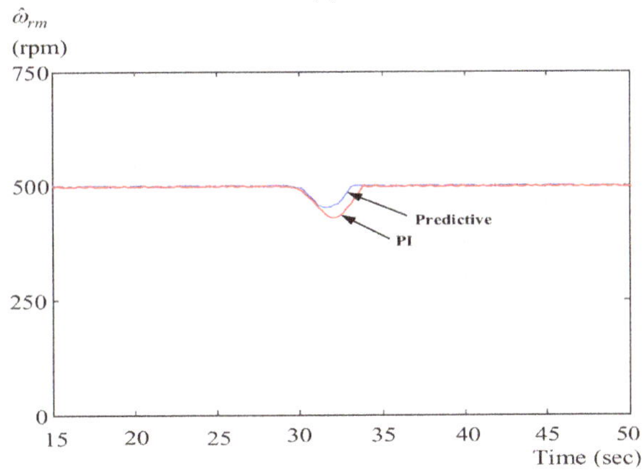

(**b**)

Figure 12. 2 N.m disturbance responses (**a**) simulated (**b**) measured.

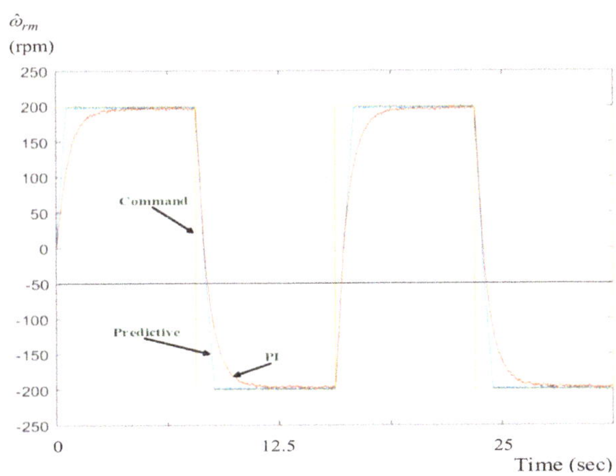

(**a**)

Figure 13. *Cont.*

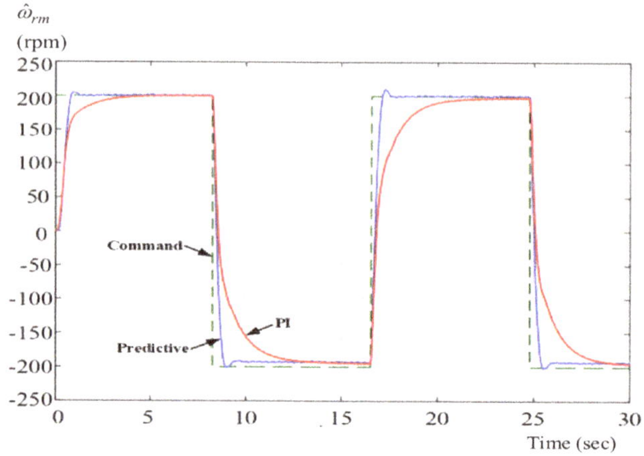

(b)

Figure 13. Square-wave tracking responses using different controllers. (a) Simulated; (b) measured.

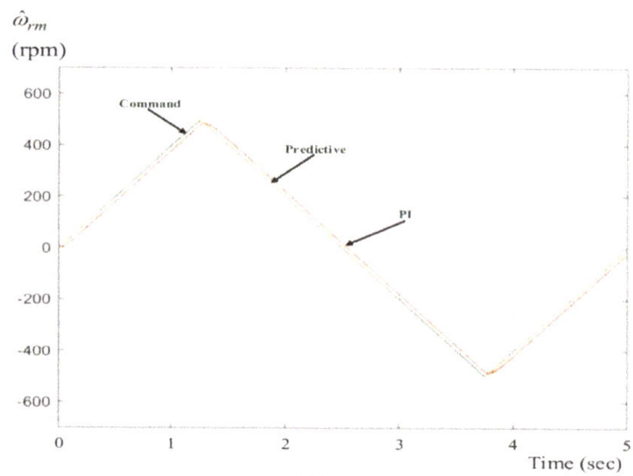

(a)

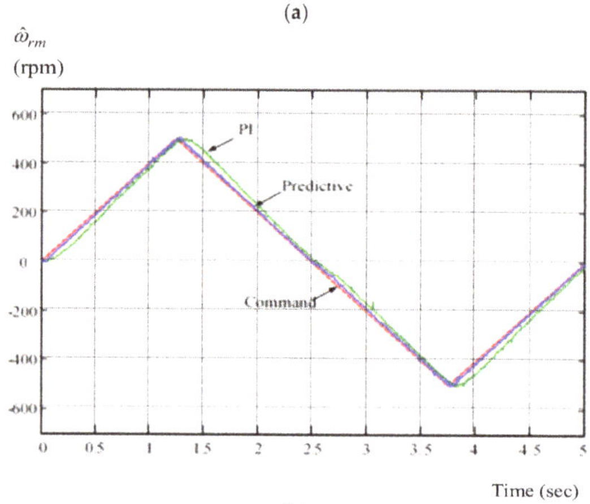

(b)

Figure 14. Triangular tracking responses using different controllers. (a) Simulated; (b) measured.

7. Conclusions

In this study, a position observer and a predictive controller for a SynRM were simulated and implemented. On the basis of the pole-assignment method, the error dynamics of the q-axis currents had similar poles and close dynamic responses. Both q-axis current errors and rotor-position observer errors could be converged in 2 s. In addition, the measured position errors of the position observer could be within four electrical degrees. Compared to previous research [11], which required a full-order observer to obtain the motor position, this paper provided a simplified method with satisfactory performance, including quick converging time and small steady-state estimation errors. In addition, the DSP computation time of the position observer in this paper was shorter than that in previous research. Furthermore, this paper proposed an original idea that combines the advanced position observer and speed-loop predictive controller for a SynRM drive system.

Experiment results demonstrated that the adjustable range of the SynRM was between 30 and 1800 rpm. Furthermore, the measured results demonstrated that the proposed rotor observer had errors below four electrical degrees, and the predictive controller had better dynamic responses than those of PI controllers. Most of the functions of the predictive controller and position observer were implemented by the DSP; therefore, the proposed synRM drive system uses simple hardware circuits that can easily be applied in industry.

Author Contributions: Conceptualization: T.-H.L. Methodology: T.-H.L. and J.–Y.C.; Software: J.-Y.C.; Validation: T.-H.L., J.-Y.C., M.S.M. and S.A.; Investigation: T.-H.L. and J.-Y.C.; Resource: T.-H.L.; Formal Analysis: T.-H.L. and J.–Y.C.; Data curation: J.–Y.C., M.S.M., and S.A.; Writing-original English draft preparation: T.-H.L. and J.–Y.C.; Writing-review & editing: J.–Y.C, M.S.M. and S.A.; Visualization: M.S.M. and S.A.; Supervision: T.-H.L.; Funding Acquisition: T.-H.L. All authors have read and agreed to the published version of the manuscript.

Funding: The paper was supported by Ministry of Science and Technology, Taiwan under grants MOST 108-2221-E-011-085 and -086.

Conflicts of Interest: The authors declare no conflicts of interest

References

1. Bolognani, S.; Ortombina, L.; Tinazzi, F.; Zigliotto, M. Model sensitivity of fundamental-frequency-based position estimators for sensorless PM and reluctance synchronous motor drives. *IEEE Trans. Ind. Electron.* **2018**, *65*, 77–85. [CrossRef]
2. Wei, M.Y.; Liu, T.H. A high-performance sensorless position control system of a synchronous reluctance motor using dual current-slope estimating technique. *IEEE Trans. Ind. Electron.* **2012**, *59*, 3411–3426.
3. Varatharajan, A.; Pellegrino, G. Sensorless synchronous reluctance motor drives: A general adaptive projection vector approach for position estimation. *IEEE Trans. Ind. Appl.* **2020**, *56*, 1495–1504. [CrossRef]
4. Tornello, L.D.; Scelba, G.; Scarcella, G.; Cacciato, M.; Testa, A.; Foti, S.; Caro, S.D.; Pulvirenti, M. Combined rotor-position estimation and temperature monitoring in sensorless, synchronous reluctance motor drives. *IEEE Trans. Ind. Appl.* **2019**, *55*, 3851–3862. [CrossRef]
5. Senjyu, T.; Shingaki, T.; Uezato, K. Sensorless vector control of synchronous reluctance motors with disturbance torque observer. *IEEE. Trans. Ind. Electron.* **2001**, *48*, 402–407. [CrossRef]
6. Lim, H.B.; Lee, J.H. The evaluation of online observer system of synchronous reluctance motor using a coupled transient FEM and Preisach model. *IEEE Trans. Magnet.* **2008**, *44*, 4139–4142.
7. Vagati, A.; Pastorelli, M.; Franceschini, G.; Drogoreanu, V. Flux-observer-based high-performance control of synchronous reluctance motors by including cross saturation. *IEEE Trans. Ind. Appl.* **1999**, *35*, 597–605. [CrossRef]
8. Tuovinen, T.; Hinkkanen, M. Adaptive full-order observer with high-frequency signal injection for synchronous reluctance motor drives. *IEEE J. Emerg. Select. Top. Power Electron.* **2014**, *2*, 181–189. [CrossRef]
9. Tuovinen, T.; Hinkkanen, M.; Luomi, J. Analysis and design of a position observer with resistance adaption for synchronous reluctance motor drives. *IEEE Trans. Ind. Appl.* **2013**, *49*, 66–73. [CrossRef]
10. Tuovinen, T.; Hinkkanen, M. Signal-injection-assisted full-order observer with parameter adaption for synchronous reluctance motor drives. *IEEE Trans. Ind. Appl.* **2014**, *50*, 3392–3402. [CrossRef]

11. Awon, H.A.A.; Tuovinen, T.; Saarakkala, S.E.; Hinkkanen, M. Discreet-time observer design for synchronous reluctance motor drives. *IEEE Trans. Ind. Appl.* **2016**, *52*, 3968–3979. [CrossRef]
12. Tuovinen, T.; Hinkkanen, M. Comparison of a reduced-order observer and a full-order observer for sensorless synchronous motor drives. *IEEE Trans. Ind. Appl.* **2012**, *48*, 1959–1967. [CrossRef]
13. Lin, C.K.; Yu, J.T.; Lai, Y.S.; Yu, H.C. Improved model-free predictive current control for synchronous reluctance motor drives. *IEEE Trans. Ind. Electron.* **2016**, *63*, 3942–3953. [CrossRef]
14. Hadla, H.; Cruz, S. Predictive stator flux and load angle control of synchronous reluctance motor drives operating in a wide speed range. *IEEE Trans. Ind. Electron.* **2017**, *64*, 6950–6959. [CrossRef]
15. Carlet, P.G.; Tinazzi, F.; Bolognani, S.; Zigliotto, M. An effective model-free predictive current control for synchronous reluctance motor drives. *IEEE Trans. Ind. Appl.* **2019**, *55*, 3781–3790. [CrossRef]
16. Antonello, R.; Carraro, M.; Peretti, L.; Zigliotto, M. Hierarchical scaled-states direct predictive control of synchronous reluctance motor drives. *IEEE Trans. Ind. Electron.* **2016**, *63*, 5176–5185. [CrossRef]
17. Lin, C.K.; Yu, J.T.; Lai, Y.S.; Yu, H.C.; Peng, C.I. Two-vector-based modeless predictive current control for four-switch inverter-fed synchronous reluctance motors emulating the six-switch inverter operation. *IEEE Trans. Electron. Lett.* **2016**, *52*, 1244–1246. [CrossRef]
18. Ioannou, P.; Fidan, B. *Adaptive Control Tutorial*; SIAM: Philadelphia, PA, USA, 2006.
19. Astrom, K.J.; Wittenmark, B. *Adaptive Control*, 2nd ed.; Addison-Wesly: New York, NY, USA, 1995.
20. Maciejowski, J.M. *Predictive Control with Constraints*; Pearson Education Limited: London, UK, 2002.
21. Rodriquez, J.; Corts, P. *Predictive Control of Power Converters and Electrical Drives*; Wiley: London, UK, 2012.
22. Wang, L.; Chai, S.; Yoo, D.; Gan, L.; Ng, K. *PID and Predictive Control of Electrical Drives and Power Converters Using MATLAB/Simulink*; Wiley: Singapore, 2015.
23. Toliyat, H.A.; Campbell, S. *DSP-Based Electromechanical Motion Control*; CRC Press: New York, NY, USA, 2003.

energies

MDPI

Article

Design and Implementation of Position Sensorless Field-Excited Flux-Switching Motor Drive Systems

Tian-Hua Liu *, Muhammad Syahril Mubarok and Yu-Hao Xu

Department of Electrical Engineering, National Taiwan University of Science and Technology, Taipei 106, Taiwan; syahril.elmubarok@gmail.com (M.S.M.); eddiechesterbn@gmail.com (Y.-H.X.)

* Correspondence: Liu@mail.ntust.edu

Received: 20 June 2020; Accepted: 15 July 2020; Published: 16 July 2020

Abstract: Field-excited flux-switching motor drive systems have become more and more popular due to their robustness and lack of need for a permanent magnet. Three different types of predictive controllers, including a single-step predictive speed controller, a multi-step predictive speed controller, and a predictive current controller are proposed for sensorless flux-switching motor drive systems in this paper. By using a 1 kHz high-frequency sinusoidal voltage injected into the field winding and by measuring the a-b-c armature currents in the stator, an estimated rotor position that is near ±2 electrical degrees is developed. To improve the dynamic responses of the field-excited flux-switching motor drive system, predictive controllers are employed. Experimental results demonstrate the proposed predictive controllers have better performance than PI controllers, including transient, load disturbance, and tracking responses. In addition, the adjustable speed range of the proposed drive system is from 4 r/min to 1500 r/min. A digital signal processor, TMS-320F-2808, is used as a control center to carry out the rotor position estimation and the predictive control algorithms. Measured results can validate the theoretical analysis to illustrate the practicability and correctness of the proposed method.

Keywords: field-excited flux-switching motor; high-frequency injection; predictive controller; digital signal processor

---

1. Introduction

The flux-switching motor is a type of double-salient structured motor with two windings in the stator, including an armature winding and a field winding, which could be replaced by a permanent magnet. The flux-switching motor has several advantages, including a robust structure, a sinusoidal back-electromotive force (back-EMF) waveform, and a reasonable torque density [1,2]. Several researchers have investigated the design of different types of flux-switching motors [3–5]. Recently, field-excited flux-switching motors have become more and more popular due to there being no need for a permanent magnet and the good flux-weakening operational characteristics [6]. Many papers have investigated field-excited flux-switching motors. For example, Ullah et al. proposed a field-excited linear flux-switching motor [7]. Gaussens proposed an analytical method of air-gap modeling for a field-excited flux-switching motor, by which the flux linkage and torque were derived [8].

Several papers have studied the control of flux-switching motor drive systems. For instance, Zhao et al. employed a model predictive controller for a flux-switching drive system to decrease its torque and flux ripples and to enhance the quality of the drive system [9]. Moreover, Zhao et al. investigated the low copper-loss control of a field-excited flux-switching drive system to improve its adjustable speed range and efficiency [10]. Zhao et al. implemented vector control of a field-excited flux-switching drive to maximize its output torque by using particle swarm optimization [11]. Yang et al. studied flux-weakening control of a hybrid flux-switching motor to increase its output torque and high-speed operating range [12]. Wu et al. investigated field-oriented control and direct torque control for a five-phase flux-switching motor drive system with a fault-tolerant capability to improve its dynamics during faulty conditions [13]. Nguyen et al. proposed rotor position sensorless control of a field-excited flux-switching motor drive system using a high-frequency square-wave voltage injecting method to replace an encoder [14]. Zhang proposed model reference adaptive control to improve the dynamics of a sensorless flux-switching motor drive system [15].

Several researchers have proposed sensorless methods for motor drives. For example, Fan et al. proposed sensorless control of a five-phase interior permanent magnet synchronous motor (IPMSM) based on a high-frequency sinusoidal voltage injection [16]. Wang et al. investigated a position-sensorless control method at low speed for PMSM based on high-frequency signal injection into a rotating reference frame [17]. Compared to previously published papers [9–17], this paper is the first to propose predictive controllers for a sensorless field-excited flux-switching motor drive system. To the authors' best knowledge, this idea is an original idea. This is the first time that a multi-step predictive speed controller is developed and compared to a single-step predictive speed controller for a flux-switching motor drive system. This is another main contribution of the paper. The implemented drive system could be applied for grass cutters and vacuum cleaners due to its robustness, lack of need for a permanent magnet, and high torque. In addition, the high torque ripple, large volume, and serious acoustic noise may not be the main issues for the applications of vacuum cleaners and grass cutters.

2. Flux-Switching Motor

This paper investigates a 3-phase, 6-slot armature winding stator, 7-salient-tooth rotor, field-excited, flux-switching motor, which is shown in Figure 1. The armature winding is located inside the motor, and the salient-tooth rotor is located outside of the motor. In Figure 1, the A1, A2, B1, B2, C1, and C2 are armature windings, and F1, F2, F3, F4, F5, and F6 are field windings. All of them are located on the teeth of the stator. Between the stator and rotor, there is a non-uniform air gap.

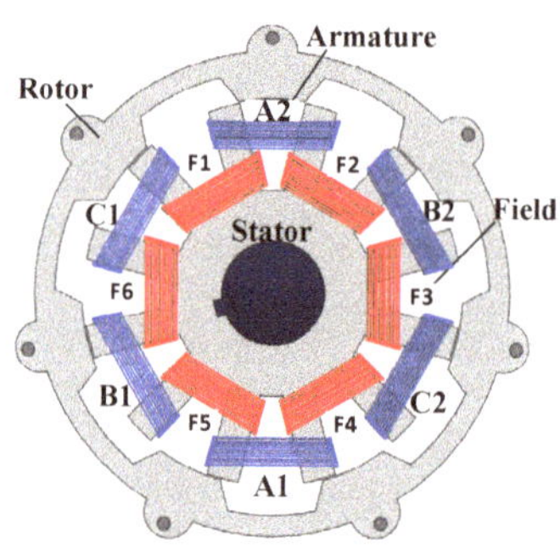

Figure 1. The structure of a flux-switching motor.

To explain the basic principle of the flux-switching motor, Figure 2a–d illustrate the different rotor positions between the rotor and the stator, in which S_1, S_2, and S_3 are the slot numbers of the stator. Figure 3 shows the induced flux linkage and its related back-EMF. Figure 2a illustrates that the flux linkage of the F1 field winding rises, but the flux linkage of the F2 field winding decreases. As a result, the total flux linkage in the A2 armature winding is zero. This situation is also indicated at the a-point in Figure 3. Here the flux linkage of the A2 armature winding is zero, but the back-EMF of the A2 armature winding reaches its maximum value. Figure 2b illustrates that both the F1 and F2 field windings provide rising flux at the A2 armature winding. As a result, the total flux of the A2 armature winding reaches its maximum value, and the back-EMF of the A2 armature winding is zero, which is shown at the b-point in Figure 3. Figure 2c illustrates that the flux linkage of the field winding F1 also rises, but the flux linkage of the field winding F2 also decreases. As a result, the total flux linkage of the A2 armature winding is zero, and the back-EMF of the A2 armature winding reaches its negative minimum value, which is shown at the c-point of Figure 3. Figure 2d illustrates that both the flux linkage of the F1 and F2 field windings decrease. As a result, the total flux linkage of the A2 armature winding reaches its negative minimum value, and the back-EMF of the A2 armature winding is zero. This is illustrated as the d-point in Figure 3. In Figure 3, ω_{re} is the electric rotor speed of the rotor and λ_m is the flux linkage of the stator when rotor rotates. According to the above analysis, it is feasible to generate a flux linkage waveform and a back-EMF waveform, which are sinusoidal and cosine waveforms, respectively, and are illustrated in Figure 3. By using the back-EMF from the armature winding, torque is generated and the motor rotates smoothly.

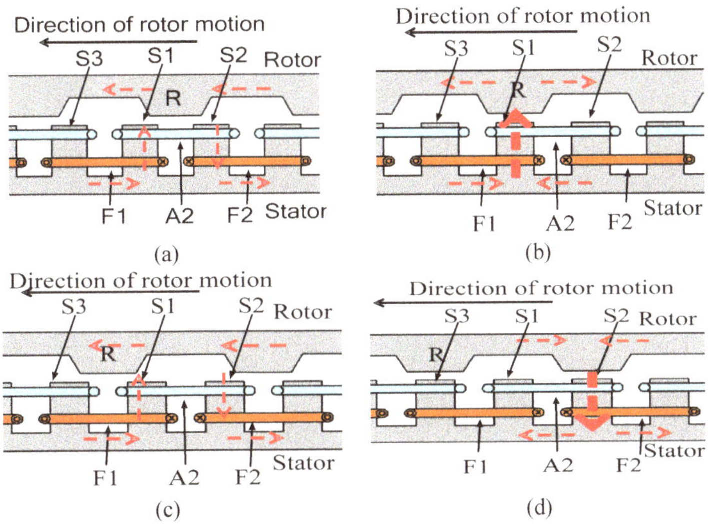

Figure 2. Rotor flux at different rotor positions. (**a**) R is aligned with the A2 axis; (**b**) R is between the F1 and A2 axes; (**c**) R is aligned with the F1 axis; (**d**) R is on the left of the F1 axis.

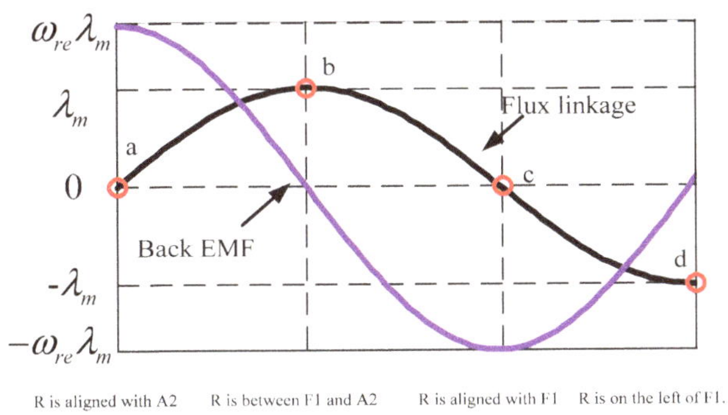

Figure 3. Flux linkage and back-EMF waveforms.

3. Mathematical Model of a Flux-Switching Motor

The synchronous frame d-q axis stator voltage of a flux-switching motor is expressed as follows:

$$\begin{bmatrix} v_d \\ v_q \end{bmatrix} = \begin{bmatrix} r_s + L_d \frac{d}{dt} & -\omega_{re} L_q & L_{df} \frac{d}{dt} \\ \omega_{re} L_d & r_s + L_q \frac{d}{dt} & \omega_{re} L_{qf} \end{bmatrix} \begin{bmatrix} i_d \\ i_q \\ i_f \end{bmatrix} \tag{1}$$

where v_d and v_q are the d- and q-axis voltages, r_s is the rotor resistance, L_d and L_q are the d- and q-axis self-inductances, L_{df} is the mutual inductance between the d-axis of the armature winding and the field winding, L_{qf} is the mutual inductance between the q-axis of the armature and the field winding, ω_{re} is the electrical rotor speed, $\frac{d}{dt}$ is the differential operator, i_d and i_q are the d- and q-axis currents, and i_f is the field current.

The flux linkage of the field winding, λ_f, can be expressed as follows:

$$\lambda_f = L_{ff} i_f + \frac{3}{2} L_{df} i_d \tag{2}$$

where λ_f is the total flux linkage of the field winding, L_{ff} is the self-inductance of the field winding, and L_{df} is the mutual inductance between the field winding and the d-axis winding. The voltage of the field winding is expressed as follows:

$$v_f = r_f i_f + L_{ff} \frac{d}{dt} i_f + \frac{3}{2} L_{df} \frac{d}{dt} i_d \tag{3}$$

where v_f is the voltage of the field winding, and r_f is the resistance of the field winding. The total torque is shown as follows:

$$T_e = \frac{3}{2} P_m \left[L_{dqf} i_f i_q + (L_d - L_q) i_d i_q \right] \tag{4}$$

The mechanical speed of the motor is

$$\frac{d}{dt} \omega_{rm} = \frac{1}{J_{st}} (T_e - T_L - B_{st} \omega_{rm}) \tag{5}$$

where ω_{rm} is the mechanical speed of the rotor, J_{st} is the inertia of the motor and load, T_e is the total output torque, T_L is external load, and B_{st} is the total viscous friction coefficient. The differential of the mechanical position of the motor is

$$\frac{d}{dt} \theta_{rm} = \omega_{rm} \tag{6}$$

4. Rotor Position Estimator Design

In this paper, assuming $L_{dh} = L_{qh} = L_{sh}$, the d-q-f axis high-frequency voltages and high- frequency currents can be described as follows:

$$\begin{bmatrix} v_{dh} \\ v_{qh} \\ v_{fh} \end{bmatrix} = j f_h \begin{bmatrix} L_{sh} & 0 & L_{msf} \\ 0 & L_{sh} & 0 \\ \frac{3}{2} L_{msf} & 0 & L_{ffh} \end{bmatrix} \begin{bmatrix} i_{dh} \\ i_{qh} \\ i_{fh} \end{bmatrix} \tag{7}$$

where v_{dh}, v_{qh}, and v_{fh} are the high-frequency d-axis, q-axis, and field winding voltages, L_{sh} is the self-inductance of the d- and q-axis, L_{msf} is the mutual inductance between the d-axis or q-axis and the field winding inductance, and L_{ffh} is the high-frequency self-inductance of the field winding. From Figure 4, the coordinate transformation between d-q-f and α-β-f can be expressed as

$$\begin{bmatrix} f_d \\ f_q \\ f_f \end{bmatrix} = \begin{bmatrix} \cos\theta_{re} & \sin\theta_{re} & 0 \\ -\sin\theta_{re} & \cos\theta_{re} & 0 \\ 0 & 0 & 1 \end{bmatrix} \begin{bmatrix} f_\alpha \\ f_\beta \\ f_f \end{bmatrix} = T(\theta_{re}) \begin{bmatrix} f_\alpha \\ f_\beta \\ f_f \end{bmatrix} \tag{8}$$

where f_d and f_q are the d- and q-axis voltages or currents, f_α and f_β are the α- and β-axis voltages or currents, and $T(\theta_{re})$ is the coordinate transformation matrix. Substituting Equation (8) into (7), one can obtain

$$\begin{bmatrix} v_{\alpha h} \\ v_{\beta h} \\ v_{fh} \end{bmatrix} = jf_h\, T(\theta_{re})^{-1} \begin{bmatrix} L_{sh} & 0 & L_{dqf} \\ 0 & L_{sh} & 0 \\ \frac{3}{2}L_{dqf} & 0 & L_{ffh} \end{bmatrix} T(\theta_{re}) \begin{bmatrix} i_{\alpha h} \\ i_{\beta h} \\ i_{fh} \end{bmatrix}$$
$$= jf_h \begin{bmatrix} \cos\theta_{re} & \sin\theta_{re} & 0 \\ -\sin\theta_{re} & \cos\theta_{re} & 0 \\ 0 & 0 & 1 \end{bmatrix}^{-1} \begin{bmatrix} L_{sh} & 0 & L_{dqf} \\ 0 & L_{sh} & 0 \\ \frac{3}{2}L_{dqf} & 0 & L_{ffh} \end{bmatrix} \begin{bmatrix} \cos\theta_{re} & \sin\theta_{re} & 0 \\ -\sin\theta_{re} & \cos\theta_{re} & 0 \\ 0 & 0 & 1 \end{bmatrix} \begin{bmatrix} i_{\alpha h} \\ i_{\beta h} \\ i_{fh} \end{bmatrix} \tag{9}$$

where $v_{\alpha h}$, $v_{\beta h}$, and v_{fh} are the $\alpha - axis$, $\beta - axis$, and field winding voltages, and $i_{\alpha h}$, $i_{\beta h}$, and i_{fh} are the $\alpha - axis$, $\beta - axis$, and field winding currents.

From Equation (9), after doing some mathematical processes, one can obtain

$$\begin{bmatrix} v_{\alpha h} \\ v_{\beta h} \\ v_{fh} \end{bmatrix} = jf_h \begin{bmatrix} L_{sh} & 0 & L_{dqf}\cos\theta_{re} \\ 0 & L_{sh} & L_{dqf}\sin\theta_{re} \\ \frac{3}{2}L_{dqf}\cos\theta_{re} & \frac{3}{2}L_{dqf}\sin\theta_{re} & L_{ffh} \end{bmatrix} \begin{bmatrix} i_{\alpha h} \\ i_{\beta h} \\ i_{fh} \end{bmatrix} \tag{10}$$

From Equation (10), one can derive the following equation:

$$\begin{bmatrix} i_{\alpha h} \\ i_{\beta h} \\ i_{fh} \end{bmatrix} = \frac{1}{jf_h(L_{sh}^2 L_{ffh} - \frac{3}{2}L_{dqf}^2)} \begin{bmatrix} L_{sh}L_{ffh} - \frac{3}{2}L_{dqf}^2 \sin\theta_{re}^2 & \frac{3}{2}L_{dqf}^2 \sin\theta_{re}\cos\theta_{re} & -\frac{3}{2}L_{sh}L_{dqf}\cos\theta_{re} \\ \frac{3}{2}L_{dqf}^2\sin\theta_{re}\cos\theta_{re} & L_{sh}L_{ffh} - \frac{3}{2}L_{dqf}^2\cos\theta_{re}^2 & -\frac{3}{2}L_{sh}L_{dqf}\sin\theta_{re} \\ -\frac{3}{2}L_{sh}L_{dqf}\cos\theta_{re} & -\frac{3}{2}L_{sh}L_{dqf}\sin\theta_{re} & L_{sh}^2 \end{bmatrix} \begin{bmatrix} v_{\alpha h} \\ v_{\beta h} \\ v_{fh} \end{bmatrix} \tag{11}$$

In this paper, a high-frequency voltage $v_{fh}(t)$, which is injected into the field winding, is expressed as follows:

$$\begin{bmatrix} v_{\alpha h} \\ v_{\beta h} \\ v_{fh} \end{bmatrix} = \begin{bmatrix} 0 \\ 0 \\ v_{fh}(t) \end{bmatrix} = \begin{bmatrix} 0 \\ 0 \\ V_{inj}\sin f_h t \end{bmatrix} \tag{12}$$

Substituting Equation (12) into (11) and doing some mathematical processes, one can obtain the high-frequency α-axis and β-axis currents as follows:

$$\begin{bmatrix} i_{\alpha h} \\ i_{\beta h} \\ i_{fh} \end{bmatrix} = \frac{V_{inj}}{jf_h(L_{sh}^2 L_{ffh} - \frac{3}{2}L_{dqf}^2)} \begin{bmatrix} -\frac{3}{2}L_{sh}L_{dqf}\cos\theta_{re} + \frac{3}{2}L_{sh}L_{dqf}\cos\theta_{re}\cos(2f_h t) \\ -\frac{3}{2}L_{sh}L_{dqf}\sin\theta_{re} - \frac{3}{2}L_{sh}L_{dqf}\cos(2f_h t) \\ L_{sh}^2 - L_{sh}^2\cos(2f_h t) \end{bmatrix} \tag{13}$$

After using a low-pass filter to remove the high-frequency components, one can obtain the following equation:

$$\begin{bmatrix} i_{\alpha h\_p} \\ i_{\beta h\_p} \\ i_{fh\_p} \end{bmatrix} = \frac{V_{inj}}{jf_h(L_{sh}^2 L_{ffh} - \frac{3}{2}L_{dqf}^2)} \begin{bmatrix} -\frac{3}{2}L_{sh}L_{dqf}\cos\theta_{re} \\ -\frac{3}{2}L_{sh}L_{dqf}\sin\theta_{re} \\ L_{sh}^2 \end{bmatrix} \tag{14}$$

where $i_{\alpha h\_p}$, $i_{\beta h\_p}$, and i_{fh_p} are the high-frequency currents after a band-pass filtering. From Equation (14), the estimated rotor position, $\widehat{\theta}_{re}$, can be derived as follows:

$$\widehat{\theta}_{re} \cong \tan^{-1}\left|\frac{i_{\beta h\_p}}{i_{\alpha h\_p}}\right| \cong \tan^{-1}\frac{\left(\dfrac{V_{inj}\frac{3}{2}L_{sh}L_{dqf}\sin\theta_{re}}{jf_h(L_{sh}^2L_{ffh}-\frac{3}{2}L_{dqf}^2)}\right)}{\left(\dfrac{V_{inj}\frac{3}{2}L_{sh}L_{dqf}\cos\theta_{re}}{jf_h(L_{sh}^2L_{ffh}-\frac{3}{2}L_{dqf}^2)}\right)} \tag{15}$$

The estimated speed can be described as follows:

$$\widehat{\omega}_{re}(k) = \frac{\widehat{\theta}_{re}(k) - \widehat{\theta}_{re}(k-1)}{\Delta T} \tag{16}$$

where $\widehat{\omega}_{re}(k)$ is the estimated speed and ΔT is the sampling interval. In this paper, a speed state estimator is constructed [18]. A high-order speed estimator is employed to avoid the high-frequency noise caused by the difference operator. Figure 4 shows the relationship between the estimated rotor position $\hat{\theta}_{re}$ and the real position θ_{re}. Figure 5 demonstrates the control block diagram of the sensorless drive system, which includes the current control of the armature winding, the current control of the field winding, the rotor position estimation, and the rotor speed control.

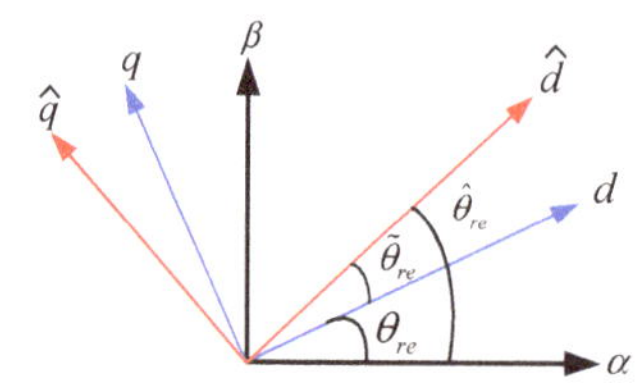

Figure 4. Estimated and real rotor positions.

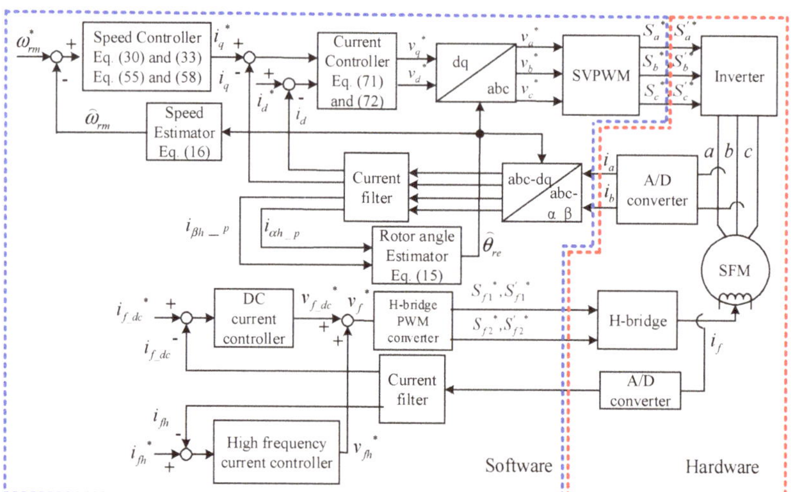

Figure 5. Control block diagram of proposed sensorless drive system.

5. Predictive Controller Design

Predictive control began in the late 1970s and has been developed significantly since then. It has been widely used in chemical processes, robotics, drying towers, and steam generators [19,20]. Recently, it has been used for power converters and motor drives [21–23]. The predictive speed-loop controllers include single-step and multi-step control inputs. The design methods are discussed as follows.

5.1. One-Step Predictive Speed Controller

By assuming that the external load T_L is zero, the discrete dynamics of the speed can be expressed as

$$\begin{aligned} \omega_{rm}(n+1) &= e^{-\frac{B_{st}}{J_{st}}T_s}\omega_{rm}(n) + \frac{1}{B_{st}}(1 - e^{-\frac{B_{st}}{J_{st}}T_s})K_{st}i_q(n) \\ &= a_{rm}\omega_{rm}(n) + b_{rm}i_q(n) \end{aligned} \tag{17}$$

and

$$a_{rm} = e^{-\frac{B_{st}}{J_{st}}T_s} \tag{18}$$

$$b_{rm} = \frac{1}{B_{st}}\left(1 - e^{-\frac{B_{st}}{J_{st}}T_s}\right)K_{st} \tag{19}$$

where K_{st} is the torque constant of the motor, a_{rm}, and b_{rm} are the discrete time parameters of the motor, and n is the step number. By defining $x_{rm}(n) = \omega_{rm}(n)$, and $u(n) = i_q(n)$, one can obtain

$$x_{rm}(n+1) = a_{rm}x_{rm}(n) + b_{rm}u(n) \tag{20}$$

The output $y_{rm}(n)$ is defined as

$$y_{rm}(n) = x_{rm}(n) \tag{21}$$

By using Equation (20) and taking one step back, one can obtain

$$x_{rm}(n) = a_{rm}x_{rm}(n-1) + b_{rm}u(n-1) \tag{22}$$

By subtracting (20) from (22), one can obtain

$$\begin{aligned} \Delta x_{rm}(n+1) &= x_{rm}(n+1) - x_{rm}(n) \\ &= a_{rm}\Delta x_{rm}(n) + b_{rm}\Delta u(n) \end{aligned} \tag{23}$$

and

$$\Delta x_{rm}(n) = x_{rm}(n) - x_{rm}(n-1) \tag{24}$$

The input difference $\Delta u(n)$ is defined as

$$\Delta u(n) = u(n) - u(n-1) \tag{25}$$

The output $\Delta y_{rm}(n+1)$ is defined as follows:

$$\Delta y_{rm}(n+1) = y_{rm}(n+1) - y_{rm}(n) = x_{rm}(n+1) - x_{rm}(n) \tag{26}$$

Then we can obtain

$$\begin{aligned} y_{rm}(n+1) &= \Delta y_{rm}(n+1) + y_{rm}(n) \\ &= \Delta x_{rm}(n+1) + y_{rm}(n) \\ &= a_{rm}\Delta x_{rm}(n) + b_{rm}\Delta u(n) + y_{rm}(n) \end{aligned} \tag{27}$$

After that, we can define a cost function as follows [19,20]:

$$\begin{aligned}\Omega_{sp}(n) &= (\omega^*_{rm}(n+1) - y_{sm}(n+1))^2 + q\Delta u^2(n) \\ &= (\Delta\omega_{rm}(n+1))^2 + q(\Delta u_n(n))^2\end{aligned} \tag{28}$$

where q is the weighting factor between the $(\Delta u_n(n))^2$ and the $(\Delta\omega_{rm}(n+1))^2$. Then, by taking the differential of the $\Omega_{sp}(n)$ to the $\Delta u(n)$ and setting its result to be zero, one can obtain

$$\Delta u(n) = \frac{b_{rm}\omega^*_{rm}(n+1) - a_{rm}b_{rm}\Delta\omega_{rm}(n) - b_{rm}\omega_{rm}(n)}{b^2_{rm} + q} \tag{29}$$

The $\Delta i_q(n)$ is used to replace the $\Delta u(n)$, and then Equation (29) can be rewritten as follows:

$$\begin{aligned}\Delta i_q(n) &= \frac{b_{rm}(\omega^*_{rm}(n+1) - \omega_{rm}(n)) - a_{rm}b_{rm}\Delta\omega_{rm}(n)}{b^2_{rm} + q} \\ &= k_1(\omega^*_{rm}(n+1) - \omega_{rm}(n)) - k_2\Delta\omega_{rm}(n)\end{aligned} \tag{30}$$

and

$$k_1 = \frac{b_{rm}}{b^2_{rm} + q} \tag{31}$$

$$k_2 = \frac{a_{rm}b_{rm}}{b^2_{rm} + q} \tag{32}$$

Finally, the q-axis current command is

$$i^*_q(n) = i^*_q(n-1) + \Delta i^*_q(n) \tag{33}$$

From Equations (30)–(33), the control block diagram of the single-step predictive speed-loop controller is demonstrated in Figure 6.

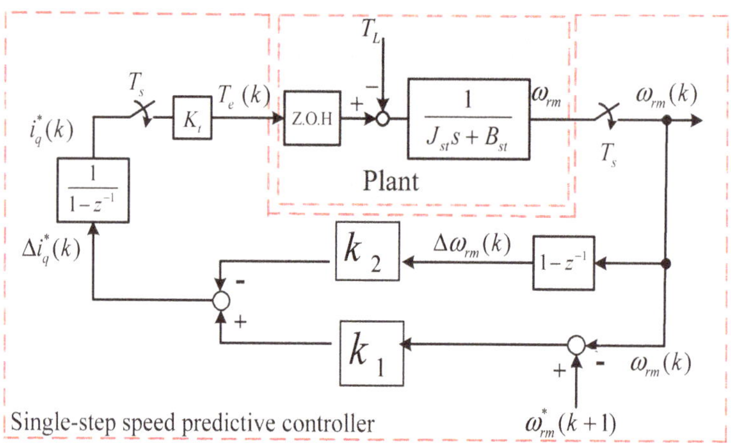

Figure 6. Single-step predictive speed controller.

5.2. Multi-Step Predictive Speed Controller

In this paper, to implement the multi-step predictive speed controller, a two-step control horizon and a two-step prediction horizon are developed. The details are discussed as follows. The $(n+1)$ augmented model is

$$X_{sm}(n+1) = A_{sm}X_{sm}(n) + B_{sm}\Delta u(n) \tag{34}$$

and

$$A_{sm} = \begin{bmatrix} a_{rm} & 0 \\ a_{rm} & 1 \end{bmatrix} \tag{35}$$

$$B_{sm} = \begin{bmatrix} b_{rm} \\ b_{rm} \end{bmatrix} \tag{36}$$

The augmented state variable for the $(n+2)$ step is

$$\begin{aligned} X_{sm}(n+2) &= A_{sm}X_{sm}(n+1) + B_{sm}\Delta u(n+1) \\ &= A_{sm}^2 X_{sm}(n) + A_{sm}B_{sm}\Delta u(n) + \Delta u(n+1) \end{aligned} \tag{37}$$

In addition, the $(n+1)$ step predictive output is

$$y_{sm}(n+1) = C_{sm}A_{sm}X_{sm}(n) + C_{sm}B_{sm}\Delta u(n) \tag{38}$$

The $(n+2)$ step predictive output is

$$y_{sm}(n+2) = C_{sm}A_{sm}^2 X_{sm}(n) + C_{sm}A_{sm}B_{sm}\Delta u(n) + C_{sm}B_{sm}\Delta u(n+1) \tag{39}$$

Then, a new predictive output vector can be defined as follows:

$$Y_{sm} = \begin{bmatrix} y_{sm}(n+1) \\ y_{sm}(n+2) \end{bmatrix} \tag{40}$$

and

$$X_{sm} = \begin{bmatrix} x_{sm}(n+1) \\ x_{sm}(n+2) \end{bmatrix} \tag{41}$$

After that, one can obtain

$$Y_{sm} = F_{sm}X_{sm} + \Theta_{sm}\Delta U_{sm} \tag{42}$$

and

$$F_{sm} = \begin{bmatrix} C_{sm}A_{sm} \\ C_{sm}A_{sm}^2 \end{bmatrix} \tag{43}$$

$$\Theta_{sm} = \begin{bmatrix} C_{sm}B_{sm} & 0 \\ C_{sm}A_{sm}B_{sm} & C_{sm}B_{sm} \end{bmatrix} \tag{44}$$

$$\Delta U_{sm} = \begin{bmatrix} \Delta u(n) \\ \Delta u(n+1) \end{bmatrix} \tag{45}$$

Now, one can define the performance index as follows [17–19]:

$$\Psi_{sp} = (R_{sm}^* - Y_{sm})^T (R_{sm}^* - Y_{sm}) + \Delta U_{sm}^T Q \Delta U_{sm} \tag{46}$$

and

$$R_{sm}^* = \begin{bmatrix} \omega_{rm}^*(n+1) \\ \omega_{rm}^*(n+2) \end{bmatrix} \tag{47}$$

$$Q = \begin{bmatrix} q & 0 \\ 0 & q \end{bmatrix} \tag{48}$$

By substituting (42) into (46) and doing the $\dfrac{\partial \Psi_{sp}}{\partial \Delta U_{sm}}$ processes, one can obtain

$$\frac{\partial \Psi_{sp}}{\partial \Delta U_{sm}} = -2\Theta_{sm}^{T}(R_{sm}^{*} - F_{sm}X_{sm}) + 2\left(\Theta_{sm}^{T}\Theta_{sm} + Q\right)\Delta U_{sm} = 0 \tag{49}$$

After that, one can obtain the optimal ΔU_{sm} as

$$\Delta U_{sm} = \left(\Theta_{sm}^{T}\Theta_{sm} + Q\right)^{-1}\left(\Theta_{sm}^{T}R_{sm}^{*} - \Theta_{sm}^{T}F_{sm}X_{sm}\right) \tag{50}$$

By substituting (41), (43)–(44), and (47)–(48) into Equation (50), one can finally derive the following equations:

$$\begin{bmatrix} \Delta u(n) \\ \Delta u(n+1) \end{bmatrix} = \begin{bmatrix} \frac{g_{rm}}{f_{rm}} \\ \frac{h_{rm}}{f_{rm}} \end{bmatrix} \tag{51}$$

and

$$f_{rm} = b_{rm}^{4} + q^{2} + b_{rm}^{2}q(a_{rm}^{2} + 2a_{rm} + 3) \tag{52}$$

$$\begin{aligned} g_{rm} &= b_{rm}^{3}(\omega_{rm}^{*}(n+1) - \omega_{rm}(n)) + a_{rm}b_{rm}q(\omega_{rm}^{*}(n+2) - \omega_{rm}(n)) \\ &+ b_{rm}q(\omega_{rm}^{*}(n+2) - 2\omega_{rm}(n)) - a_{rm}b_{rm}^{3}\Delta\omega_{rm}(n) \\ &- b_{rm}q\Delta\omega_{rm}(n)(a_{rm}^{3} + 2a_{rm}^{2} + 2a_{rm}) \end{aligned} \tag{53}$$

$$\begin{aligned} h_{rm} &= b_{rm}^{3}(\omega_{rm}^{*}(n+2) - a_{rm}\omega_{rm}^{*}(n+1) - \omega_{rm}^{*}(n+1) + a_{rm}\omega_{rm}(n)) \\ &+ b_{rm}q(\omega_{rm}^{*}(n+2) - \omega_{rm}(n)) - a_{rm}b_{rm}q\Delta\omega_{rm}(n)(a_{rm} + 1) \end{aligned} \tag{54}$$

The control output in this paper can be rearranged as follows:

$$\begin{bmatrix} \Delta i_{q}(n) \\ \Delta i_{q}(n+1) \end{bmatrix} = \begin{bmatrix} \frac{g_{rm}}{f_{rm}} \\ \frac{h_{rm}}{f_{rm}} \end{bmatrix} \tag{55}$$

$$i_{q}^{*}(n) = i_{q}^{*}(n-1) + \Delta i_{q}^{*}(n) \tag{56}$$

$$i_{q}^{*}(n+1) = i_{q}^{*}(n) + \Delta i_{q}^{*}(n+1) \tag{57}$$

Combining (56) and (57), one can provide the control input q-axis current command as follows:

$$i_{qc}^{*}(n) = \rho i_{q}^{*}(n) + (1-\rho)i_{q}^{*}(n+1) \tag{58}$$

where ρ is the weighting factor of the control input power. From Equations (51)–(58), one can obtain the multi-step predictive speed controller, which is displayed in Figure 7.

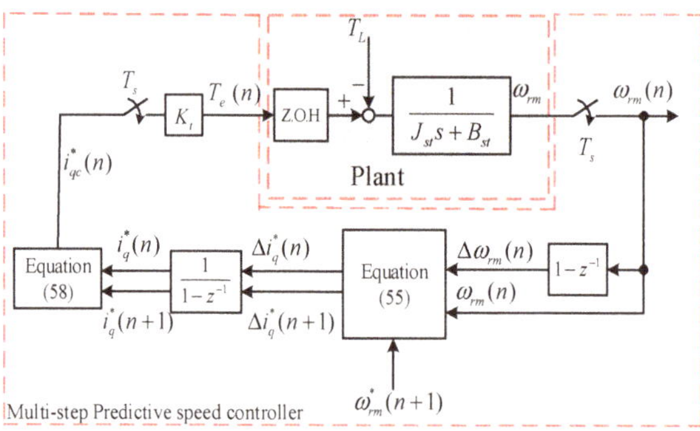

Figure 7. Multi-step predictive speed controller.

5.3. Predictive Current Loop Controller

From Equation (1), one can easily derive the following d-q axis current equations:

$$\frac{d}{dt}i_d = \frac{1}{L_d}\left(v_d - r_s i_d + \omega_{re} L_q i_q\right) \tag{59}$$

and

$$\frac{d}{dt}i_q = \frac{1}{L_q}\left(v_q - r_s i_q - \omega_{re}(L_d i_d + \lambda_m)\right) \tag{60}$$

From Equations (59) and (60), and by inserting a zero-order-hold and taking the discrete form, one can obtain

$$i_d(n+1) = e^{-\frac{r_s}{L_d}T_{cu}} i_d(n) + \frac{1 - e^{-\frac{r_s}{L_d}T_{cu}}}{r_s}\left[v_d(n) + \omega_{re}(n) L_q i_q(n)\right] \tag{61}$$

and

$$i_q(n+1) = e^{-\frac{r_s}{L_q}T_{cu}} i_q(n) + \frac{1 - e^{-\frac{r_s}{L_q}T_{cu}}}{r_s}\left[v_q(n) - \omega_{re}(n)(L_d i_d(k) + \lambda_m)\right] \tag{62}$$

$$u_d(n) = v_d(n) + \omega_{re}(n) L_q i_q(n) \tag{63}$$

and

$$u_q(n) = v_q(n) - \omega_{re}(n)(L_d i_d(n) + \lambda_m) \tag{64}$$

where $u_d(n)$ and $u_q(n)$ are the d- and q-axis current loop control inputs, and by finally substituting (63)–(64) into (61)–(62), one can obtain

$$i_d(n+1) = a_{cd} i_d(n) + b_{cd} u_d(n) \tag{65}$$

and

$$i_q(n+1) = a_{cq} i_q(n) + b_{cq} u_q(n) \tag{66}$$

In Equations (65) and (66), a_{cd}, b_{cd}, a_{cq}, and b_{cq}, which are the parameters of the IPMSM, can be expressed as follows:

$$a_{cd} = e^{-\frac{r_s}{L_d}T_c} \tag{67}$$

$$b_{cd} = \frac{1 - e^{-\frac{r_s}{L_d}T_c}}{r_s} \tag{68}$$

$$a_{cq} = e^{-\frac{r_s}{L_q}T_c} \tag{69}$$

and

$$b_{cq} = \frac{1 - e^{-\frac{r_s}{L_q}T_c}}{r_s} \tag{70}$$

By using similar processes [22], one can obtain

$$v_d^*(n) = \frac{z}{z-1}\left(\frac{b_{cd} q (id^*(n+1) - i_d(n))}{b_{cd}^2 q + r} - \frac{a_{cd} b_{cd} q \Delta i_d(n)}{b_{cd}^2 q + r}\right) - \omega_{re}(n)\left(L_q i_q(n)\right) \tag{71}$$

and

$$v_q{}^*(n) = \frac{z}{z-1}\left(\frac{b_{cq}q\left(i_q^*(n+1) - i_q(n)\right)}{b_q^2 q + r} - \frac{a_{cq}b_{cq}q\Delta i_q(n)}{b_{cq}^2 q + r}\right) + \omega_{re}(n)(L_d i_d(n) + \lambda_m) \quad (72)$$

Then, from Equations (71) and (72), the block diagram of the proposed predictive current control can be obtained as shown in Figure 8.

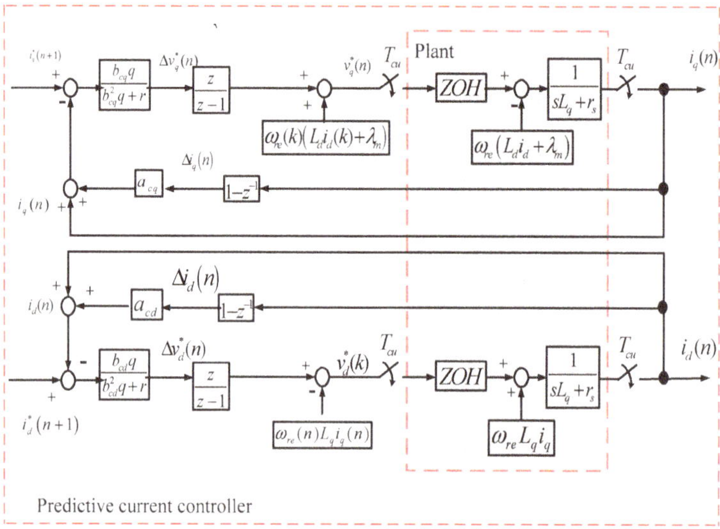

Figure 8. Current loop predictive controller.

6. Implementation

To evaluate the correctness and feasibility of the proposed method, a field-excited flux-switching sensorless flux-switching motor drive system is implemented. The implemented block diagram of the drive system is displayed in Figure 9a, which includes a voltage-source inverter, a field-excited flux-switching motor, an H-bridge circuit that controls the excited field winding, a digital signal processor, Hall-effect sensors, and A/D converters. The switching frequency of the inverter is 10 kHz, and the switching frequency of the H-bridge circuit is 20 kHz. In addition, the sampling time of the current control is 100 μs and the sampling time of the speed control is 1 ms. The DC voltages of the inverter and H-bridge are both 250 V. The digital signal processor (DSP) is manufactured by Texas Instruments, type TMS-320-F2808 [23]. To obtain a closed-loop high performance drive system, the DSP reads the a-phase current, b-phase current, and field winding current via A/D converters. After that, the DSP executes the rotor position estimation and all of the predictive control algorithms, and it also determines the inverter and H-bridge triggering signals.

A photograph of the implemented hardware circuit is displayed in Figure 9b, including a DSP, two drivers, two Hall-effect current sensors, an inverter, and an H-bridge circuit. The insulated gate bipolar transistor (IGBT) modules are made by the Mitsubishi Company, type CM200dy-12NF. The photo-couple gate drivers are manufactured by Avago Company, type HCPL-3120. The Hall-effect sensors are manufactured by the Swiss LEM company, type LA25-NP with a 100 kHz bandwidth. The A/D converters with fast conversion time are made by Analog Devices, type AD7655. Figure 9c displays the photograph of the flux-switching motor, which is connected to the dynamometer, in which a DC generator is used as an external load. The field-excited flux-switching motor is a seven stator salient teeth motor, with rated specifications of a power of 540 W, a speed of 600 r/min, and a current of 6 A. Its parameters include a stator resistance of 1.3 Ω, a d-axis inductance of 18.9 mH, a q-axis inductance of 23 mH, an inertia of 0.0143 kg-m^2, and a friction coefficient of 0.0047 N.m.s/rad. The motor was assembled in our laboratory because it is not available on the market.

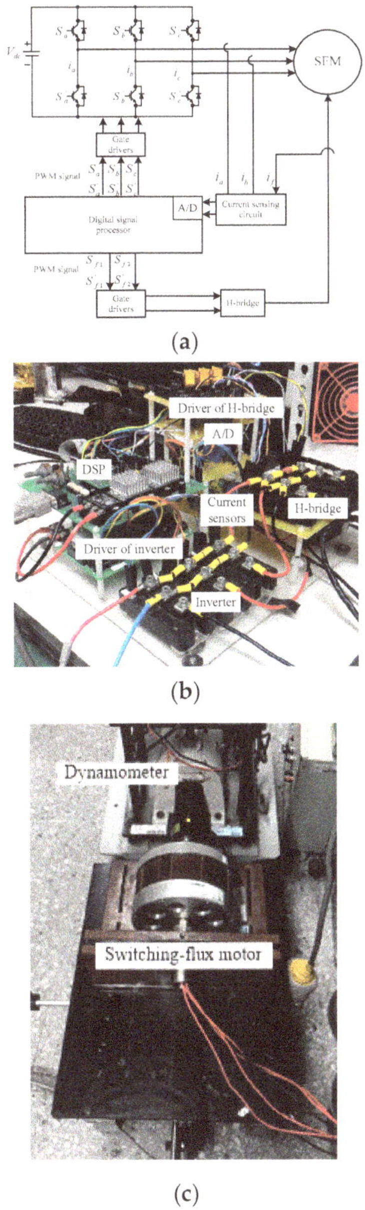

Figure 9. Photographs of the implementation system. (**a**) Block diagram, (**b**) hardware circuits, and (**c**) motor and dynamometer.

The details of the control algorithms are shown in Figure 10a–d, which are the flowcharts of the DSP. Figure 10a is the flowchart of the main program, which waits for the zero-voltage vector to detect currents. Figure 10b is the speed-loop interrupt service routine, which uses Equation (16) to compute

$\hat{\omega}_{rm}$ and Equations (30) and (33) to determine i_q^*. Figure 10c is the current loop interrupt service routine, which uses Equation (8) to execute the coordinate transformation, the band-pass filtering to obtain $i_{\alpha h}$, and finally computes $\hat{\theta}_{re}$. Figure 10d is the flowchart of the H-bridge field current control. The principle is measuring the current and comparing it with the field current command. Then, proportional-integral (PI) controller is used as the current controller to obtain the field voltage. Finally, the high-frequency voltage is injected and pulse-width modulation (PWM) is executed.

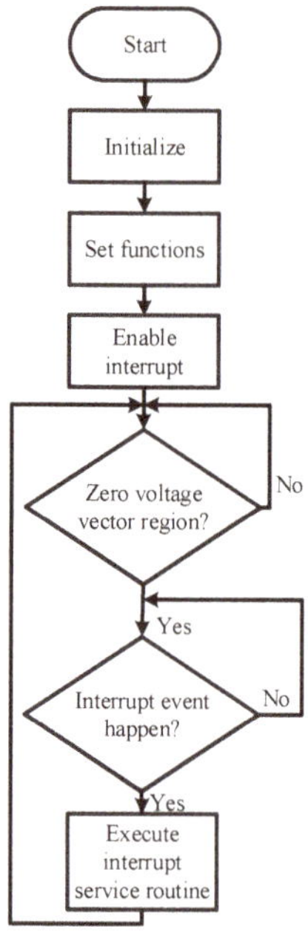

(a)

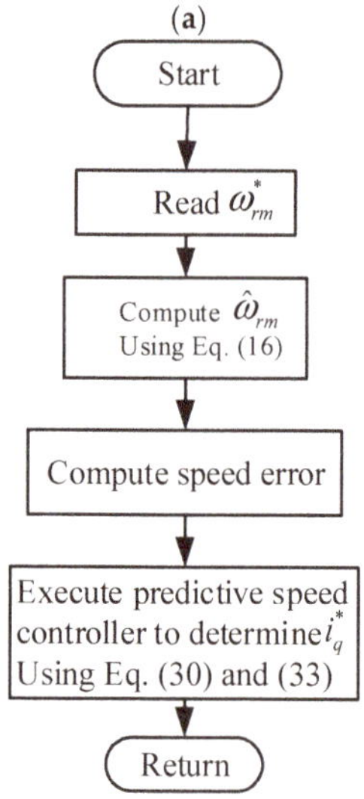

(b)

Figure 10. *Cont.*

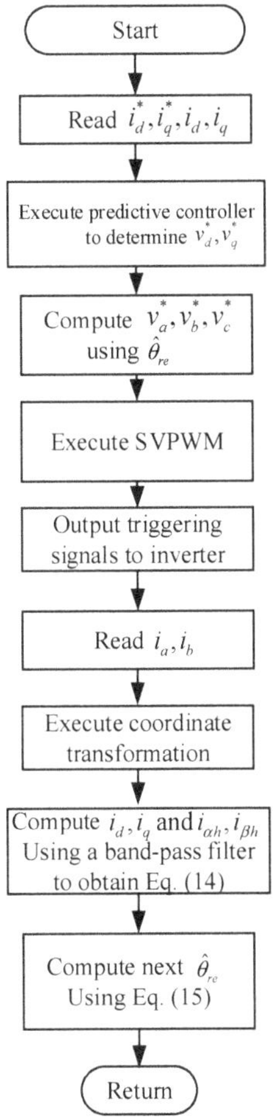

(**c**)

Figure 10. *Cont.*

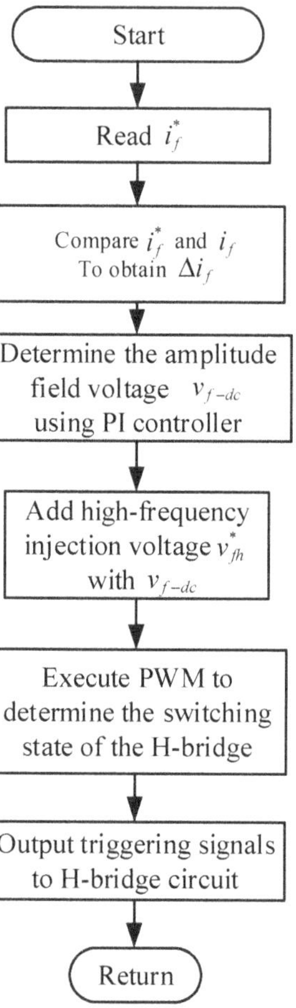

(**d**)

Figure 10. Digital signal processor (DSP) flowcharts. (**a**) Main program, (**b**) current loop interrupt service routine, (**c**) speed-loop interrupt service routine, and (**d**) H-bridge field current control.

7. Experimental Results

To verify the proposed methods, our experimental results are illustrated in this section. The input DC voltage of the inverter and the H-bridge are both 250 V. The switching frequency of the inverter is 10 kHz, and the switching frequency of the H-bridge is 20 kHz. The sampling time of the a-b-c axis current control is 100 μs, and the sampling time of the speed control is 1 ms. The selection of the sampling intervals is based on the required computation time of the current loop and the speed loop. In addition, a synchronous relationship, in which the current loop executes 10 times and then the speed loop executes one time, is employed in this paper. The injection high-frequency voltage of the field winding is ±25V, which is near 10% of the DC bus voltage of the H-bridge.

Figure 11a demonstrates the measured a-phase current at 300 r/min and a 2 N.m load. The current is influenced by the different switching frequencies of the inverter and field winding. The measured a-phase current is a near sinusoidal waveform with high-frequency harmonics. The b-phase and c-phase are similar to the a-phase with a 120 degree phase shift. Figure 11b demonstrates the measured current of the field winding, which is clearly influenced by an injection of high-frequency voltage, which is 1 kHz. Figure 11c demonstrates the measured 1 kHz and 25 V high-frequency voltage, which is used to inject into the field winding.

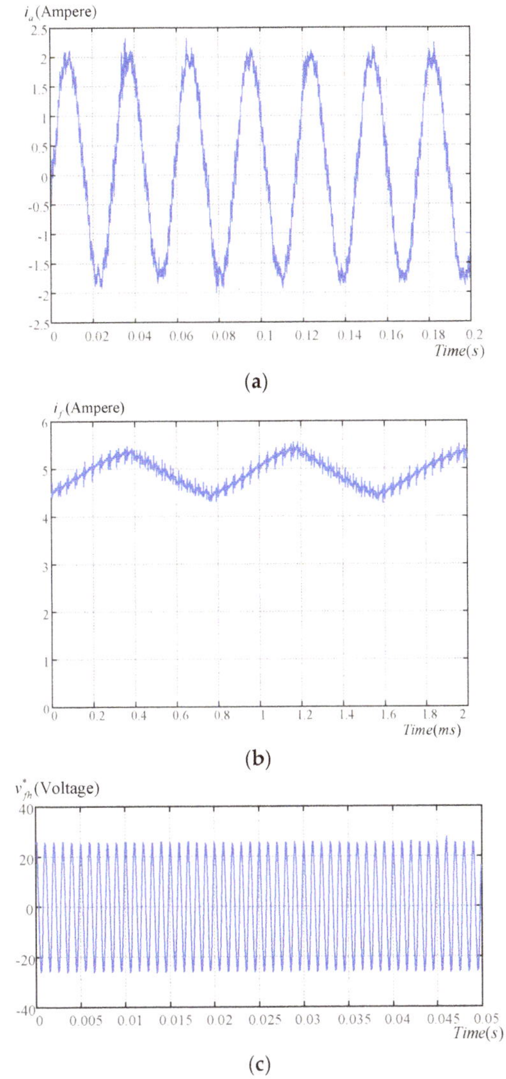

Figure 11. Measured signals at 300 r/min and 2 N.m external load; (**a**) a-phase current, (**b**) field-excited winding current, and (**c**) high-frequency voltage.

Figure 12a demonstrates the measured speed responses of step input at 300 r/min. The rise time is 0.23 s, and the speed reaches its steady-state condition at 0.41 s. In addition, the estimated speed $\widehat{\omega}_{rm}$ is very close to the speed ω_{rm}. Figure 12b demonstrates the estimated and real rotor positions and both of them are very close. Figure 12c demonstrates the errors of the estimated position and they are near ±2 electrical degrees.

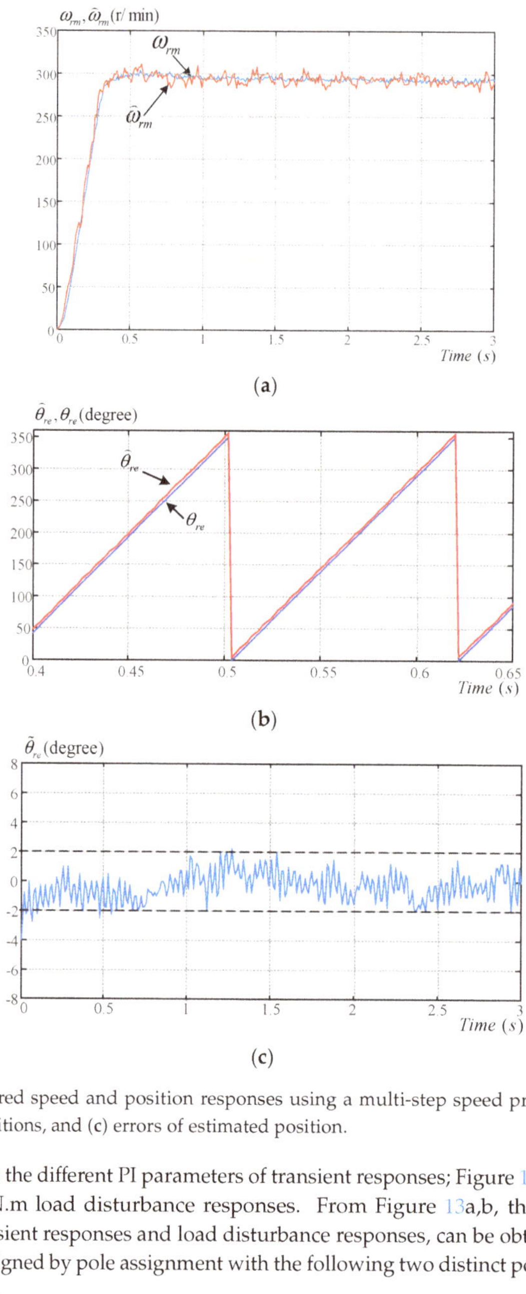

Figure 12. Measured speed and position responses using a multi-step speed predictive controller. (**a**) Speeds, (**b**) positions, and (**c**) errors of estimated position.

Figure 13a shows the different PI parameters of transient responses; Figure 13b shows the different PI parameters of 2 N.m load disturbance responses. From Figure 13a,b, the best PI parameters, considering both transient responses and load disturbance responses, can be obtained as $K_p = 0.1$ and $K_I = 4.5$, which is designed by pole assignment with the following two distinct poles, $P_1 = -1.9 + j12.4$ and $P_2 = -1.9 - j12.4$.

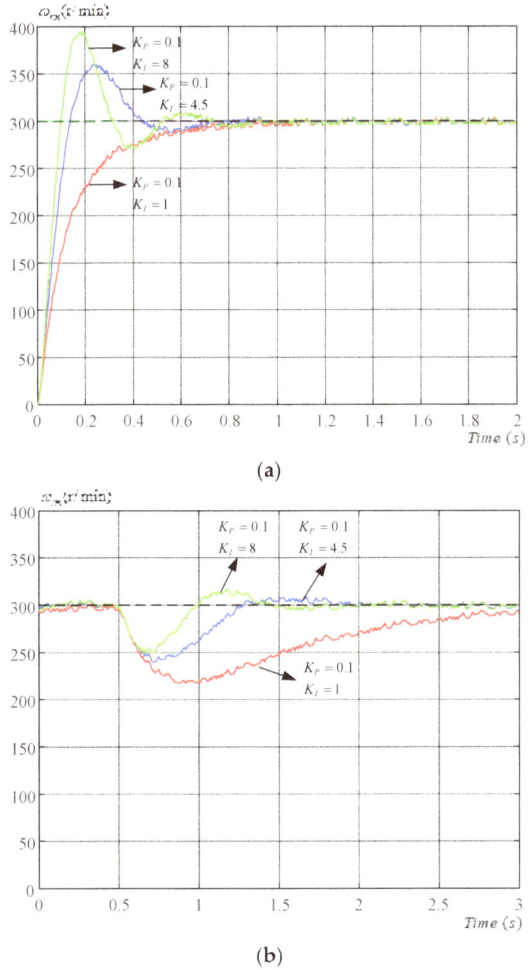

Figure 13. Measured speed responses with different proportional-integral (PI) control parameters. (**a**) Transient and (**b**) load disturbance.

Figure 14a demonstrates the measured speed responses using different controllers, including a multi-step predictive speed controller ($H_P = 2$), a single-step predictive speed controller ($H_P = 1$), and a PI speed controller. As one can observe, the multi-step predictive speed controller has a very smooth response, but the single-step controller has a quicker response with larger overshoot. The PI controller performs the worst, with the slowest response and largest overshoot. Figure 14b demonstrates the measured load disturbance responses. The multi-step controller has a very similar response to the single-step controller. The PI controller, however, has the largest speed drop and the slowest recovery time.

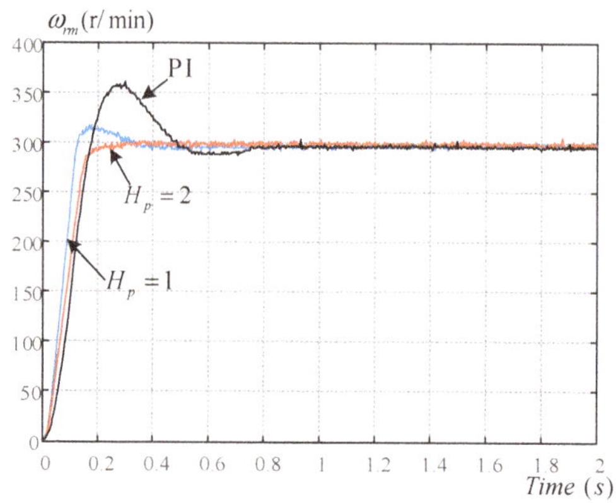

(a)

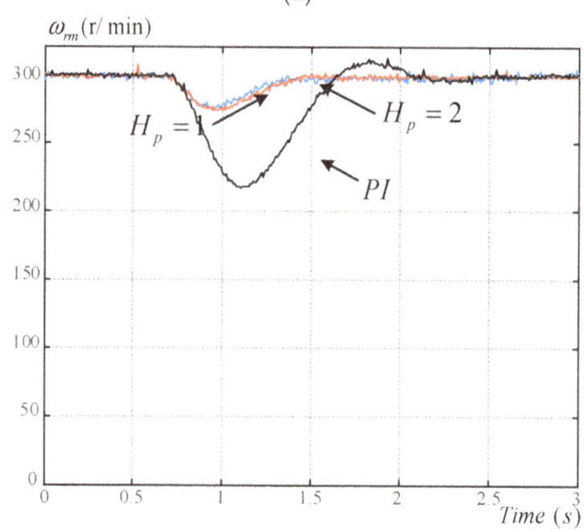

(b)

Figure 14. Measured speed responses. (**a**) Transient and (**b**) load disturbance.

Figure 15a demonstrates the measured speed responses of a triangular command by using an encoder for comparison. The multi-step controller has a similar response to the single-step controller. Figure 15b demonstrates the measured speed responses of a triangular command by using the proposed sensorless method. The speed ripples are increased when compared with using an encoder. Figure 16a demonstrates the measured speed responses of a trapezoidal command using an encoder. The multi-step controller has similar responses to the single-step controller. Figure 16b demonstrates the measured speed responses of a trapezoidal command using the proposed sensorless method. Figure 16c shows the measured responses of a step-input command using the proposed sensorless method. The multi-step controller with a small weighting factor has a slower but smoother response than the multi-step controller, which has a weighting factor of 1.

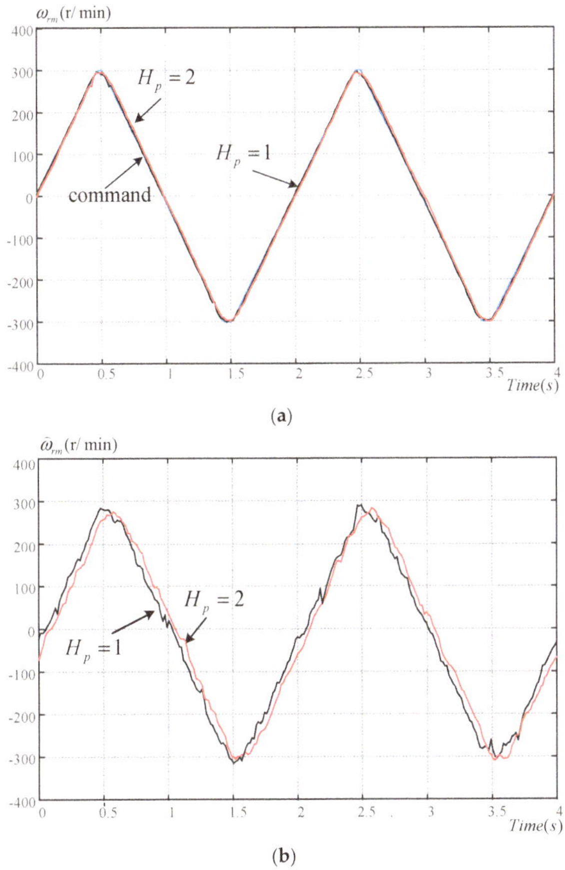

Figure 15. Measured triangular speed tracking responses. (**a**) Using an encoder and (**b**) using the proposed senseless method.

Figure 17a–c demonstrate the comparison of different speed operations. Figure 17a demonstrates the highest speed by using a field-weakening control, which is 1550 r/min with an 1 N.m external load. The control method effectively reduces the d-axis flux to extend the operational speed range. In addition, Figure 17a also demonstrates the highest operational speed without using a field-weakening control, which is 1080 r/min. Figure 17b demonstrates the mid-speed operational range from 100 to 600 r/min with a 2 N.m external load. All of the different speeds have linear responses. Figure 17c demonstrates the lowest operational speed, which is near 4 r/min with a 0.5 N.m external load. The lowest speed has obvious speed ripples. Figure 18a demonstrates the measured line-to-line voltage, v_{ab}, which has high-frequency PWM modulation pulses. Figure 18b demonstrates the excited winding voltage, v_f, which is controlled by an H-bridge circuit. Figure 18c demonstrates the excited winding current, i_f, during the transient time interval of the field current regulated control, which is controlled by an H-bridge circuit. The additional high-frequency flux on the core created core loss, which increases core loss and reduces the efficiency of the motor. Roughly speaking, this increased high frequency core loss is near 2% of the total losses of the motor. Comparing the torque and efficiency for analogously dimension PM and induction motors, for a 0.5 KW motor, the efficiency of a PM is near 90%, the efficiency of an induction motor is near 87%, and the efficiency of a field-excited flux-switching motor is 87%. For a 0.5 kW motor, the torque of a PM is 100% as a reference, the torque of an induction motor is 90%, and the torque of a field-excited flux-switching motor is 93%.

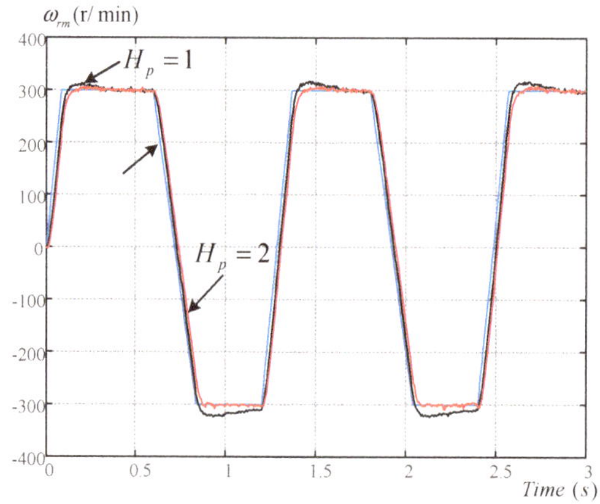

(a)

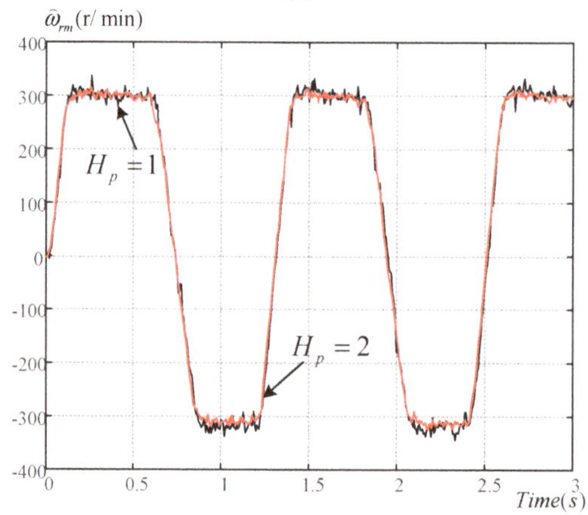

(b)

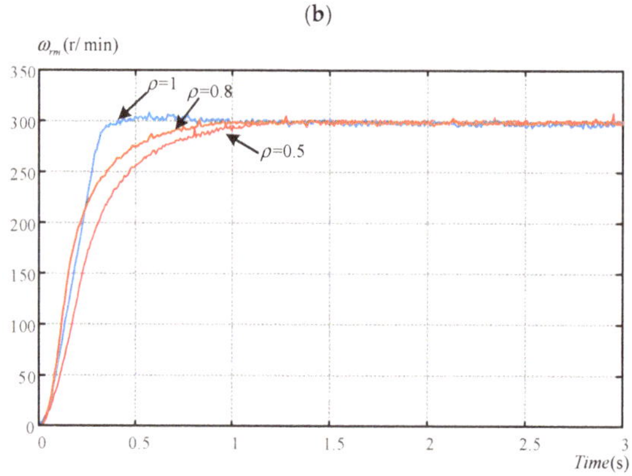

(c)

Figure 16. Measured speed tracking responses. (**a**) Trapezoidal response using an encoder, (**b**) trapezoidal response using the proposed sensorless method, and (**c**) step input using the proposed sensorless method.

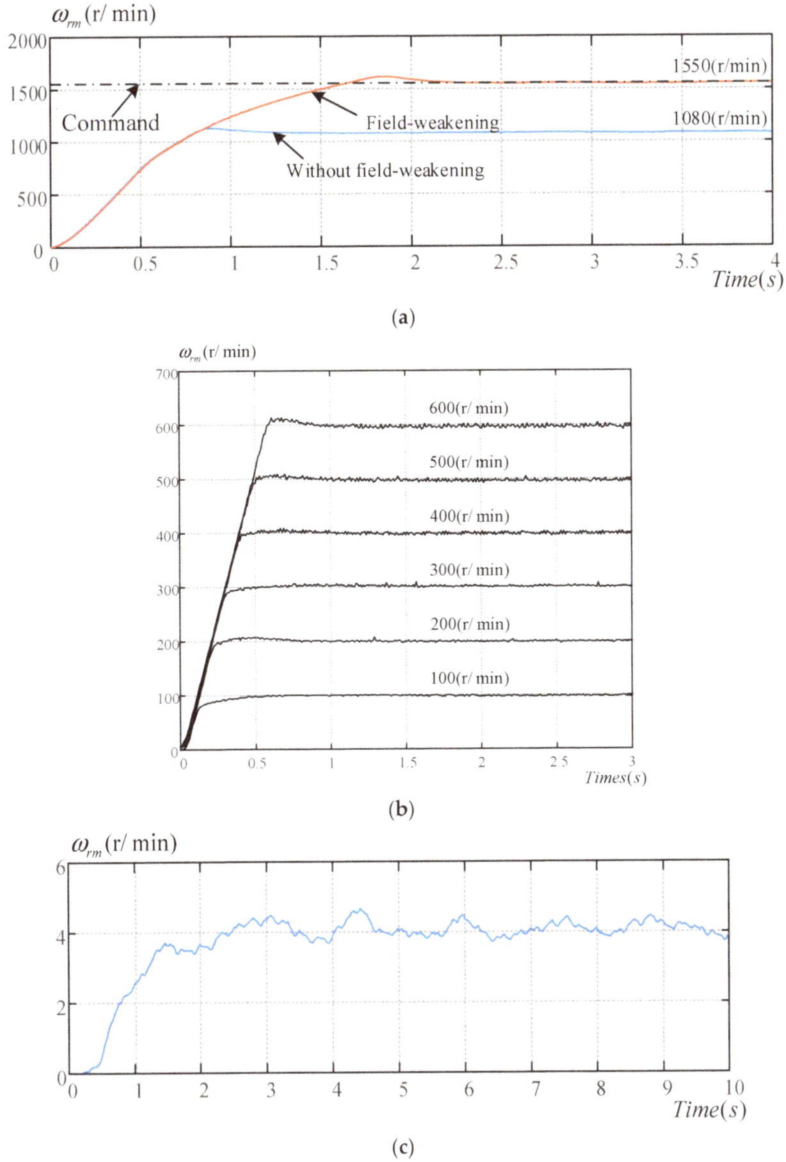

Figure 17. Measured highest, middle, and lowest operational speeds. (**a**) Highest speed, (**b**) middle speeds, and (**c**) lowest speed.

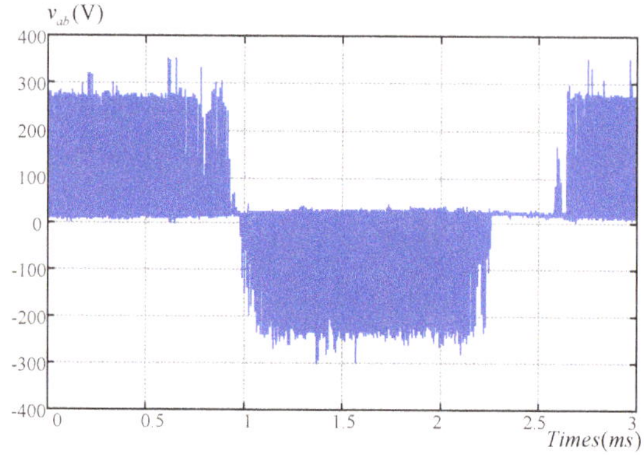

(a)

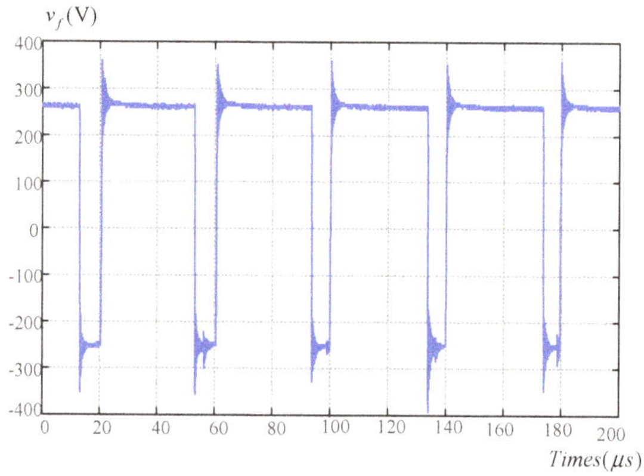

(b)

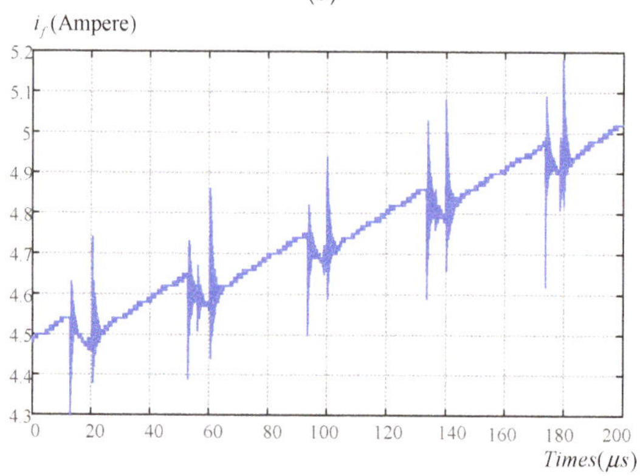

(c)

Figure 18. Measured voltages and currents at 300 r/min. (**a**)-line voltage v_{ab}, (**b**) field-excited voltage v_f, and (**c**) field-excited current i_f.

8. Conclusions

A rotor position estimator and three predictive controllers, which include a multi-step speed predictive controller, a single-step speed predictive controller, and a predictive current controller for a field-excited flux-switching motor drive system have been designed and implemented for this paper.

A 1 kHz, high-frequency voltage is injected into the field winding to obtain the estimated rotor position. In addition, the high-frequency injection voltage does not occupy any available output voltages in the inverter. The methods in this paper provide more available voltage for the PWM modulation. The two different predictive controllers implemented in this research enhance the transient responses, decrease the speed drops of load disturbances, and improve tracking responses more effectively than previously published control methods for flux-switching drive systems. In addition, the multiple-step predictive controller provides smoother speed responses than the single-step predictive controller. These are the main contributions of this paper versus other similar research.

Experimental results show that the controllable speed range of the motor drive system discussed in this paper is from 4 r/min to 1500 r/min. Furthermore, the measured results demonstrate that the errors of the rotor position estimations are below ±2 electrical degrees. Moreover, the proposed predictive controllers provide better performance than PI controllers at different speeds. However, all the considerations in the paper are based on a circuit model with fixed parameters. In the future, methods using field methods or online measuring methods will be thoroughly investigated.

Most functions of the rotor position estimators and predictive controllers are implemented by the DSP, and only simple hardware circuits are employed. This proposed flux-switching drive system can be easily applied to meet the high dynamic requirements for applications in household appliances, such as vacuum cleaners and lawn mowers.

Author Contributions: Conceptualization, T.-H.L.; Methodology, Y.-H.X.; Software, Y.-H.X.; Validation, T.-H.L., M.S.M., and Y.-H.X.; Formal Analysis, T.-H.L.; Investigation, T.-H.L. and Y.-H.X.; Resources, T.-H.L.; Data Curation, M.S.M.; Writing—Original Draft Preparation, T.-H.L.; Writing—Review & Editing, T.-H.L., and M.S.M.; Visualization, M.S.M.; Supervision, T.-H.L.; Project Administration, T.-H.L.; Funding Acquisition, T.-H.L. All authors have read and agreed to the published version of the manuscript.

Funding: This research was funded by Ministry of Science and Technology under grant number MOST 108-2221-E-011-085 and -086.

Acknowledgments: The paper was supported by the MOST, Taiwan under Grant MOST 108-2221-E-011-085 and -086.

Conflicts of Interest: The authors declare no conflict of interest.

References

1. Tang, Y.; Paulides, J.J.H.; Lomonova, E.A. Energy conversion in DC excited flux-switching machines. *IEEE Trans. Magn.* **2014**, *50*, 8105004. [CrossRef]
2. Evans, D.J.; Zhu, Z.Q.; Zhan, H.L.; Wu, Z.Z. Flux-weakening control performance of partitioned stator-switched flux PM machines. *IEEE Trans. Ind. Appl.* **2016**, *52*, 2350–2359. [CrossRef]
3. Deodhar, R.P.; Pride, A.; Iwasaki, S.; Bremner, J.J. Performance Improvement in Flux-Switching PM Machines Using Flux Diverters. *IEEE Trans. Ind. Appl.* **2014**, *50*, 973–978. [CrossRef]
4. Hua, H.; Zhu, Z.Q. Novel Hybrid-Excited Switched-Flux Machine Having Separate Field Winding Stator. *IEEE Trans. Magn.* **2016**, *52*, 8104004. [CrossRef]
5. Yang, H.; Lin, H.; Zhu, Z.Q.; Wang, D.; Fang, S.; Huang, Y. A variable-flux hybrid-PM switched-flux memory machine for EV/HEV applications. *IEEE. Trans. Ind. Appl.* **2016**, *52*, 2203–2214. [CrossRef]
6. Ahmed, N.; Khan, F.; Ali, H.; Ishaq, S.; Sulaiman, E. Outer rotor wound field flux switching machine for in-wheel direct drive application. *IET Electr. Power Appl.* **2019**, *13*, 757–765. [CrossRef]
7. Hua, W.; Cheng, M.; Zhu, Z.Q.; Howe, D. Analysis and optimization of back EMF waveform of a flux-switching permanent magnet motor. *IEEE Trans. Energy Convers.* **2008**, *23*, 727–733. [CrossRef]
8. Gaussens, B.; Hoang, E.; Barriere, O.D.L.; Saint-Michel, J.; Lecrivain, M.; Gabsi, M. Analytical approach for air-gap modeling of field-excited flux-switching machine: No-load operation. *IEEE Trans. Magn.* **2012**, *48*, 2505–2517. [CrossRef]
9. Zhao, J.; Quan, X.; Lin, M. Model predictive torque control of a hybrid excited axial field flux-switching permanent magnet machine. *IEEE Access* **2020**, *8*, 33703–33712. [CrossRef]
10. Zhao, J.; Lin, M.; Xu, D. Minimum-copper-loss control of hybrid excited axial field flux-switching machine. *IET Electr. Power Appl.* **2016**, *10*, 82–90. [CrossRef]

11. Zhao, J.; Lin, M.; Xu, D.; Hao, L.; Zhang, W. Vector control of a hybrid axial field flux-switching permanent magnet machine based on particle swarm optimization. *IEEE Trans. Magn.* **2015**, *51*, 8204004. [CrossRef]
12. Yang, G.; Lin, M.; Li, N.; Tan, G.; Zhang, B. Flux-weakening control combined with magnetization state manipulation of hybrid permanent magnet axial field flux-switching memory machine. *IEEE Trans. Energy Convers.* **2018**, *33*, 2210–2219. [CrossRef]
13. Wu, B.; Xu, D.; Ji, J.; Zhao, W.; Jiang, Q. Field-oriented control and direct torque control for a five-phase fault-tolerant flux-switching permanent-magnet motor. *Chin. J. Electr. Eng.* **2018**, *4*, 48–56. [CrossRef]
14. Nguyen, H.Q.; Yang, S.M. Rotor position sensorless control of wound-field flux-switching machine based on high frequency square-wave voltage injection. *IEEE Access* **2018**, *6*, 48776–48784. [CrossRef]
15. Zhang, W.; Yang, Z.; Zhai, L.; Wang, J. Speed sensorless control of hybrid excitation axial field flux-switching permanent-magnet machine based on model reference adaptive system. *IEEE Access* **2020**, *8*, 22013–22024. [CrossRef]
16. Fan, Y.; Wang, R.; Zhang, L. Sensorless control of five-phase IPM motor based on high-frequency sinusoidal voltage injection. In Proceedings of the 2017 20th International Conference on Electrical Machines and Systems (ICEMS), Sydney, NSW, Australia, 11–14 August 2017; pp. 1–5. [CrossRef]
17. Wang, S.; Yang, K.; Chen, K. An Improved Position-Sensorless Control Method at Low Speed for PMSM Based on High-Frequency Signal Injection into a Rotating Reference Frame. *IEEE Access* **2019**, *7*, 86510–86521. [CrossRef]
18. Chen, C.G.; Liu, T.H.; Lin, M.T.; Tai, C.A. Position control of a sensorless synchronous reluctance motor. *IEEE Trans. Ind. Electron.* **2004**, *51*, 15–25. [CrossRef]
19. Maciejowski, J.M. *Predictive Control with Constraints*; Pearson Education Limited: Harlow, UK, 2002.
20. Rodriquez, J.; Corts, P. *Predictive Control of Power Converters and Electrical Drives*; Wiley: New York, NY, USA, 2012.
21. Wang, L.; Chai, S.; Yoo, D.; Gan, L.; Ng, K. *PID and Predictive Control of Electrical Drives and Power Converters Using MATLAB/Simulink*; Wiley: Singapore, 2015.
22. Wang, L. *Model Predictive Control System Design and Implementation Using MATLAB*; Springer: London, UK, 2009.
23. Toliyat, H.A.; Campbell, S. *DSP-Based Electromechanical Motion Control*; CRC Press: New York, NY, USA, 2003.

energies

MDPI

Article

A Wide-Adjustable Sensorless IPMSM Speed Drive Based on Current Deviation Detection under Space-Vector Modulation

Muhammad Syahril Mubarok, Tian-Hua Liu *, Chung-Yuan Tsai and Zuo-Ying Wei

Department of Electrical Engineering, National Taiwan University of Science and Technology, Taipei 106, Taiwan; syahril.elmubarok@gmail.com (M.S.M.); zhoan831020@gmail.com (C.-Y.T.); a75334798@gmail.com (Z.-Y.W.)
* Correspondence: Liu@mail.ntust.edu.tw

Received: 1 August 2020; Accepted: 26 August 2020; Published: 27 August 2020

Abstract: This paper investigates the implementation of a wide-adjustable sensorless interior permanent magnet synchronous motor drive based on current deviation detection under space-vector modulation. A hybrid method that includes a zero voltage vector current deviation and an active voltage vector current deviation under space-vector pulse-width modulation is proposed to determine the rotor position. In addition, the linear transition algorithm between the two current deviation methods is investigated to obtain smooth speed responses at various operational ranges, including at a standstill and at different operating speeds, from 0 to 3000 rpm. A predictive speed-loop controller is proposed to improve the transient, load disturbance, and tracking responses for the sensorless interior permanent magnet synchronous motor (IPMSM) drive system. The computations of the position estimator and control algorithms are implemented by using a digital signal processor (DSP), TMS-320F-2808. Several experimental results are provided to validate the theoretical analysis.

Keywords: current deviation; SVPWM; IPMSM; wide-adjustable speed; sensorless drive

1. Introduction

Interior permanent magnet synchronous motors (IPMSMs) have better performance than any other motors because of their robustness, high efficiency, and high ratio of torque to ampere characteristics [1]. They have been used in industry and household appliances, including machine tools, rolling mills, high-speed trains, electric vehicles, and elevators [2,3]. A sensorless IPMSM drive system can save space, reduce costs, and prevent the noise interference of high-frequency pulse-width modulation (PWM) switching. The major sensorless technologies for an IPMSM include three methods. The first method uses the extended back-electromotive force (back-EMF) estimation to obtain the estimated rotor position [4–9]. The extended back-EMF estimation method, however, cannot effectively determine the rotor position at low-speed. The second method uses a high-frequency injection voltage to produce its related high-frequency currents [10–15]. The high-frequency injection method, however, requires some extra hardware or software to implement a high-frequency generator and also produces audible noise and electromagnetic interference [10–15]. To overcome these issues, the third method uses a stator current deviation to estimate the rotor position. By detecting the current deviation of the stator current, the position estimator has been successfully applied. For example, Bui et al. studied a modified sensorless scheme by using a PWM excitation signal [16]. Wei et al. presented a dual current deviation estimating method to obtain better accuracy of the position estimator [17]. Raute et al. investigated a sensorless technique with the analysis of the inverter nonlinearity effect [18]. Hosogay et al. implemented a position sensorless technique for low-speed ranges based on the d-axis and q-axis current derivative method [19]. These studies have not required any high-frequency injecting signals or a back-EMF estimation [16–19]. These methods [16–19], however, have not been applied for all ranges of different speeds.

To solve this problem, a current deviation of the stator current using the active voltage vector (AVV) is developed in [20–23]. These papers, however, have only focused on middle- to high-speed ranges. In this paper, two novel ideas are proposed as follows:

- a zero voltage vector (ZVV) rotor estimation method is originally proposed to obtain a high-performance sensorless drive system at a zero-speed with full load conditions.
- a linear combination method, including a zero voltage vector (ZVV) algorithm and an AVV algorithm, is proposed with predictive control to operate the sensorless IPMSM drive system from 0 rpm to 3000 rpm. This linear combination method is easily implemented when compared to fuzzy-logic combination methods [24].

2. Mathematical Model of an IPMSM

2.1. The d-q Axis Synchronous Frame Model

According to the rotor flux-linkage coordinate system, the mathematical model of the IPMSM in the d-q axis synchronous frame is expressed as follows:

$$\begin{bmatrix} v_d \\ v_q \end{bmatrix} = \begin{bmatrix} R_s + \frac{d}{dt}L_d & -\omega_e L_q \\ \omega_e L_d & R_s + \frac{d}{dt}L_q \end{bmatrix} \begin{bmatrix} i_d \\ i_q \end{bmatrix} + \begin{bmatrix} 0 \\ \omega_e \lambda_m \end{bmatrix} \quad (1)$$

where v_d and v_q are the d-q axis voltages, R_s is the stator resistance, i_d and i_q are the d-q axis currents, L_d and L_q are the d-q axis inductances, ω_e is the electrical rotor speed, and λ_m is the flux linkage generated by the permanent magnet material, which is placed in the rotor. The addition of the reluctance torque and electromagnetic torque can be obtained as

$$T_e = \frac{3}{2}\frac{P}{2}\left[(L_d - L_q)i_d + \lambda_m\right]i_q \quad (2)$$

The dynamic mechanical equations of the rotor speed and rotor position are expressed as

$$\frac{d}{dt}\omega_m = \frac{1}{J_m}(T_e - T_L - B_m\omega_m) \quad (3)$$

and

$$\frac{d}{dt}\theta_m = \omega_m \quad (4)$$

The electrical rotor position and speed are independently expressed as follows:

$$\theta_e = \frac{P}{2}\theta_m \quad (5)$$

and

$$\omega_e = \frac{P}{2}\omega_m \quad (6)$$

where T_L is the external load, J_m is the inertia, B_m is the viscous coefficient, ω_m is the mechanical angular speed, θ_m is the mechanical position, and θ_e is the electrical position.

2.2. The a-b-c Axis Synchronous Frame Model

Considering the three-phase voltage balanced and Y-connected windings of an IPMSM, the a-b-c-axis stator voltage equation can be stated as follows:

$$\begin{bmatrix} v_{as} \\ v_{bs} \\ v_{cs} \end{bmatrix} = \begin{bmatrix} r_a & 0 & 0 \\ 0 & r_b & 0 \\ 0 & 0 & r_c \end{bmatrix} \begin{bmatrix} i_{as} \\ i_{bs} \\ i_{cs} \end{bmatrix} + \frac{d}{dt} \begin{bmatrix} \lambda_{as} \\ \lambda_{bs} \\ \lambda_{cs} \end{bmatrix} \tag{7}$$

where v_{as}, v_{bs}, and v_{cs} are the a-, b-, and c-phase voltages, r_a, r_b, and r_c are the a-, b-, and c-phase resistances, and λ_{as}, λ_{bs}, and λ_{cs} are the a-, b-, and c-phase flux linkages. In Equation (7), the neutral point voltage is assumed to be zero when the neutral point is free because the fundamental components of the v_{as}, v_{bs}, and v_{cs} are balanced. The relationship between the magnetic flux linkages of three-phase stator windings and the three-phase self-inductances, stator currents, and rotor position can be stated as follows:

$$\begin{bmatrix} \lambda_{as} \\ \lambda_{bs} \\ \lambda_{cs} \end{bmatrix} = \begin{bmatrix} L_{aa} & M_{ab} & M_{ac} \\ M_{ba} & L_{bb} & M_{bc} \\ M_{ca} & M_{cb} & L_{cc} \end{bmatrix} \begin{bmatrix} i_{as} \\ i_{bs} \\ i_{cs} \end{bmatrix} + \lambda_m \begin{bmatrix} \cos\theta_e \\ cos(\theta_e - \frac{2\pi}{3}) \\ cos(\theta_e + \frac{2\pi}{3}) \end{bmatrix} \tag{8}$$

where L_{aa}, L_{bb}, and L_{cc} are the three-phase self-inductances, and M_{ab}, M_{ac}, M_{ba}, M_{bc}, M_{ca}, and M_{cb} are the three-phase mutual inductances. The self-inductances are expressed as

$$L_{aa} = L_{ls} + L_{AA} - L_{BB}\cos(2\theta_e) \tag{9}$$

$$L_{bb} = L_{ls} + L_{AA} - L_{BB}\cos(2\theta_e + \frac{2\pi}{3}) \tag{10}$$

and

$$L_{cc} = L_{ls} + L_{AA} - L_{BB}\cos(2\theta_e - \frac{2\pi}{3}) \tag{11}$$

The mutual inductances are expressed as

$$M_{ab} = M_{ba} == -\frac{1}{2}L_{AA} - L_{BB}\cos(2\theta_e - \frac{2\pi}{3}) \tag{12}$$

$$M_{bc} = M_{cb} = -\frac{1}{2}L_{AA} - L_{BB}\cos(2\theta_e) \tag{13}$$

and

$$M_{ca} = M_{ac} = -\frac{1}{2}L_{AA} - L_{BB}\cos(2\theta_e + \frac{2\pi}{3}) \tag{14}$$

The L_{AA} and L_{BB} are constant parameters.

3. Zero Voltage Vector-Based Current Deviation Rotor Position Estimator

3.1. Basic Principle

The ZVV dominates the duty cycle of a PWM at a standstill and low-speed operating ranges; as a result, one can substitute $v_d = 0$ and $v_q = 0$ into Equation (1) and obtain

$$\begin{bmatrix} \frac{di_d}{dt} \\ \frac{di_q}{dt} \end{bmatrix} = \begin{bmatrix} -\frac{R_s}{L_d} & \frac{L_q}{L_d}\omega_e \\ -\frac{L_d}{L_q}\omega_e & -\frac{R_s}{L_q} \end{bmatrix} \begin{bmatrix} i_d \\ i_q \end{bmatrix} - \begin{bmatrix} 0 \\ \frac{\lambda_m \omega_e}{L_q} \end{bmatrix} \tag{15}$$

From Figure 1, the estimated position error $\tilde{\theta}_e$ is first defined as follows:

$$\tilde{\theta}_e = \theta_e - \hat{\theta}_e \tag{16}$$

where $\tilde{\theta}_e$ and $\hat{\theta}_e$ are the estimated rotor position error and estimated rotor position, respectively. Second, the estimated d- and -q axis currents are expressed as follows:

$$\begin{bmatrix} \hat{i}_d \\ \hat{i}_q \end{bmatrix} = \begin{bmatrix} \cos\widetilde{\theta}_e & -\sin\widetilde{\theta}_e \\ \sin\widetilde{\theta}_e & \cos\widetilde{\theta}_e \end{bmatrix} \begin{bmatrix} i_d \\ i_q \end{bmatrix} = T\left(\widetilde{\theta}_e\right) \begin{bmatrix} i_d \\ i_q \end{bmatrix} \tag{17}$$

where $\hat{i}_d$ and $\hat{i}_q$ are the estimated d-axis and q-axis currents, and $T(\widetilde{\theta}_e)$ is the coordinate transformation matrix. By substituting Equation (17) into Equation (15), the dynamic equation of the ZVV-based estimated q-axis current is

$$\begin{aligned} \frac{d\hat{i}_q}{dt} \cong\ & \frac{1}{L_d L_q}\left\{R_s\left[\left(L_q - L_d\right)\cos^2\left(\widetilde{\theta}_e\right) - L_q\right] - \frac{1}{2}\omega_e\left(L_q^2 - L_d^2\right)\sin\left(2\widetilde{\theta}_e\right)\right\}\hat{i}_q \\ & + \frac{1}{L_d L_q}\left\{\omega_e\left[\left(L_q^2 - L_d^2\right)\cos^2\left(\widetilde{\theta}_e\right) - L_q^2\right] - \frac{R_s}{2}\left(L_q - L_d\right)\sin\left(2\widetilde{\theta}_e\right)\right\}\hat{i}_d \\ & - \frac{\lambda_m \omega_e}{L_q}\cos\left(\widetilde{\theta}_e\right) \end{aligned} \tag{18}$$

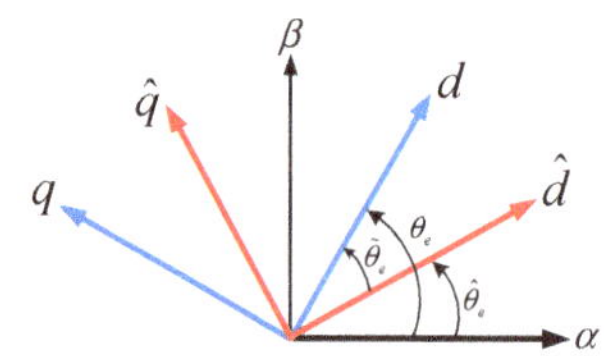

Figure 1. The coordinate transformation relationship.

The values of the L_d and L_q are only a few *mHs* and are lower than 1 Henry. In consequence, L_d is much greater than L_d^2, and L_q is much greater than L_q^2. By omitting the L_d^2, L_q^2, and related items that are multiplied by the ω_e, the dynamic equation of the ZVV-based estimated q-axis current can be derived as follows:

$$\begin{aligned} \frac{d\hat{i}_q}{dt} \cong\ & \frac{R_s}{L_d L_q}\left\{-\left[\left(L_d - L_q\right)\cos^2\left(\widetilde{\theta}_e\right) + L_q\right]\hat{i}_q \right. \\ & \left. + \frac{1}{2}(L_d - L_q)\sin\left(2\widetilde{\theta}_e\right)\hat{i}_d\right\} - \frac{\lambda_m \omega_e}{L_q}\cos\left(\widetilde{\theta}_e\right) \end{aligned} \tag{19}$$

Because the estimated position error $\widetilde{\theta}_e$ is close by zero at steady-state, it is feasible to assume that $\sin\left(2\widetilde{\theta}_e\right) \cong 2\widetilde{\theta}_e$, $\sin\left(\widetilde{\theta}_e\right) \cong \widetilde{\theta}_e$, $\cos^2\left(\widetilde{\theta}_e\right) \cong 1$, and $\cos\left(\widetilde{\theta}_e\right) \cong 1$. By using these guestimates into Equation (19), the following equation can be obtained

$$\frac{d\hat{i}_q}{dt} \cong -\frac{R_s \hat{i}_q}{L_q} + \frac{R_s\left(L_d - L_q\right)\hat{i}_d}{L_d L_q}\widetilde{\theta}_e - \frac{\lambda_m \omega_e}{L_q} \tag{20}$$

3.2. The ZVV Rotor Position Estimating Scheme

In fact, the estimated speed $\hat{\omega}_e$ is employed to replace the real speed ω_e. The estimated rotor position $\widetilde{\theta}_e$ can be assumed to be zero under steady-state conditions. Equation (20), as a result, is rearranged as follows:

$$\frac{d\hat{i}_q}{dt} + \frac{R\hat{i}_q + \lambda_m \hat{\omega}_e}{L_q} \cong \frac{R\left(L_d - L_q\right)i_d^*}{L_d L_q}\widetilde{\theta}_e \tag{21}$$

Then, a new variable, $D\hat{i}_q$, can be defined as follows:

$$D\hat{i}_q \triangleq \frac{d\hat{i}_q}{dt} + \frac{R_s \hat{i}_q + \lambda_m \hat{\omega}_e}{L_q} \tag{22}$$

Substituting (22) into Equation (21), one can obtain

$$D\hat{i}_q \cong K_q \widetilde{\theta}_e \tag{23}$$

and

$$K_q \triangleq \frac{R_s\left(L_d - L_q\right)i_d^*}{L_d L_q} \tag{24}$$

From Equation (23), the $D\hat{i}_q$ is nearly proportional to the estimated position error $\widetilde{\theta}_e$. Figure 2 illustrates the proposed ZVV-based estimator. First, by using the coordinate transformation, one can transfer the a-, b-, and c-phase current deviations into the estimated d-q axis current deviations. Then, one can derive the current deviation $d\hat{i}_q/dt$. By summation of the $d\hat{i}_q/dt$, $\lambda_m\hat{\omega}_e/L_q$, and $R_s\hat{i}_q/L_q$, one can obtain $D\hat{i}_q$. Next, one can compute the value of $\widetilde{\theta}_e$ after dividing $D\hat{i}_q$ by K_q and, thus, obtain the estimated position error. After that, one can use a proportional-integral (PI) controller to acquire the estimated rotor speed and obtain the estimated position by using an integral operation.

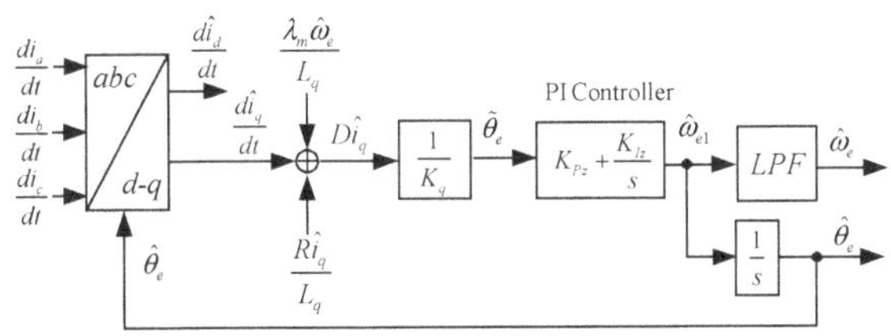

Figure 2. Block diagram of the zero voltage vector (ZVV)-based estimator.

4. Active Voltage Vector-Based Current Deviation Rotor Position Estimator

4.1. The AVV Rotor Position Estimation Scheme

At middle and high speeds, the current deviation of the AVV in the a-b-c-stationary frame is used to estimate the position. From Figure 3, when the a-phase upper leg turns on and the b-phase and c-phase lower legs turn on, one can obtain

$$V_{dc} = v_{as} - v_{bs} \tag{25}$$

By substituting Equation (7) into (25), one can derive

$$V_{dc} = r_s i_{as} + \frac{d\lambda_a}{dt} - r_s i_{bs} - \frac{d\lambda_b}{dt} \tag{26}$$

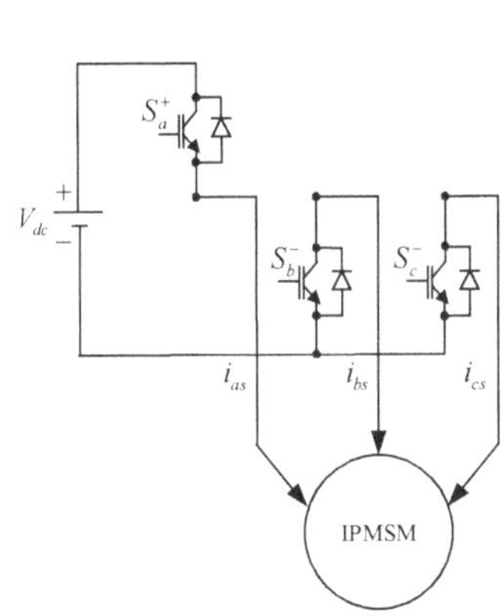

Figure 3. The main circuit of the mode A+.

Substituting Equation (8) into (26), one can derive

$$\begin{aligned} V_{dc} &= r_s i_{as} + \tfrac{d}{dt}(L_{aa} i_{as} + L_{ab} i_{bs} + L_{ac} i_{cs} + \lambda_m \cos\theta_e) - r_s i_{bs} \\ &\quad - \tfrac{d}{dt}(L_{ba} i_{as} + L_{bb} i_{bs} + L_{bc} i_{cs} + \lambda_m \cos(\theta_e - \tfrac{2\pi}{3})) \end{aligned} \tag{27}$$

By substituting Equations (9)–(14) into Equation (27) and doing some mathematical processes, one can obtain

$$\begin{aligned} V_{dc} = \; & \Big[\Big(\tfrac{1}{2}(3L_{AA} + 2L_{ls}) + \sqrt{3}\sin\big(2\theta_e - \tfrac{\pi}{3}\big)L_{BB}\Big)\tfrac{di_{as}}{dt} + \Big(-\tfrac{1}{2}(3L_{AA} + 2L_{ls}) + \sqrt{3}\sin(2\theta_e + \pi)L_{BB}\Big)\tfrac{di_{bs}}{dt} \\ & + \Big(\sqrt{3}\sin\big(2\theta_e + \tfrac{\pi}{3}\big)L_{BB}\Big)\tfrac{di_{cs}}{dt}\Big] + 2\sqrt{3}\omega_e L_{BB}\Big[\sin\big(2\theta_e + \tfrac{\pi}{6}\big)i_{as} + \sin\big(2\theta_e - \tfrac{\pi}{2}\big)i_{bs} + \sin\big(2\theta_e + \tfrac{5\pi}{6}\big)i_{cs}\Big] \\ & - \lambda_m \omega_e\Big(\tfrac{3}{2}\sin\theta_e + \tfrac{\sqrt{3}}{2}\cos\theta_e\Big) + r_s(i_{as} - i_{bs}) \end{aligned} \tag{28}$$

In addition, from Figure 3, when the a-phase upper leg turns on and the b-phase and c-phase lower legs turn on, one can obtain the voltage between the b-phase and c-phase as follows:

$$0 = v_{bs} - v_{cs} \tag{29}$$

By using the similar processes shown in Equations (26)–(29), one can derive

$$\begin{aligned} 0 = \; & \Big[\Big(\sqrt{3}\sin(2\theta_e + \pi)L_{BB}\Big)\tfrac{di_{as}}{dt} + \Big(\tfrac{1}{2}(3L_{AA} + 2L_{ls}) + \sqrt{3}\sin\big(2\theta_e + \tfrac{\pi}{3}\big)L_{BB}\Big)\tfrac{di_{bs}}{dt} + \Big(-\tfrac{1}{2}(3L_{AA} + 2L_{ls}) + \sqrt{3}\sin\big(2\theta_e - \tfrac{\pi}{3}\big)L_{BB}\Big)\tfrac{di_{cs}}{dt}\Big] \\ & + 2\sqrt{3}\omega_e L_{BB}\Big[\sin\big(2\theta_e - \tfrac{\pi}{2}\big)i_{as} + \sin\big(2\theta_e + \tfrac{5\pi}{6}\big)i_{bs} + \sin\big(2\theta_e + \tfrac{\pi}{6}\big)i_{cs}\Big] + \sqrt{3}\lambda_m\omega_e\cos\theta_e + r_s(i_{bs} - i_{cs}) \end{aligned} \tag{30}$$

Moreover, from Figure 3, when the a-phase upper leg turns on and the b-phase and c-phase lower legs turn on, one can obtain

$$i_{as} = -(i_{bs} + i_{cs}) \tag{31}$$

Taking the differential of both sides in Equation (31), one can obtain

$$\frac{di_{as}}{dt} = -\Big(\frac{di_{bs}}{dt} + \frac{di_{cs}}{dt}\Big) \tag{32}$$

From Equations (28), (30), and (32), one can obtain the following equation:

$$\begin{aligned} \tfrac{di_{as}}{dt}\_{mode\ A+} = \; & \Big\{4V_{dc}\Big(\tfrac{2}{3}L_{ls} + L_{AA} + L_{BB}\cos(2\theta_e)\Big) + 2\omega_e L_{BB} \\ & \Big[\Big(-3i_{as}\sin(2\theta_e) + \sqrt{3}(i_{bs} - i_{cs})(3L_{BB} + \cos(2\theta_e))\Big) \\ & (3L_{AA} + 2L_{ls}) + K_a\Big] + R_{sa}\Big\} / [(2L_{ls} + 3L_{\Delta})(2L_{ls} + 3L_{\Sigma})] \end{aligned} \tag{33}$$

where

$$L_{\Delta} = L_{AA} - L_{BB} \tag{34}$$

and

$$L_{\Sigma} = L_{AA} + L_{BB} \tag{35}$$

When the switching state of the inverter is at a zero voltage state, the motor short circuits. Consequently, the direct-current (DC) voltage V_{dc} is equal to zero. By substituting $V_{dc} = 0$ into Equation (33), one can obtain

$$\begin{aligned} \tfrac{di_{as}}{dt}\_{mode\ 0} = \; & \Big\{2\omega_e L_{BB}\Big[\Big(-3i_{as}\sin(2\theta_e) + \sqrt{3}(i_{bs} - i_{cs})(3L_{BB} + \cos(2\theta_e))\Big) \\ & (3L_{AA} + 2L_{ls}) + K_a\Big] + R_{sa}\Big\} / [(2L_{ls} + 3L_{\Delta})(2L_{ls} + 3L_{\Sigma})] \end{aligned} \tag{36}$$

From Equation (36), the current deviation $di_{as}/dt_{\text{\_mode 0}}$ includes a resistance influencing part, a back-EMF influencing part, and an inductance influencing part. To eject the influences of the back-EMF and resistance parts, the a-phase compensated current deviation $Di_{as\_\text{mode A+}}$ is computed as

$$\begin{aligned} Di_{as\_\text{mode A+}} &= \frac{di_{as}}{dt}_{\text{\_mode A+}} - \frac{di_{as}}{dt}_{\text{\_mode 0}} \\ &= \frac{4V_{dc}\left[\frac{2}{3}L_{ls}+L_{AA}+L_{BB}\cos(2\theta_e)\right]}{(2L_{ls}+3L_{\Delta})(2L_{ls}+3L_{\Sigma})} \end{aligned} \tag{37}$$

From Equation (37), one can observe that the $Di_{as\_\text{mode A+}}$ is only related to the inductance and input DC voltage V_{dc}. By using a similar method, one can derive the b-phase compensation current deviation and the c-phase compensation current deviation.

By transferring the a-, b-, and c-axis current deviations into the $\alpha-$ and $\beta-$ axis current-deviations, one can obtain the current deviation Di_{α} and Di_{β} as follows:

$$\begin{bmatrix} Di_{\alpha} \\ Di_{\beta} \end{bmatrix} = \begin{bmatrix} 1 & -\frac{1}{2} & -\frac{1}{2} \\ 0 & \frac{\sqrt{3}}{2} & -\frac{\sqrt{3}}{2} \end{bmatrix} \begin{bmatrix} Di_{as\_\text{mode A+}} \\ Di_{bs\_\text{mode B+}} \\ Di_{cs\_\text{mode C+}} \end{bmatrix} \tag{38}$$

By substituting the a-b-c-phase compensated current deviation into Equation (38), the Di_{α} can be simplified as

$$Di_{\alpha} = \frac{6V_{dc}L_B\cos(2\theta_e)}{(2L_{ls}+3L_{\Delta})(2L_{ls}+3L_{\Sigma})} \tag{39}$$

and the Di_{β} can be simplified as

$$Di_{\beta} = \frac{-6V_{dc}L_B\sin(2\theta_e)}{(2L_{ls}+3L_{\Delta})(2L_{ls}+3L_{\Sigma})} \tag{40}$$

By using the following $\tan^{-1}$ mathematical process, one can obtain the AVV-based estimated electrical rotor position as follows:

$$\tan^{-1}\left(\frac{-Di_{\beta}}{Di_{\alpha}}\right) = 2\theta_e \tag{41}$$

4.2. Space-Vector Extension And Compensation

The current deviation is used to estimate the position and is discussed in this section. To obtain the current deviation, two different currents are sampled for each switching state, and then the current deviation can be computed. In the real world, the time interval between the two-sampling intervals should be large enough to obtain an accurate current deviation. However, some turn-on or turn-off intervals are reduced as the motor speed increases. In addition, the switching interval is also varied when the voltage vector moves to different positions. To solve this issue, an extension with a compensation method is used [13]. When the digital signal processor (DSP) detects that the time interval of the switching state is too short, an extension time is automatically provided to make the switching time maintain a minimum required switching time, $T_{\min}$. In this paper, the minimum required switching time interval $T_{\min}$ is set as 20 μs.

However, the extension time of the switching state causes DC bias and harmonics. Therefore, a compensation time of the whole switching interval is required. Figure 4a,b show the space-vector pulse width modulation (SVPWM) switching states used in this paper. In each switching interval, which is 100 μs, three switching states are generated. Figure 4a shows the T_1 switching state when it is too short. When this occurs, the switching state "100" is extended, and its complementary state "011" is compensated for at the end because the voltage in the whole-time interval T_s should be balanced. As a result, the DC bias is reduced to zero, and the ac harmonics are compensated. Figure 4b shows the T_2 switching state when it is too short. When this occurs, the switching state "110" is extended. Then, "001" is compensated for at the end of the entire switching interval T_s.

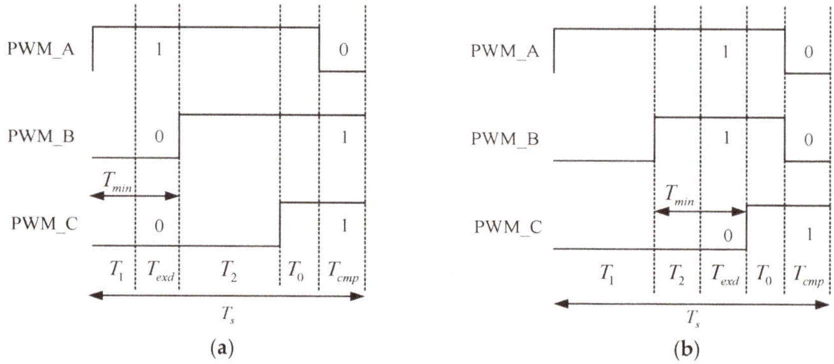

Figure 4. Space-vector pulse width modulation extension and compensation. (**a**) T1 voltage vector; (**b**) T2 voltage vector.

5. Current Deviation Detection Technique

In this paper, the current deviation is used to estimate the position of IPMSM. As a result, the precise detection of the current deviation at AVV and ZVV is very important. As one can observe in the waveform, shown in Figure 5, the current deviation is obtained for every sampling time, T_s. The current spike can be avoided by carefully selecting the sampling instance. The first current sampling instance of each switching state is delayed 10 μs after the power device turns on. The second current sampling instance is 5 μs before the next switching state occurs. After that, the current deviation can be computed by subtracting the second captured current with the first captured current over the time difference.

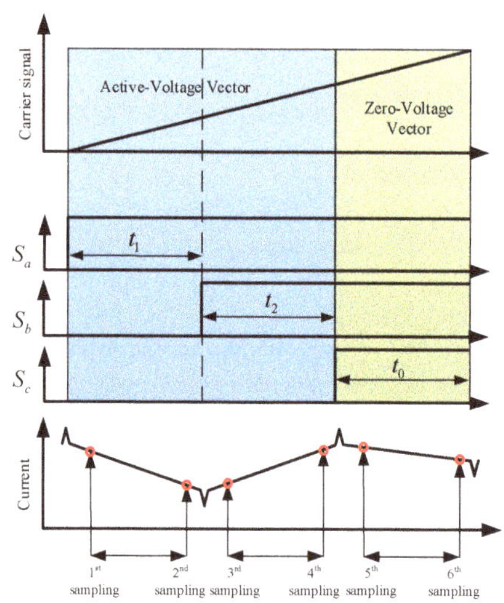

Figure 5. The modulation of inverter and current sampling waveform.

The current deviation can be precisely obtained since the time period of AVV is extended if it is too narrow when the motor is operated at middle- and high-speeds. The current-deviation of the ZVV can be obtained because of the large duty cycle of the ZVV when the motor is operated at a standstill and low-speed ranges.

6. Linear Transition from Standstill and Low-Speed to High-Speed

A linear transferring method between the ZVV and AVV algorithms is displayed in Figure 6. This linear transition method is easy to implement and can achieve better performance than other advanced transition methods, such as fuzzy logic methods. Even though fuzzy logic methods have better performance and faster responses, they need more complex computations [24]. A lower bound transition speed, which is 60 rpm, is defined as ω_{s1}. In addition, a higher bound transition speed, which is 100 rpm, is defined as ω_{s2}. A weighting factor β is utilized in the subsequent equations:

$$\hat{\theta}_e = \beta\hat{\theta}_{zvv} + (1-\beta)\hat{\theta}_{avv} \tag{42}$$

and

$$\hat{\omega}_e = \beta\hat{\omega}_{zvv} + (1-\beta)\hat{\omega}_{avv} \tag{43}$$

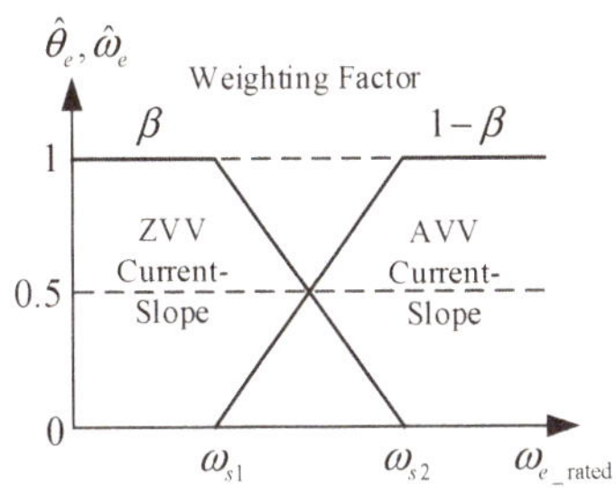

Figure 6. The linear transfer between ZVV current deviation and active voltage vector (AVV) current deviation.

The weighting factor β in (42) and (43) is defined as follows:

$$\beta = \begin{cases} 1 & \text{, when } \hat{\omega}_e \le \omega_{s1} \\ \frac{\omega_{s2}-\hat{\omega}_e}{\omega_{s2}-\omega_{s1}} & \text{, when } \omega_{s1} < \hat{\omega}_e < \omega_{s2} \\ 0 & \text{, when } \hat{\omega}_e \ge \omega_{s2} \end{cases} \tag{44}$$

where $\hat{\theta}_{zvv}$ and $\hat{\theta}_{avv}$ are the estimated positions by using the ZVV and the AVV algorithms, respectively. The $\hat{\omega}_{zvv}$ and $\hat{\omega}_{avv}$ are the estimated speeds after using the ZVV and AVV algorithms. By using this method, the estimated positions and estimated speeds can be accurately obtained.

7. Predictive Speed Controller Design

Predictive controllers have been applied in chemical process industries, robotic controls, and other multivariable systems [25–27]. Recently, predictive controllers have been successfully employed in motor drives and power electronics [28–30]. A predictive speed controller is designed to improve the responses of the drive systems. By omitting the external load T_L from (3), when the d-axis is zero, it is possible to derive the transfer function $G_p(s)$ of the IPMSM, which is shown in Figure 7, as follows:

$$G_p(s) = \frac{\omega_m(s)}{i_q(s)} = \frac{K_t/J_t}{s+(B_t/J_t)} \tag{45}$$

By inserting a zero-order hold device and taking the z-transformation, one can obtain

$$G_p(z) = \frac{\omega_m(z)}{i_q(z)} = \frac{K_t}{B_t}\frac{(1-e^{-\frac{B_t}{J_t}T_s})}{(z-e^{-\frac{B_t}{J_t}T_s})} \tag{46}$$

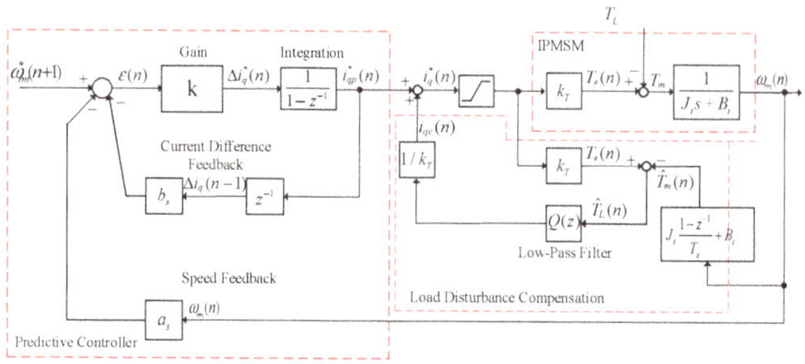

Figure 7. Block diagram of the proposed speed-loop predictive controller.

From (46), it is straightforward to derive the speed prediction as

$$\omega_m(n+1) = a_s\omega_m(n) + b_s i_q(n) \tag{47}$$

where

$$a_s = e^{-\frac{B_t}{J_t}T_s} \tag{48}$$

and

$$b_s = \frac{K_t}{B_t}(1 - e^{-\frac{B_t}{J_t}T_s}) \tag{49}$$

The predictive speed equation is shown as

$$\begin{aligned}\widehat{\omega}_m(n+1) &= a_s\omega_m(n) + b_s i_{qp}{}^*(n) \\ &= a_s\omega_m(n) + b_s i_{qp}(n-1) + b_s\Delta i_q{}^*(n)\end{aligned} \tag{50}$$

The performance index can be defined as [28–30]

$$J_p(n) = \alpha\left[\widehat{\omega}_m(n+1) - \omega_m{}^*(n+1)\right]^2 + \left[\Delta i_q{}^*(n)\right]^2 \tag{51}$$

After taking $\partial J_p(n)/\partial\Delta i_q{}^*(n) = 0$, one can obtain

$$2\alpha b_s\left[a_s\omega_m(n) + b_s\left(i_{qp}(n-1) + \Delta i_q{}^*(n)\right) - \omega_m{}^*(n+1)\right] + 2\Delta i_q{}^*(n) = 0 \tag{52}$$

Next, one can obtain

$$\Delta i_q{}^*(n) = \frac{\alpha b_s}{\alpha b_s{}^2 + 1}\left[\omega_m{}^*(n+1) - a_s\omega_m(n) - b_s i_{qp}(n-1)\right] \tag{53}$$

and

$$k = \frac{\alpha b_s}{\alpha b_s{}^2 + 1} \tag{54}$$

Finally, the q-axis current command is

$$i_{qp}{}^*(n) = i_{qp}(n-1) + \Delta i_q{}^*(n) \tag{55}$$

Load disturbance compensation is shown in Figure 7. By computing the difference between $k_t i_q^*(n)$ and $J_t(\Delta\omega_m/\Delta t) + B_t\omega_m$, the estimated mechanical load can be obtained. Next, the external load $\hat{T}_m(n)$ is estimated by using a low-pass filter. After that, the compensation current $i_{qc}(n)$ can be obtained. Finally, the q-axis current command $i_q{}^*(n)$, which is the summation of the $i_{qp}{}^*(n)$ and the compensation current $i_{qc}(n)$, can be computed, shown in Figure 7.

8. Implementation

The implemented circuit of the sensorless IPMSM drive system is discussed here. Figure 8a demonstrates the closed-loop block diagram of the implemented drive system. The digital signal processor (DSP) executes the position estimation and predictive control algorithms. As a result, the DSP is the control center of the IPMSM drive system. Figure 8b displays the implemented circuit, including an inverter, a 3-phase driving circuit, a DSP, two A/D converters, and two Hall-effect current sensing circuits. Figure 8c demonstrates the drive system with a dynamometer, which is driven by a DC permanent magnet motor to equip the external load. By suitably adjusting the input voltage of the DC motor, a varied load can be obtained. The IPMSM is an 8-pole, 7.7 A, 2000 rpm, 2 kW IPMSM. The motor has the following parameters: stator resistance = 0.32 Ω, d-axis self-inductance = 0.0049 H, q-axis self-inductance = 0.0078 H, flux linkage = 0.16 V.s/rad, inertia = 0.00455 kg-m^2, viscous coefficient = 0.003 N.m.s/rad, switching frequency of the inverter = 10 kHz, DC bus voltage = 300 V, current-loop sampling interval = 100 μs, and the speed-loop sampling interval = 1 ms.

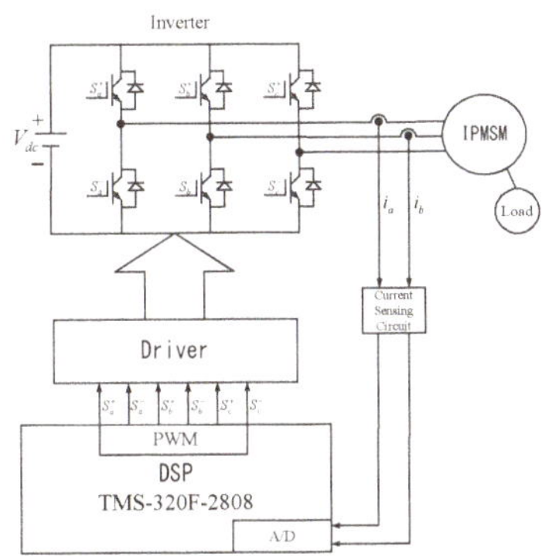

(a)

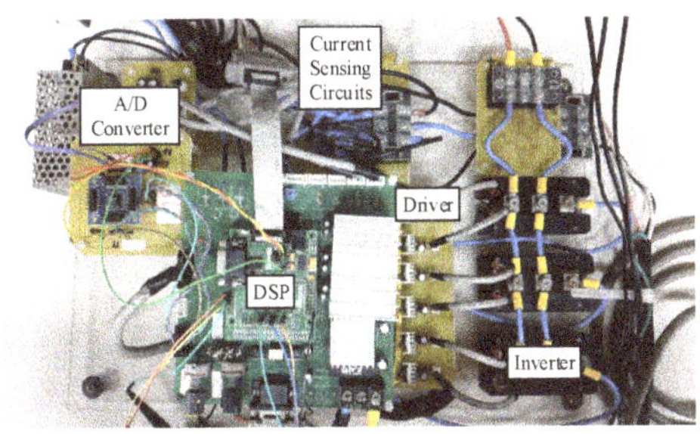

(b)

(c)

Figure 8. Implemented system. (**a**) block diagram; (**b**) circuit; (**c**) dynamometer.

9. Experimental Results

To verify the theoretical analysis, several measured results are demonstrated in this section. The speed-loop PI controller is designed by a pole assignment technique. Figure 9 demonstrates the relation between the q-axis current deviations and estimated position errors at 0 rpm. When the d-axis current is too low, the amplitude of the q-axis current deviation is also very low. In order to enhance the amplitude of the current deviation to reduce the estimated rotor position error, a higher d-axis current is selected for the rotor position estimation at a standstill condition. From Figure 9, the relationship between the q-axis current deviation and the estimated position error is nearly a sinusoidal curve. In addition, the current deviation slope and the estimated position error have the same polarity. Therefore, it is reasonable to use the q-axis current deviation slope to estimate the position, as explained in Section 3. Figure 10a demonstrates the estimated and real positions, and they are similar. However, the estimated rotor position varies above and below the real positions. Figure 10b demonstrates the estimated and real speeds at 5 rpm. Both of them have obvious speed ripples, which are near ±1 rpm. In addition, the speed computed from rotor estimation has larger speed ripples than the speed measured by an encoder, which provides more accurate speed information. Figure 10c demonstrates the estimated position error with only nearby ±2 electrical degrees. This result shows that the estimated d-axis moves above and below the real d-axis. Figure 11a demonstrates the rotor speed and estimated speed at 0 rpm under an 11 N·m external load. According to this figure, the estimated speed follows the real speed well, even though a heavy load is added. However, the estimated speed has a larger speed ripples than the speed obtained from the encoder. Figure 11b demonstrates the estimated and real positions using an encoder. Both of them finally reach 0 electrical degrees at steady-states, at which the motor provides maximum holding torque. Figure 11c shows the position error at 0 rpm under an 11 N·m load. The estimated error is at a near −2 degrees at no load; however, it reaches 2 degrees under an 11 N·m load. According to Figure 11a–c, one can conclude that the proposed method can provide a high-performance speed control at a standstill with a heavy load.

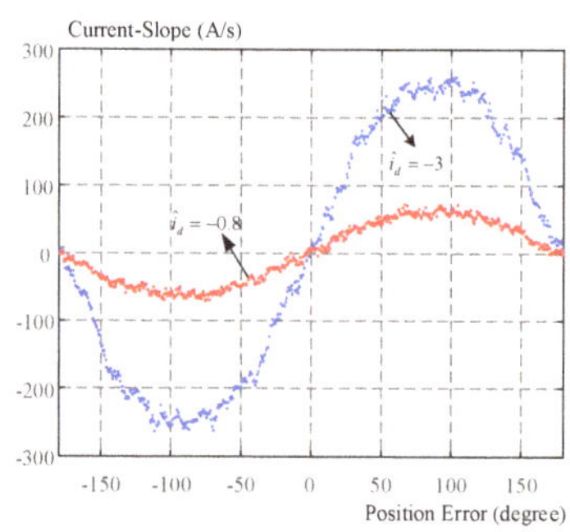

Figure 9. Measured q-axis current deviations to estimated position errors.

Figure 12a,b show the comparison of the a-phase output current of the hysteresis current control and the proposed extension and compensation SVPWM when the motor is running at 600 rpm under a 1.5 N·m external load. As we can observe, the output current with the proposed method provides lower current harmonics than the hysteresis current control method. The major reason is that hysteresis control uses an infinite gain for a current-loop and then creates high current harmonics. Figure 13a shows the a-phase current and its related sampling signal. The current is detected twice for each switching interval in order to compute the current deviation. Figure 13b shows the extended and compensated SVPWM switching states. In that figure, it can be seen that the sampling interval T_s is 100 μs. The first and second applied voltage vectors are "110" and "010" and are both less than 20 μs. Therefore, these voltage vectors need to be extended at the first stage and then be compensated for at the next stage. Besides that, the switching time for the zero voltage vector reaches 20 μs, and it does not need to be extended. The extended voltage vectors "110" and "010" also need to be compensated for in the opposite direction, which is "001" and "101", to keep the total time of switching states from changing. Hence, the compensation for switching states "001" and "101" are performed before the end of the switching interval T_S. Figure 13c shows the related switching points (broken lines) and sampling points (solid lines). As can be observed, two current sampling points are required for the AVV "110" and "010" and the ZVV "000".

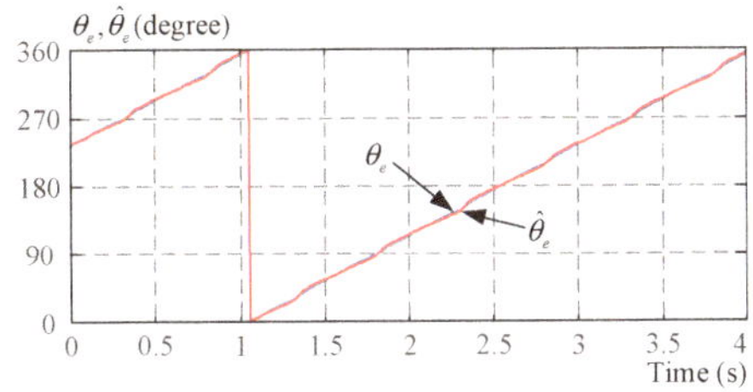

(**a**)

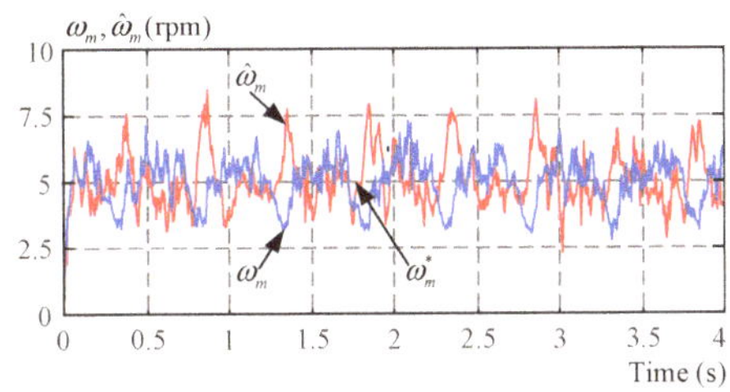

(**b**)

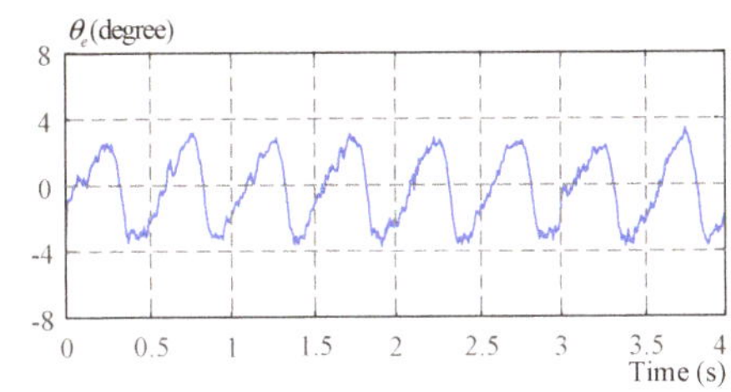

(**c**)

Figure 10. Measured 5 rpm responses. (**a**) estimated and real positions; (**b**) estimated and real speeds; (**c**) estimated position errors.

Figure 14a–d show the α and β current responses using the AVV and ZVV algorithms at 60 rpm under a 2 N·m load. Figure 14a shows the i_α and i_β using the AVV and the blue line is i_α and the red line is i_β. Figure 14b shows the i_α and i_β using the ZVV and the blue line is i_α and the red line is i_β. As we can observe, at low speeds, the ZVV performs better than the AVV with fewer current ripples and current distortions. The reason is that the ZVV uses a lot of zero vectors, but the AVV does not. In addition, the $\alpha\beta$ trajectory for the ZVV is closer to a circle. This is why the ZVV is employed for low-speed ranges. Figure 15a–c illustrate the transitional responses from the ZVV algorithm to the AVV algorithm. Figure 15a shows the estimated and real speeds at about 150 rpm. According to the transition rules in Equations (48)–(50), when the speed is below 60 rpm, the ZVV estimation algorithm is used. However, when the motor is at more than 100 rpm, the AVV rotor estimation algorithm is used. If the speeds are between those two limitations, a weighting factor is considered, which allows the two estimators to transition smoothly to obtain estimated speeds and positions. Figure 15b compares the estimated and real positions, and they are close at a standstill and low-speed ranges. Moreover, during the transitional interval, the estimated and real positions can provide smooth transient responses to demonstrate that the proposed method is practical and useful. Figure 15c demonstrates the estimated position errors. The errors are near six electrical degrees during transitional intervals and then are reduced to two electrical degrees at steady-states. Figure 16a demonstrates the reverse speed responses from 600 rpm to −600 rpm by using a predictive controller. Figure 16b demonstrates the real and estimated position responses, and both of them are similar. The results show that the proposed rotor position estimator can track the real position very well in transient responses and at steady-states. Figure 16c demonstrates the measured estimated rotor position errors, which are near ±4 electrical degrees during the transient responses. The estimated motor position errors in these transient responses are larger than the estimated rotor position errors at a steady-state.

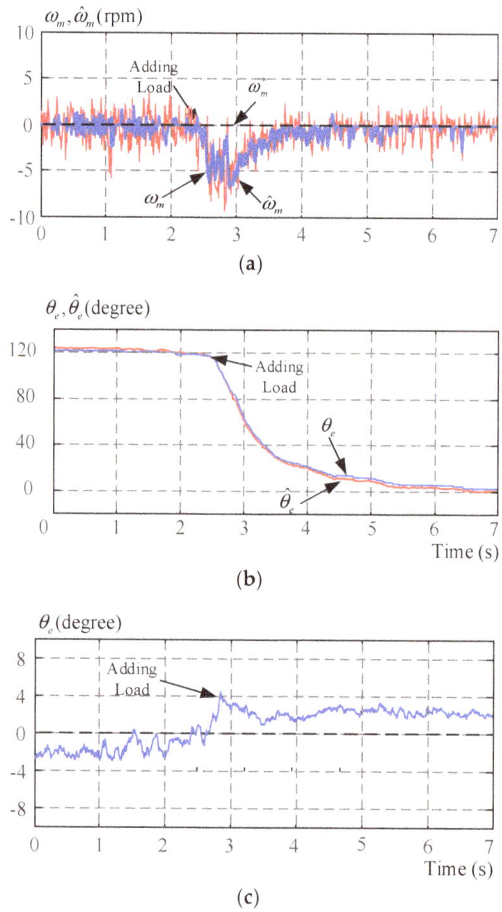

Figure 11. Measured waveforms at 0 rpm and 11 N-m load. (**a**) speeds; (**b**) positions; (**c**) estimated position errors.

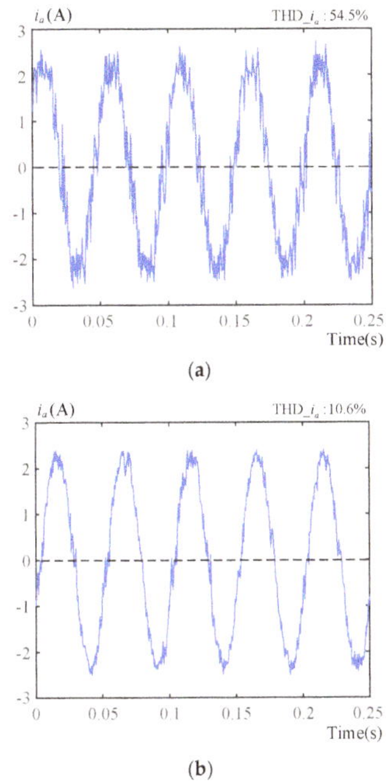

Figure 12. Measured phase current at 600 rpm and 1.5 N-m external load. (**a**) hysteresis current control; (**b**) proposed SVPWM.

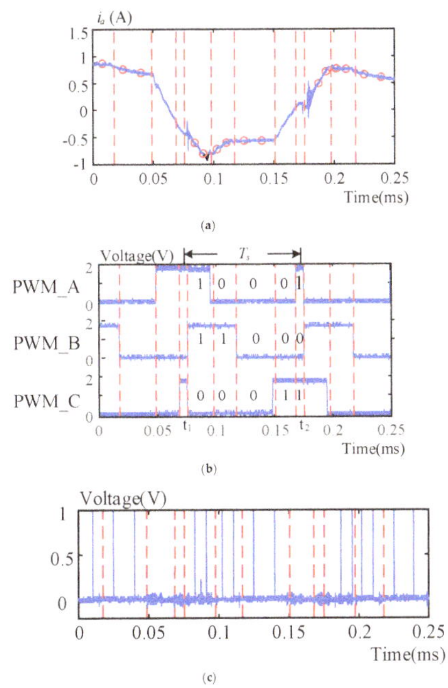

Figure 13. Measured waveforms. (**a**) a-phase current and its sampling points; (**b**) SVPWM; (**c**) sampling and switching points.

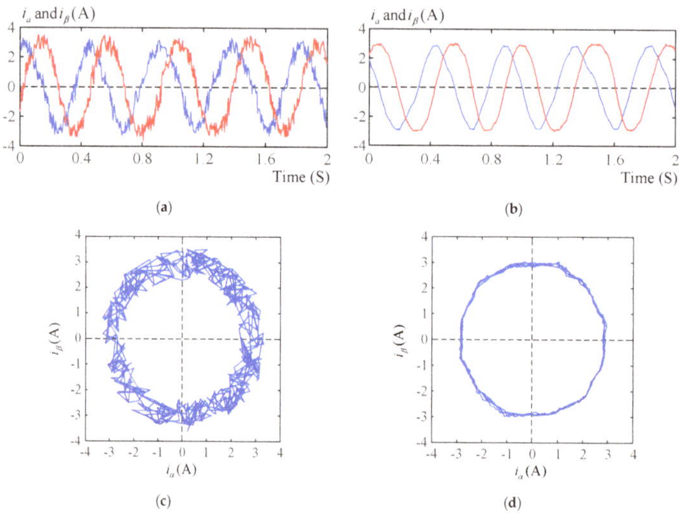

Figure 14. Measured current waveforms at 60 rpm and 2 N-m load. (**a**) axis current using AVV; (**b**) axis current using ZVV; (**c**) axis trajectory using AVV; (**d**) axis trajectory using ZVV.

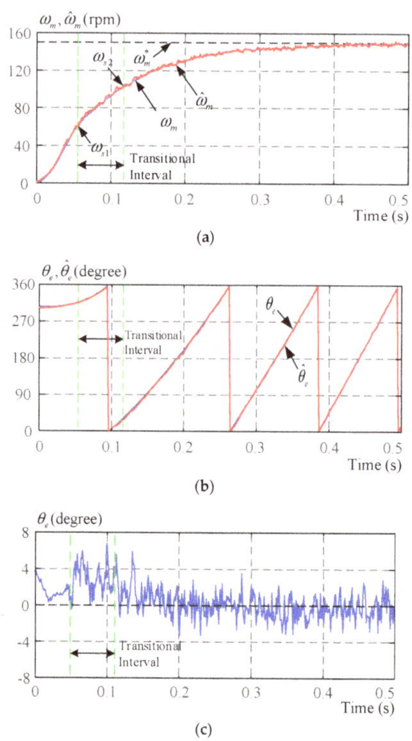

Figure 15. Measured responses during transition intervals. (**a**) speeds; (**b**) positions; (**c**) position estimated errors.

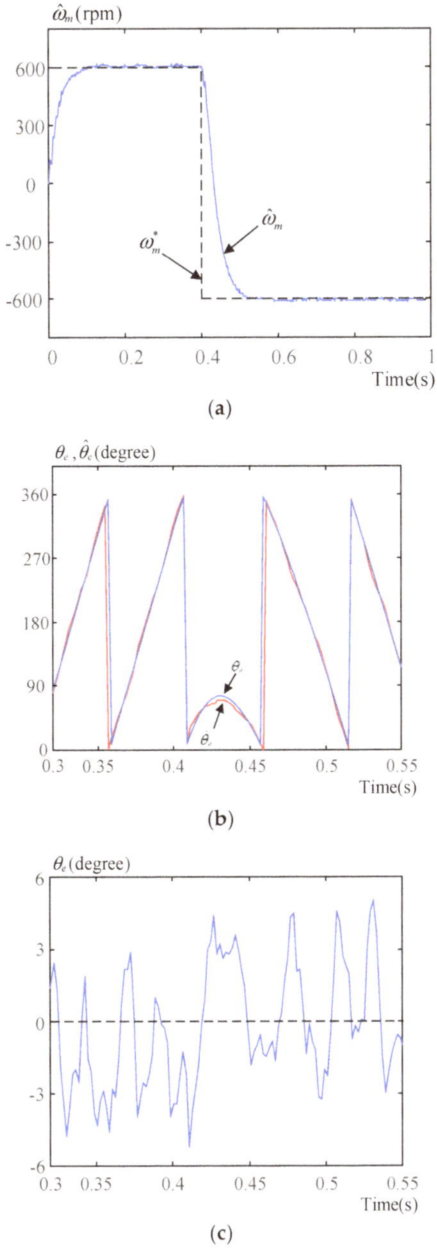

Figure 16. Measured responses from 0 rpm to 600 rpm to −600 rpm. (**a**) speeds; (**b**) positions; (**c**) position estimated errors.

Figure 17a,b show the measured responses of the drive system with different values of external loads. Figure 17a shows the responses when a 1 N·m load is added. As we can observe, the estimated rotor position and estimated rotor speed can follow measured rotor position and rotor speed well. The position error is only ±2 electrical degrees. Figure 17b shows the responses when a 4 N·m load is added, and the estimated position error is near ±3 electrical degrees. Based on these experimental results, the performance of the sensorless drive system can work well at 1 N·m load and 4 N·m load.

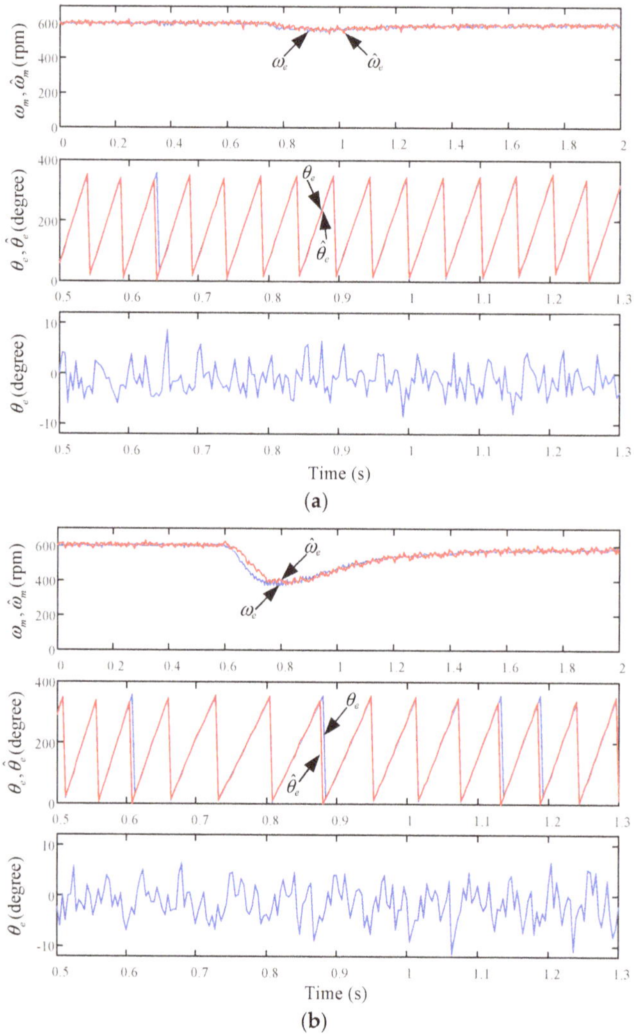

Figure 17. Measured responses of different loads at 600 rpm. (**a**) 1 N·m; (**b**) 4 N·m.

Figure 18a demonstrates a comparison of the speed responses by using different controllers. The PI controller has near 10% overshoot, which requires 0.3 s to reach a steady-state. However, the predictive controller only has a 3% overshoot and can reach a steady-state quickly. Figure 18b demonstrates the load disturbance responses at 600 rpm with a 2 N·m load. The PI controller drops by 150 rpm, but the predictive controller drops by only 60 rpm. In addition, the predictive controller has a quicker recovery time than the PI controller. Figure 18c demonstrates the step-input responses at different speeds by using the proposed predictive controller. All of them are linear responses. The reason is that the predictive controller uses more state variables than the PI controller. This demonstrates that the proposed drive system has wide and adjustable ranges, which include different speed ranges.

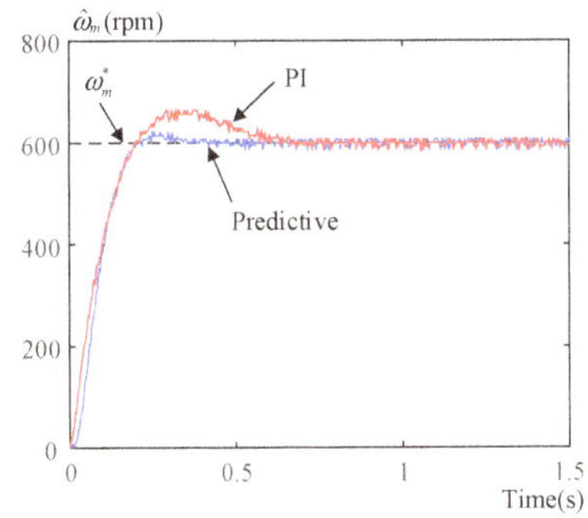

(**a**)

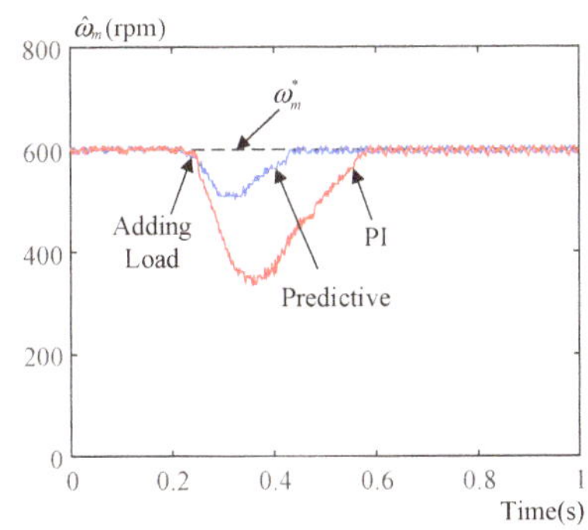

(**b**)

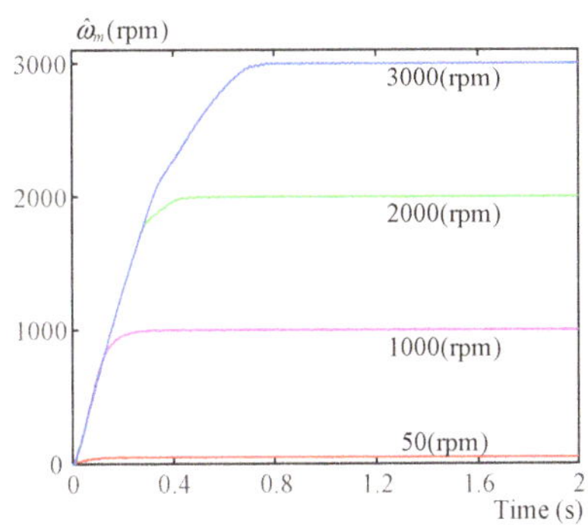

(**c**)

Figure 18. Measured responses at different speeds. (**a**) 600 rpm transient responses; (**b**) 600 rpm 2 N·m load responses; (**c**) different speed responses.

10. Conclusions

A sensorless DSP-based IPMSM drive system using current deviation detection is implemented in this paper. The ZVV rotor estimating method, which is very suitable for IPMSMs operating at zero-speeds and low-speeds, is originally investigated. Experimental results show that this ZVV estimating method can achieve high-performance sensorless speed control at zero-speed under full load conditions. In addition, a linear combination method, which is the simplest method to combine a ZVV algorithm and an AVV algorithm, is also proposed in this paper. As a result, the IPMSM drive system can be operated from 0 r/min to 3000 r/min. This linear combination method is more easily implemented than fuzzy-logic combination methods.

Author Contributions: Conceptualization, T.-H.L.; methodology, T.-H.L.; validation, C.-Y.T. and Z.-Y.W.; resources, T.-H.L.; writing—original draft preparation, T.-H.L. and M.S.M.; writing—review and editing, T.-H.L. and M.S.M.; supervision, T.-H.L.; funding acquisition, T.-H.L. All authors have read and agreed to the published version of the manuscript.

Funding: This research wan funded by the Ministry of Science and Technology, Taiwan, grant number MOST 109-2221-E011-050-, and the APC was funded by the Ministry of Science and Technology, Taiwan.

Conflicts of Interest: The authors declare no conflict of interest.

References

1. Piippo, A.; Hinkkanen, M.; Luomi, J. Analysis of an Adaptive Observer for Sensorless Control of Interior Permanent Magnet Synchronous Motors. *IEEE Trans. Ind. Electron.* **2008**, *55*, 570–576. [CrossRef]
2. Sheng, L.; Sheng, L.; Wang, Y.; Fan, M.; Yang, X. Sensorless Control of a Shearer Short-Range Cutting Interior Permanent Magnet Synchronous Motor Based on a New Sliding Mode Observer. *IEEE Access* **2017**, *5*, 18439–18450. [CrossRef]
3. Batzel, T.; Lee, K. Electric Propulsion With the Sensorless Permanent Magnet Synchronous Motor: Model and Approach. *IEEE Trans. Energy Convers.* **2005**, *20*, 818–825. [CrossRef]
4. Zhang, G.; Wang, G.; Xu, D.G.; Zhao, N. ADALINE-Network-Based PLL for Position Sensorless Interior Permanent Magnet Synchronous Motor Drives. *IEEE Trans. Power Electron.* **2015**, *31*, 1450–1460. [CrossRef]
5. Wu, X.; Huang, S.; Liu, K.; Lu, K.; Hu, Y.; Pan, W.; Peng, X. Enhanced Position Sensorless Control Using Bilinear Recursive Least Squares Adaptive Filter for Interior Permanent Magnet Synchronous Motor. *IEEE Trans. Power Electron.* **2020**, *35*, 681–698. [CrossRef]
6. Colli, V.D.; Di Stefano, R.L.; Marignetti, F. A System-on-Chip Sensorless Control for a Permanent-Magnet Synchronous Motor. *IEEE Trans. Ind. Electron.* **2010**, *57*, 3822–3829. [CrossRef]
7. Chen, Z.; Tomita, M.; Doki, S.; Okuma, S. An extended electromotive force model for sensorless control of interior permanent-magnet synchronous motors. *IEEE Trans. Ind. Electron.* **2003**, *50*, 288–295. [CrossRef]
8. Wang, G.; Ding, L.; Li, Z.; Xu, J.; Zhang, G.; Zhan, H.; Ni, R.; Xu, D.G. Enhanced Position Observer Using Second-Order Generalized Integrator for Sensorless Interior Permanent Magnet Synchronous Motor Drives. *IEEE Trans. Energy Convers.* **2014**, *29*, 486–495. [CrossRef]
9. An, Q.-T.; Zhang, J.; An, Q.; Liu, X.; Shamekov, A.; Bi, K. Frequency-Adaptive Complex-Coefficient Filter-Based Enhanced Sliding Mode Observer for Sensorless Control of Permanent Magnet Synchronous Motor Drives. *IEEE Trans. Ind. Appl.* **2020**, *56*, 335–343. [CrossRef]
10. Nguyen, D.; Dutta, R.; Rahman, M.F.; Fletcher, J.E. Performance of a Sensorless Controlled Concentrated-Wound Interior Permanent-Magnet Synchronous Machine at Low and Zero Speed. *IEEE Trans. Ind. Electron.* **2015**, *63*, 2016–2026. [CrossRef]
11. Sun, X.; Cao, J.; Lei, G.; Guo, Y.; Zhu, J. Speed Sensorless Control for Permanent Magnet Synchronous Motors Based on Finite Position Set. *IEEE Trans. Ind. Electron.* **2020**, *67*, 6089–6100. [CrossRef]
12. Barcaro, M.; Morandin, M.; Pradella, T.; Bianchi, N.; Furlan, I. Iron Saturation Impact on High-Frequency Sensorless Control of Synchronous Permanent-Magnet Motor. *IEEE Trans. Ind. Appl.* **2017**, *53*, 5470–5478. [CrossRef]
13. Shinnaka, S. A New Speed-Varying Ellipse Voltage Injection Method for Sensorless Drive of Permanent-Magnet Synchronous Motors With Pole Saliency—New PLL Method Using High-Frequency Current Component Multiplied Signal. *IEEE Trans. Ind. Appl.* **2008**, *44*, 777–788. [CrossRef]
14. Park, N.-C.; Kim, S.-H. Simple sensorless algorithm for interior permanent magnet synchronous motors based on high-frequency voltage injection method. *IET Electron. Power Appl.* **2014**, *8*, 68–75. [CrossRef]
15. Haque, M.; Zhong, L.; Rahman, M. A sensorless initial rotor position estimation scheme for a direct torque controlled interior permanent magnet synchronous motor drive. *IEEE Trans. Energy Convers.* **2003**, *18*, 1376–1383. [CrossRef]
16. Bui, M.X.; Guan, D.; Xiao, D.; Rahman, M.F. A Modified Sensorless Control Scheme for Interior Permanent Magnet Synchronous Motor Over Zero to Rated Speed Range Using Current Derivative Measurements. *IEEE Trans. Ind. Electron.* **2018**, *66*, 102–113. [CrossRef]

17. Wei, M.Y.; Liu, T.H. A high-performance sensorless position control system of a synchronous reluctance motor using dual current-deviation estimating technique. *IEEE Trans. Ind. Electron.* **2012**, *59*, 3411–3426.
18. Raute, R.; Caruana, C.; Staines, C.S.; Cilia, J.; Sumner, M.; Asher, G. Analysis and Compensation of Inverter Nonlinearity Effect on a Sensorless PMSM Drive at Very Low and Zero Speed Operation. *IEEE Trans. Ind. Electron.* **2010**, *57*, 4065–4074. [CrossRef]
19. Hosogaya, Y.; Kubota, H. Position estimating method of IPMSM at low speed region using dq-axis current derivative without high frequency component. In Proceedings of the 2013 IEEE 10th International Conference on Power Electronics and Drive Systems (PEDS), Kitakyushu, Japan, 22–25 April 2013; pp. 1306–1311.
20. Hua, Y.; Sumner, M.; Asher, G.; Gao, Q.; Saleh, K. Improved sensorless control of a permanent magnet machine using fundamental pulse width modulation excitation. *IET Electron. Power Appl.* **2011**, *5*, 359–370. [CrossRef]
21. Zhang, H.; Liu, W.; Chen, Z.; Mao, S.; Meng, T.; Peng, J.; Jiao, N. A Time-Delay Compensation Method for IPMSM Hybrid Sensorless Drives in Rail Transit Applications. *IEEE Trans. Ind. Electron.* **2018**, *66*, 6715–6726. [CrossRef]
22. Gu, M.; Ogasawara, S.; Takemoto, M. Novel PWM Schemes With Multi SVPWM of Sensorless IPMSM Drives for Reducing Current Ripple. *IEEE Trans. Power Electron.* **2015**, *31*, 6461–6475. [CrossRef]
23. Zhang, Z.; Guo, H.; Liu, Y.; Zhang, Q.; Zhu, P.; Iqbal, R. An Improved Sensorless Control Strategy of Ship IPMSM at Full Speed Range. *IEEE Access* **2019**, *7*, 178652–178661. [CrossRef]
24. Fan, W.; Luo, S.; Zou, J.; Zheng, G. A hybrid speed sensorless control strategy for PMSM based on MRAS and fuzzy control. In Proceedings of the 7th International Power Electronics and Motion Control Conference, Harbin, China, 2–5 June 2012; pp. 2976–2980.
25. Maciejowski, J.M. *Predictive Control with Constraints*; Prentice Hall: New York, NY, USA, 2002.
26. Soeterboek, R. *Predictive Control—A Unified Approach*; Prentice Hall: New York, NY, USA, 1992.
27. Camacho, E.F.; Bordons, C. *Modern Predictive Control*, 2nd ed.; Springer: Berlin/Heidelberg, Germany, 2003.
28. Wang, L. *Model Predictive Control System Design and Implementation Using MATLAB*; Springer: Berlin/Heidelberg, Germany, 2009.
29. Wang, S.; Chai, L.; Yoo, D.; Gan, L.; Ng, K. *PID and Predictive Control of Electrical Drives and Power Converters Using MATLAB/Simulink*; Wiley: Hoboken, NJ, USA, 2015.
30. Rodriguez, J.; Cortes, P. *Predictive Control of Power Converters and Electrical Drives*; Wiley: Hoboken, NJ, USA, 2012.

energies

MDPI

Article

Permanent-Magnet Synchronous Motor Drive System Using Backstepping Control with Three Adaptive Rules and Revised Recurring Sieved Pollaczek Polynomials Neural Network with Reformed Grey Wolf Optimization and Recouped Controller

Chih-Hong Lin

Department of Electrical Engineering, National United University, Miaoli 360, Taiwan; jhlin@nuu.edu.tw; Tel.: +886-3-7382464

Received: 1 October 2020; Accepted: 8 November 2020; Published: 10 November 2020

Abstract: Owing to some nonlinear characteristics in the permanent-magnet synchronous motor (SM), such as nonlinear friction, cogging torque, wind stray torque, external load torque, and unmodeled systems, fine control performances cannot be accomplished by utilizing the general linear controllers. Thereby, the backstepping approach adopting three adaptive rules and a swapping function is brought forward for controlling the rotor motion in the permanent-magnet SM drive system to reduce nonlinear uncertainties effects. To improve the chattering phenomenon, the backstepping control with three adaptive rules using a revised recurring sieved Pollaczek polynomials neural network (RRSPPNN) with reformed grey wolf optimization (RGWO) and a recouped controller is proposed to estimate the internal collection and external collection torque uncertainties, and to recoup the smallest fabricated error of the appraised rule. In the light of the Lyapunov stability, the on-line parametric training method of the RRSPPNN can be derived through an adaptive rule. Furthermore, to obtain a beneficial learning rate and improve the convergence of the weights, the RGWO algorithm adopting two exponential-functional adjustable factors is applied to adjust the two learning rates of the weights. Then, the efficiency of the used controller is validated by test results.

Keywords: backstepping control; Lyapunov stability theorem; grey wolf optimization; permanent-magnet synchronous motor; Sieved-Pollaczek polynomials neural network

---

1. Introduction

The permanent-magnet synchronous motors (SMs) [1] with many merits are superior to the switched reluctance motors (SRMs) and induction motors (IMs). The permanent-magnet SMs [2] can offer higher efficiency, higher power density, lower power loss, and higher robustness in comparsion with the SRMs and the IMs at the same volume. The permanent-magnet SMs have mostly adopted the field-oriented control technique owing to their easy implementation. Thereby, output torque can result in lower ripple torque in comparison with the SRMs and IMs at the same output torque. On the other hand, the permanent-magnet SMs controlled by field-oriented control [1,2], which can achieve fast four-quadrant operation, are much less sensitive to the parameter variations of the motor. Therefore, they have been widely used in many industrial applications such as robotics [1], computer numerical control (CNC) tools [2], and other mechatronics [3].

The backstepping designs [4] are befitting for a large type of linearizable nonlinear systems. Each backstepping phase can produce a novel fictitious-control design denoted by previous design phases. When the procedure ends, a feedback design can achieve the primitive design aim by utilizing the last Lyapunov function that made up by adding into the Lyapunov functions regarding all individual design phases [4]. Moreover, the backstepping control with an adaptive law has been applied in a microgyroscope [5], automatic train operation [6], aircraft flight control [7], power switcher [8], and synchronous generator [9]. Further, the backstepping control by using the modified recurrent Rogers–Szego polynomials neural network with decorated gray wolf optimization (DGWO) has been used in the permanent-magnet synchronous linear motor drive system [10]. The recouped mechanisms in these methods were absent for the estimated uncertainty. Therefore, the main aim of this paper is to improve control performances by using the proposed backstepping control with three adaptive rules using a revised recurring sieved Pollaczek polynomials neural network (RRSPPNN) with reformed grey wolf optimization (RGWO) and recouped controller.

Neural networks (NNs) have better approximation behavior in modeling [11], identification [12] and control [13] of systems. These NNs were the feedforward network structures with static mapping functions. They may not exactly respond the dynamic behavior in real time because of absent feedback loop. The recurrent NNs with feedback loop have been broadly used in the prediction of photovoltaic power output [14], an accurate electricity spot price prediction scheme [15], a photovoltaic power forecasting approach [16], and an adaptive energy management control [17] as result of higher certification and finer control performance. The primary significant property of the recurrent NN is to recollect feedback message of the foretime effect in the same neuron via its self-link. Moreover, in the general recurrent NNs, the specific self-link feedback of the hidden neuron or output neuron is in charge of recollecting the designated preceding activation of the hidden neuron or output neuron and provender to itself only. Therefore, the outputs of the other neurons have no capacity to infect the designated neuron. However, in a complex nonlinear dynamic system such as the permanent-magnet SM with nonlinear wind stray torque, flux saturation torque, cogging torque, external load torque, and interference of time-varying uncertainties, in general, seriously effect system performances. Hence, if each neuron in the recurring neural networks is considered as a state in the nonlinear dynamic systems, the self-connection feedback type is unable to approximate the dynamic systems efficiently. Due to the recurring neurons, it has certain dynamical merits over static NN and it also has been proverbially applied in photovoltaic power forecasting and electricity spot price prediction. However, these NNs take a longer time to process the online training procedure. Hence, some functional-type NNs, such as the amended recurrent Gegenbauer-functional-expansions NN [18], reformed recurrent Hermite polynomial NN [19], and mended recurrent Romanovski polynomials NN [20], have been broadly applied in the control and identification of nonlinear systems as a result of less calculation complexity. However, the adjustment mechanics of weights were not discussed in these control methods that combined with NNs. It is leads to larger error in control and identification for system. Moreover, the sieved Pollaczek polynomials [21] belong to the sieved orthogonal polynomials, according to Ismail [21]. However, the sieved Pollaczek polynomials combined with the NN have never presented in any control of nonlinear systems. Although the feedforward sieved Pollaczek polynomials neural network (SPPNN) can approximate nonlinear function, it may not be an approximated dynamic act of nonlinear uncertainties as a result of lacking a reflect loop. Because of the many benefits compared to the feedforward SPPNN, the revised recurring sieved Pollaczek polynomials neural network (RRSPPNN) control is not introduced yet for controlling the permanent-magnet SM drive system to improve the performances of the nonlinear system and computation complexity. However, the backstepping technique utilizing the RRSPPNN with error recouped agency to decrease uncertainties is thus the main motivation in this topic. Additionally, these learning rates, by utilizing acceleration factors, did not present that the convergent speed of weights is tardy.

A multi-objective grey wolf optimization (GWO) proposed by Emary et al. [22] was used to attribute the reduction of system. A GWO and conventional NN training method proposed by Mosavi et al. [23] was used in a sonar dataset category. Khandelwal et al. [24] proposed to track the programming question of transmitting network by utilizing the modified GWO. Mirjalili et al. [25] put forward a hunting mechanism of GWO to mimic the social behavior. Even though these algorithms are highly competitive and have been used in certain fields, such as distribution system [26], melanoma detection [27], and feature selection in classification [28], they have poor exploration capability and suffer from local optima stagnation. So, to improve the explorative abilities of GWO, a reformed grey wolf optimization (RGWO) algorithm adopting two exponential-functional adjustable factors is put forward as the novel method in this paper. This newly proposed algorithm makes up two revisions: Firstly, it can explore new regions in the look for space because of diverse locations assigned to the leaders. This can increase the exploration and avoid local perfect stagnation problem. Secondly, an opposition-based learning method has been used in the initial half of iterations to provide diversity among the search agents. To speed up the convergence of weights in the RRSPPNN, the RGWO with two exponential-functional adjustable factors, that is the novel method, is used to adjust the two learning rates of the weights. This novel method can prevent premature convergence and to acquire optimal learning rates with fast convergence.

The better control performance of the permanent-magnet SM drive system cannot be reached by utilizing the linear controller due to the influences of these uncertainties. To heighten robustness, the backstepping approach with three adaptive rules and a swapping function is proposed to control the permanent-magnet SM drive system to trace different periodical references. With the backstepping approach with three adaptive rules and a swapping function, the rotor position of the permanent-magnet SM drive system preserves the merits of fine transient control performance and robustness to uncertainties for the tracedifferent periodical references. Moreover, to improve the large chattering influence under uncertainties, the backstepping control with three adaptive rules by utilizing RRSPPNN with RGWO is proposed to estimate the internal bunched uncertainty and external bunched force uncertainty and the recouped controller to recoup the smallest fabricated error of the appraised rule.

Furthermore, the RGWO algorithm by using two exponential-functional adjustable factors that is applied for regulating two learning rates of the weights in the RRSPPNN is a novel method to speed up the convergence of weights in this paper. Finally, the efficiency of the backstepping control with three adaptive rules using RRSPPNN with RGWO and recouped controller is validated by some test results.

The important issue in this paper is described below. Section 2 presents the models and conformation of the permanent-magnet SM drive. Section 3 describes the backstepping control with three adaptive rules using RRSPPNN with RGWO and the recouped controller. Section 4 is the examination consequences for the permanent-magnet SM utilizing three control methods at five tested events. Section 5 is the conclusion.

2. Models and Conformation of Permanent-Magnet SM Drive

2.1. Models of Permanent-Magnet SM

For simplicity, in the three-phase $as-bs-cs$ axis coordinate frames via the Clarke and Park transformations, the voltage equations in the coordinate frames transformation from the three-phase $as-bs-cs$ axis to the $qs-ds$ axis in the permanent-magnet SM [1] are typified by

$$u_{qs} = r_s i_{qs} + L_{qs} di_{qs}/dt + \omega_{es}(L_{ds} i_{ds} + \lambda_{ps}) \quad (1)$$

$$u_{ds} = r_s i_{ds} + (L_{ds} di_{ds}/dt + \lambda_{ps}) - \omega_{es} L_{qs} i_{qs} \quad (2)$$

$$\omega_{es} = p_s \omega_{r1} \quad (3)$$

$$\theta_{es} = p_s \theta_{r1} \quad (4)$$

The electromagnetic power P_{es} in the air gap as well as the electromagnetic torque d_{e1} [1] can be typified by:

$$P_{es} = d_{e1}\omega_{es} = 3p_s[\lambda_{ps} i_{qs} + (L_{ds} - L_{qs}) i_{ds} i_{qs}]\omega_{es}/2 \quad (5)$$

$$d_{e1} = 3p_s[\lambda_{ps} i_{qs} + (L_{ds} - L_{qs}) i_{ds} i_{qs}]/2 \quad (6)$$

The dynamic equation can be typified by:

$$H_s d\omega_{r1}/dt + U_s \omega_{r1} = d_{e1} - d_{l1} - d_{w1} - d_{f1} - d_{c1} \quad (7)$$

where $u_{ds} - u_{qs}$, $i_{ds} - i_{qs}$ and $L_{ds} - L_{qs}$ stand for the $qs - ds$ axis voltages, the $qs - ds$ axis currents, and the $qs - ds$ axis inductances; r_s and p_s stand for the phase winding resistance and the number of pole pairs; λ_{ps}, ω_{es}, and ω_{r1} stand for the permanent-magnet flux linkage, the electric angular speed, and the angular speed of the rotor; θ_{es} and θ_{r1} stand for the electrical angular position and the mechanical angular position of the rotor; d_{e1}, d_{l1}, d_{f1}, d_{c1}, and d_{w1} stand for the electromagnetic torque, the external load torque, the flux saturation torque, the cogging torque, and the wind stray torque; H_s and U_s stand for total moment of inertia of permanent-magnet SM and the total viscous frictional coefficient of permanent-magnet SM.

2.2. Conformation of Permanent-Magnet SM Drive

The decoupled control technology of the permanent-magnet SM drive system is general adopting the field-oriented control (FOC) [1,2] to raise dynamic performance. The electromagnetic torque is produced by the $qs-$ axis current based on the FOC and the rotor flux is generated by the $ds-$ axis current only. When i_{ds} is equal to zero and λ_{ps} is equal to a constant in Equations (5) and (6), then the electromagnetic torque d_{e1} will be direct ratio to i_{qs} for a permanent-magnet SM drive in the closed-loop control. When the generated torque is linearly direct ratio to the $qs-$ axis current, and the $ds-$ axis rotor flux is a fixed value, the larger torque per ampere can be reached. The electromagnetic torque Equation (6) can be typified by

$$d_{e1} = 3\lambda_{ps} i_{qs}^{*}/2 = k_s i_{qs}^{*} \quad (8)$$

where $k_s = 3p_s\lambda_{ps}/2$ stands for the propulsion coefficient and i_{qs}^{*} stands for the mandate of control current. The permanent-magnet SM drive system can be predigested as $W_t(s) = 1/(sH_s + U_s)$.

Figure 1 is the conformation of FOC permanent-magnet SM drive system, which makes up an encoder and three Hall sensors, a permanent-magnet SM, a sinusoidal pulse-width-modulation (SPWM) current control, a coordinate transformation including inverse Park and Clarke coordinate's transformations, cos/sin generator, a speed control loop and a position control loop. The control technologies in the real-time realization are realized by utilizing the digital signal processor (DSP) controller. Rotor of permanent-magnet SM is equipped on magnet force brake that is mounted with different sizes of iron disks to change the total moment of inertia and the total viscous frictional coefficient, and to add load torque.

The FOC was realized by a digital signal processor (DSP) controller. The nominal values of used permanent-magnet SM are given as 3-phase, 2-poles, 60 Hz, 220 V, 1 kW, 2.8 A, and 3600 r/min. For the convenience of controller design, the position and speed signals in the control loop are set at 1 V = 50 rad and 1 V = 50 rad/s, respectively. The mechanical and electrical parameters of the permanent-magnet SM are given as $H_s = 2.142 \times 10^{-3}$ Nm sec^2 $= 0.1071$ Nm sec rad/V, $U_s = 5.86 \times 10^{-3}$ Nms/rad $= 0.293$ Nm/V, $r_s = 2.5\ \Omega$, $L_{ds} = L_{qs} = 4.62$ mH, $k_s = 0.947$ Nm/A by means of an open circuit test, short circuit test, blocked rotor test, and added load test. With the fulfillment of FOC [1–3], the permanent-magnet SM drive system can be predigested as the control block diagram portrayed in Figure 2. The perfect electromagnetic property for the drive system is hence implemented by controlling the torque current distributions to lie in the $qs-$ axis current when the d-axis current is equal to zero. Then, the torque per amp property for the drive system will generate.

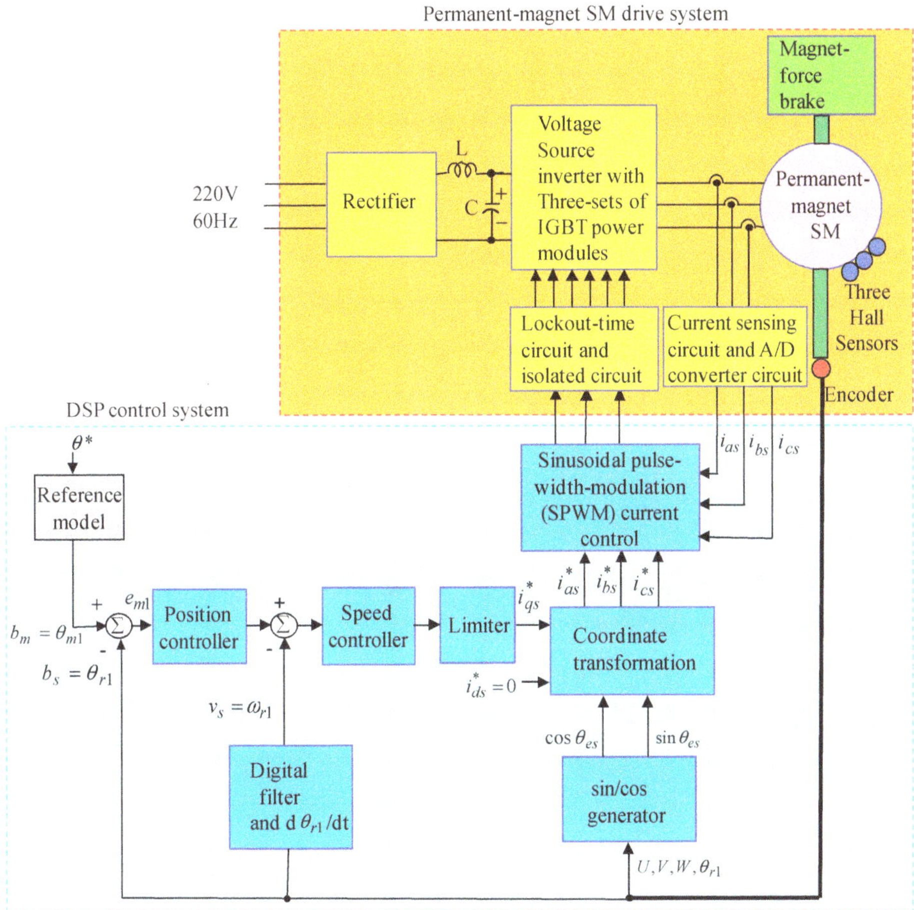

Figure 1. Conformation of FOC permanent-magnet SM drive system with DSP controller.

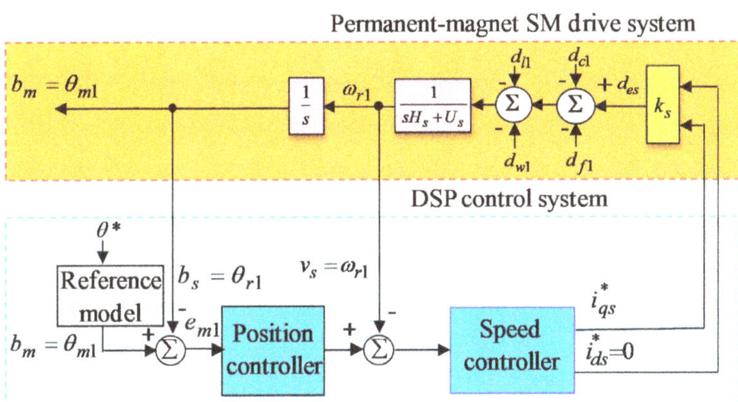

Figure 2. Predigested controller with control block diagram.

3. Design of the Controller

When the permanent-magnet SM drive system with the electromagnetic torque, the wind stray torque, the flux saturation torque, the cogging torque, the parametric variations, and the external load torque disturbance is enacted, then Equation (7) is typified by

$$dv_s/dt = (e_s + \Delta e_s)v_s + (f_s + \Delta f_s)l_s + g_s(d_{l1} + d_{w1} + d_{f1} + d_{c1}) \quad (9)$$

$$d\theta_{r1}/dt = \omega_{r1} = v_s \quad (10)$$

$$b_s = \theta_{r1} \quad (11)$$

where θ_{r1} and v_s stand for the rotor position and rotor speed of the SM to be presumed bounded; Δe_s and Δf_s stand for two parametric uncertainties from H_s and U_s to be presumed bounded; $e_s = -U_s/H_s$; $f_s = k_s/H_s > 0$; $g_s = -1/H_s$ stand for three real numbers to be presumed bounded; $l_s = i_{qs}$ is the control propulsion of the permanent-magnet SM drive system, i.e., the propulsion current. Equation (9) can be typified by

$$dv_s/dt = e_s v_s + f_s l_s + d_{l1} + d_{w1} + d_{f1} + d_{c1} \quad (12)$$

where $w_1 = \Delta e_s v_s$ and $w_2 = \Delta f_s d_s$ stand for two parametric variation that are to be presumed bounded; $w_3 = g_s(d_{f1} + d_{c1})$ and $w_4 = g_s(d_{l1} + d_{w1})$ stand for the internal bunched uncertainty and external bunched force uncertainty to be presumed bounded.

The trace reference locus $b_m = \theta_{m1}$ is the control goal. The design procedure is as below.

The trace error is typified by

$$e_{m1} = b_m - b_s = \theta_{m1} - \theta_{r1} \quad (13)$$

Take differential of (13) by

$$de_{m1}/dt = d\theta_{m1}/dt - d\theta_{r1}/dt = d\theta_{m1}/dt - v_s \quad (14)$$

The stabilizing function is typified by:

$$\varepsilon_{m1} = k_{m1}e_{m1} + d\theta_{m1}/dt + k_{m2}\mu_m \quad (15)$$

where k_{m1} and k_{m2} stand for two positive constants; $\mu_m = \int e_{m1}(t)dt$ stands for the integral factor.

The fictitious trace error is typified by:

$$e_{m2} = v_s - \varepsilon_{m1} \quad (16)$$

Take the differential of (16) by

$$de_{m2}/dt = dv_s/dt - d\varepsilon_{m1}/dt = (e_s v_s + f_s l_s + w_1 + w_2 + w_3 + w_4) - d\varepsilon_{m1}/dt \quad (17)$$

where w_1, w_2, and w_3 stand for three unknown parameters. The estimated errors are typified by

$$e_{w1} = \hat{w}_1 - w_1 \quad (18)$$

$$e_{w2} = \hat{w}_2 - w_2 \quad (19)$$

$$e_{w3} = \hat{w}_3 - w_3 \quad (20)$$

where e_{w1}, e_{w2}, and e_{w3} are the estimated errors; $\hat{w}_1$, $\hat{w}_2$, and $\hat{w}_3$ are the estimated values of w_1, w_2 and w_3. The internal bunched uncertainty and external bunched force uncertainty w_4 satisfies the condition $|w_4| \le \beta_1$ and is to be presumed bounded.

3.1. Design of the Backstepping Control with Three Adaptive Rules and a Swapping Function

So, to design the backstepping control with three adaptive rules and a swapping function, the Lyapunov function can be typified by:

$$y_{m1} = e_{m1}^2/2 + e_{m2}^2/2 + e_{w1}^2/(2\varsigma_1) + e_{w2}^2/(2\varsigma_2) + e_{w3}^2/(2\varsigma_3) + c_{m2}\mu_m^2/2 \tag{21}$$

By taking the differential of y_{m1} and by utilizing Equations (14)–(20) and the integral factor $\mu_m = \int e_{m1}(t)dt$, Equation (21) can be typified by

$$\begin{aligned} dy_{m1}/dt &= e_{m1}de_{m1}/dt + e_{m2}de_{m2}/dt + e_{w1}\varsigma_1^{-1}de_{w1}/dt + e_{w2}\varsigma_2^{-1}de_{w2}/dt \\ &\quad + e_{w3}\varsigma_3^{-1}de_{w3}/dt + k_{m2}\mu_m d\mu_m/dt \\ &= e_{m1}(d\theta_{m1}/dt - v_s) + e_{m2}((e_s v_s + f_s l_s + w_1 + w_2 + w_3 + w_4) - d\varepsilon_{m1}/dt) \\ &\quad + e_{w1}\varsigma_1^{-1}de_{w1}/dt + e_{w2}\varsigma_2^{-1}de_{w2}/dt + e_{w3}\varsigma_3^{-1}de_{w3}/dt + k_{m2}\mu_m d\mu_m/dt \\ &= e_{m1}(-k_{m1}e_{m1} + \varepsilon_{m1} - v_s) + e_{m2}(e_s v_s + f_s l_s + w_1 + w_2 + w_3 + w_4) - d\varepsilon_{m1}/dt) + e_{w1}\varsigma_1^{-1}de_{w1}/dt \\ &\quad + e_{w2}\varsigma_2^{-1}de_{w2}/dt + e_{w3}\varsigma_3^{-1}de_{w3}/dt + k_{m2}\mu_m d\mu_m/dt \\ &= e_{m1}(-k_{m1}e_{m1} - e_{m2}) + e_{m2}(e_s v_s + f_s l_s + w_4 - d\varepsilon_{m1}/dt) + \hat{w}_1\varsigma_1^{-1}d\hat{w}_1/dt + (e_{m2}w_1 - w_1\varsigma_1^{-1}d\hat{w}_1/dt) \\ &\quad + \hat{w}_2\varsigma_1^{-1}d\hat{w}_2/dt + (e_{m2}w_2 - w_2\varsigma_1^{-1}d\hat{w}_2/dt) + \hat{w}_3\varsigma_1^{-1}d\hat{w}_3/dt + (e_{m2}w_3 - w_3\varsigma_1^{-1}d\hat{w}_3/dt) \end{aligned} \tag{22}$$

In accordance with Equation (22), the control propulsion l_s in the backstepping control with three adaptive rules and a swapping function can be typified by

$$l_s = i_{qs}^* = f_s^{-1}[e_{m1} - k_{m3}e_{m2} - e_s v_s + d\varepsilon_{m1}/dt - \beta_1 sgn(e_{m2}) - (\hat{w}_1 + \hat{w}_2 + \hat{w}_3)] \tag{23}$$

where k_{m3} stands for a positive constant; β_1 stands for upper bound that is a constant, and $f_s^{-1}\beta_1\mathrm{sgn}(e_{m2})$ stands for a swapping function. By utilizing Equation (23), Equation (22) can be typified by:

$$\begin{aligned} dy_{m1}/dt &= -k_{m1}e_{m1}^2 - c_{m3}e_{m2}^2 - e_{m2}[\beta_1 sgn\ (e_{m2}) - w_4] - e_{m2}[\hat{w}_1 + \hat{w}_2 + \hat{w}_3] + \hat{w}_1\varsigma_1^{-1}d\hat{w}_1/dt \\ &+ w_1(e_{m2} - \varsigma_1^{-1}d\hat{w}_1/dt) + \hat{w}_2\varsigma_1^{-1}d\hat{w}_2/dt + w_2(e_{m2} - \varsigma_1^{-1}d\hat{w}_2/dt) + \hat{w}_3\varsigma_1^{-1}d\hat{w}_3/dt + w_3(e_{m2} - \varsigma_1^{-1}d\hat{w}_3/dt) \end{aligned} \tag{24}$$

For reaching $dy_{m1}(t)/dt \le 0$, three adaptive rules $d\hat{w}_1/dt$, $d\hat{w}_2/dt$, and $d\hat{w}_3/dt$ can be typified by

$$d\hat{w}_1/dt = \varsigma_1 e_{m2} \tag{25}$$

$$d\hat{w}_2/dt = \varsigma_2 e_{m2} \tag{26}$$

$$d\hat{w}_3/dt = \varsigma_3 e_{m2} \tag{27}$$

By utilizing Equations (25)–(27), and $|w_4| \le \beta_1$, then Equation (24) can be typified by

$$\begin{aligned} dy_{m1}/dt &= -k_{m1}e_{m1}^2 - k_{m3}e_{m2}^2 - e_{m2}[\beta_1 sgn\ (e_{m2}) - w_4] \\ &\le -k_{m1}e_{m1}^2 - k_{m3}e_{m2}^2 - |e_{m2}|[\beta_1 - |w|] \\ &\le -k_{m1}e_{m1}^2 - k_{m3}e_{m2}^2 \le 0 \end{aligned} \tag{28}$$

Equation (28) shows $dy_{m1}(t)/dt$ to be negative semi-definite (i.e., $y_{m1}(t) \le y_{m1}(0)$), meaning that e_{m1} and e_{m2} are bounded. The following term is typified by

$$h_{m1}(t) = k_{m1}e_{m1}^2 + k_{m3}\ e_{m2}^2 = -dq_{m1}/dt \tag{29}$$

The integration of Equation (29) is typified by:

$$\int_0^t h_{m1}(v)dv = q_{m1}(e_{m1}(0),\ e_{m2}(0)) - q_{m1}(e_{m1}(t), e_{m2}(t)) \tag{30}$$

Because $q_{m1}(e_{m1}(0),\ e_{m2}(0))$ is bounded and $q_{m2}(e_{m1}(t),\ e_{m2}(t))$ is nonincreasing and presumed bounded, then $\lim_{t\to\infty}\int_0^t q_{m1}(\mathrm{v})d\tau < \infty$. Moreover, $dh_{m1}(t)/dt$ is presumed bounded, hence $h_{m1}(t)$ is a uniformly continuous function. By utilizing the Barbalat's lemma [29,30], it can be portrayed that $\lim_{t\to\infty} h_{m1}(t) = 0$. That is, e_{m1} and e_{m2} will converge to zero when $t \to \infty$. Furthermore, $\lim_{t\to\infty} \theta_{r1}(t) = \theta_{m1}$ and $\lim_{t\to\infty} v_s = d\theta_{m1}/dt$. The stability of the backstepping control with three adaptive rules and a swapping function can be guaranteed, and consequently, the control block diagram is portrayed in Figure 3.

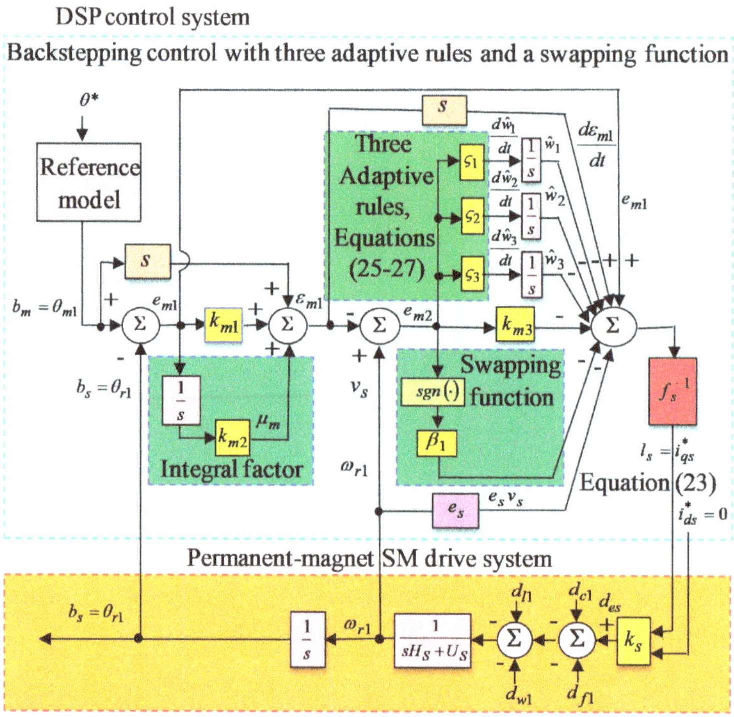

Figure 3. Block diagram by using the backstepping control with three adaptive rules and a swapping function.

3.2. Design of the Backstepping Control with Three Adaptive Rules Using RRSPPNN with RGWO and Recouped Controller

Because the internal bunched uncertainty and external bunched force uncertainty w_4 is unknown, and its upper bound is troublesome to be decided. The appraised value $\hat{w}_4$ of the internal bunched uncertainty and external bunched force uncertainty w_4 is not easy to be estimated, and consequently, the revised recurring sieved Pollaczek polynomials neural network (RRSPPNN) is proposed to adapt the real value of the internal bunched uncertainty and external bunched force uncertainty w_4.

3.2.1. Constitution of RRSPPNN

The RRSPPNN has a three-layer constitution, with the first layer (input layer), the second layer (hidden layer 1), and the third layer (output layer) portrayed in Figure 4. The semaphore intentions in each node for each layer are explained in the following expression.

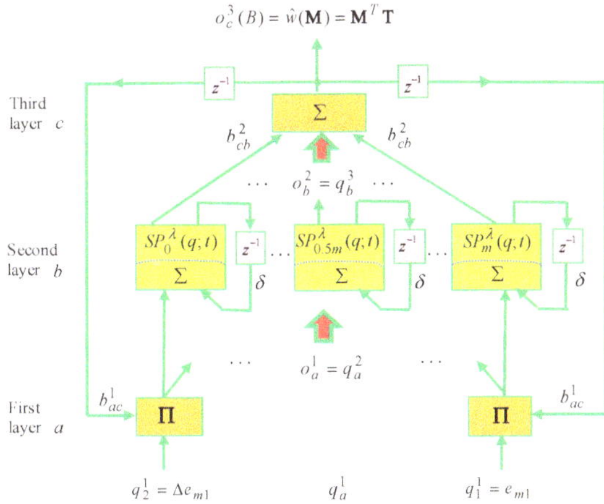

Figure 4. Constitution of the RRRSPNN.

At the first layer, input semaphore and output semaphore are typified by

$$ne_a^1 = \prod_c q_a^1(B) b_{ac}^1 o_c^3(B-1), o_a^1 = g_a^1(ne_a^1) = ne_a^1,\ a = 1,\ 2 \quad (31)$$

where $q_1^1 = \theta_{m1} - \theta_{r1} = e_{m1}$ and $q_2^1 = e_{m1}(1 - z^{-1}) = \Delta e_{m1}$ stand for the speed discrepancy and speed discrepancy alteration, respectively. B is the iteration count. b_{ac}^1 stands for the recurring weight through the third layer and the first layer. o_c^3 stands for the output of node at the third layer. The symbol $\prod$ stands for a multiply factor.

At the second layer, input semaphore and output semaphore are typified by

$$ne_b^2(B) = \sum_{a=1}^{2} o_a^1(B) + \delta\, o_b^2(B-1),\ o_b^2 = g_b^2(ne_b^2) = SP_b^\lambda(ne_b^2; q),\ b = 0,\ 1, \cdots,\ m-1 \quad (32)$$

where δ stands for the recurring gain at the second layer. Sieved-Pollaczek polynomials function [21,31] is adopted as the activation function. $SP_j^\lambda(q;t)$ stands for the sieved Pollaczek polynomials in the interval $[-1, 1]$. $SP_0^\lambda(q;t) = 1$, $SP_1^\lambda(q;t) = 2q$ and $SP_2^\lambda(q;t) = 4q^2 - 1$ stand for the zero-order, first order and second order sieved Pollaczek polynomials, respectively. The sieved Pollaczek polynomials may be generated by the recurrence relation [21,31] $SP_{n+1}^\lambda(q;t) = 2qSP_n^\lambda(q;t) - SP_{n-1}^\lambda(q;t)$. The symbol $\sum$ stands for a summation factor.

At the third layer, semaphore and output semaphore are typified by

$$ne_o^3 = \sum_{b=0}^{m-1} b_{cb}^2 o_b^2(B),\ \ o_c^3 = g_c^3(ne_c^3) = ne_c^3,\ c = 1 \quad (33)$$

where b_{cb}^2 stands for the connecting weight through the second layer and the third layer. g_c^3 stands for the linear activation function. The output $o_c^3(B)$ at the third layer of the RRSPPNN can be typified by:

$$o_c^3(B) = \hat{w}_4(\mathbf{M}) = \mathbf{M}^T\mathbf{T} \quad (34)$$

where $\mathbf{M} = \begin{bmatrix} b_{10}^2 & \cdots & b_{1,\,m-1}^2 \end{bmatrix}^T$ and $\mathbf{T} = \begin{bmatrix} o_0^2 & \cdots & o_{m-1}^2 \end{bmatrix}^T$ stands for the weight vector at the third layer and the input vector at the third layer, respectively. The smallest fabricated error φ_{w4} is typified by

$$\varphi_w = w_4 - w_4(\mathbf{M}^*) = w_4 - (\mathbf{M}^*)^T\mathbf{T} \quad (35)$$

where $\mathbf{M}^*$ stands for an ideal weight vector that reaches the smallest fabricated error. So as to make up the smallest fabricated error φ_w, the recouped controller x_c with an appraised rule is proposed. It is presumed that the small positive number e_{w4} stands for greater than absolute value of φ_w, i.e., $e_{w4} \geq |\varphi_w|$. The Lyapunov function is typified by

$$y_{m2} = y_{m1} + (\widetilde{e}_{w4})^2/(2\sigma_1) + (\mathbf{M} - \mathbf{M}^*)^T(\mathbf{M} - \mathbf{M}^*)/(2\gamma_1) \tag{36}$$

where σ_1 stands for an adaptive gain. $\widetilde{e}_{w4} = \hat{e}_{w4} - e_{w4}$ stands for the appraised error to be presumed bounded. By taking the derivative of y_{m2} utilizing Equations (14)–(20) and the integral factor $\mu_m = \int e_{m1}(t)dt$, then Equation (36) is typified by

$$\begin{aligned} dy_{m2}/dt = e_{m1}(-k_{m1}e_{m1} - e_{m2}) + e_{m2}(e_s v_s + f_s l_s + w_4 - d\varepsilon_{m1}/dt) + \hat{w}_1\varsigma_1^{-1}d\hat{w}_1/dt \\ + w_1(e_{m2} - \varsigma_1^{-1}d\hat{w}_1/dt) + \widetilde{e}_{w4}\sigma_1^{-1}d\widetilde{e}_{w4}/dt + \hat{w}_2\varsigma_1^{-1}d\hat{w}_2/dt + w_2(e_{m2} - \varsigma_1^{-1}d\hat{w}_2/dt) \\ + \hat{w}_3\varsigma_1^{-1}d\hat{w}_3/dt + w_3(e_{m2} - \varsigma_1^{-1}d\hat{w}_3/dt) + \gamma_1^{-1}(\mathbf{M} - \mathbf{M}^*)^T d\,\mathbf{M}/dt \end{aligned} \tag{37}$$

In accordance with Equation (37), the control propulsion $l_s = \hat{l}_s$ in the backstepping control with three adaptive rules by using RRSPPNN with RGWO and recouped controller can be typified by

$$l_s = \hat{l}_s = i^*_{qs} = f_s^{-1}[e_{m1} - c_{m3}e_{m2} - e_s v_s + d\varepsilon_{m1}/dt - (\hat{w}_1 + \hat{w}_2 + \hat{w}_3) - \hat{w}_4(\mathbf{M}) - x_c] \tag{38}$$

By utilizing Equation (38), then Equation (37) can be typified by

$$\begin{aligned} dy_{m2}/dt = -k_{m1}e_{m1}^2 - k_{m3}e_{m2}^2 + e_{m2}[w_4 - \hat{w}_4(\mathbf{M}) - x_c - (\hat{w}_1 + \hat{w}_2 + \hat{w}_3)] \\ + \hat{w}_1\varsigma_1^{-1}d\hat{w}_1/dt + w_1(e_{m2} - \varsigma_1^{-1}d\hat{w}_1/dt) + \widetilde{e}_{w4}\sigma_1^{-1}d\widetilde{e}_{w4}/dt + \hat{w}_2\varsigma_1^{-1}d\hat{w}_2/dt + w_2(e_{m2} - \varsigma_1^{-1}d\hat{w}_2/dt) \\ + \hat{w}_3\varsigma_1^{-1}d\hat{w}_3/dt + w_3(e_{m2} - \varsigma_1^{-1}d\hat{w}_3/dt) + \gamma_1^{-1}(\mathbf{M} - \mathbf{M}^*)^T d\,\mathbf{M}/dt \end{aligned} \tag{39}$$

By utilizing Equations (25)–(27) and $\widetilde{e}_{w4} = \hat{e}_{w4} - e_{w4}$, then Equation (39) can be typified by

$$\begin{aligned} dy_{m2}/dt &= -k_{m1}e_{m1}^2 - k_{m3}e_{m2}^2 + e_{m2}[w_4 - w_4(\mathbf{M}^*)] - e_{m2}[\hat{w}_4(\mathbf{M}) - w_4(\mathbf{M}^*)] - e_{m2}x_c \\ &\quad + \widetilde{e}_{w4}\sigma_1^{-1}d\,\widetilde{e}_{w4}/dt + \gamma_1^{-1}(\mathbf{M} - \mathbf{M}^*)^T d\,\mathbf{M}/dt \\ &= -k_{m1}e_{m1}^2 - k_{m3}e_{m2}^2 + e_{m2}\varphi_w - e_{m2}[\hat{w}_4(\mathbf{M}) - \hat{w}_4(\mathbf{M}^*)] \\ &\quad - e_{m2}x_c + (\hat{e}_{w4} - e_{w4})\sigma_1^{-1}d\hat{e}_{w4}/dt + \gamma_1^{-1}(\mathbf{M} - \mathbf{M}^*)^T d\,\mathbf{M}/dt \end{aligned} \tag{40}$$

3.2.2. Recouped Controller with an Adaptive Rule

To reach $dy_{m2}/dt \leq 0$, the adaptive rule $d\,\mathbf{M}/dt$, the recouped controller x_c, and the appraised rule $d\hat{e}_{w4}/dt$ to reduce uncertainties influences can be typified by:

$$d\mathbf{M}/dt = \gamma_1\, e_{m2}\,\mathbf{T} \tag{41}$$

$$x_c = \hat{e}_{w4}\mathrm{sgn}(e_{m2}) \tag{42}$$

$$d\,\hat{e}_{w4}/dt = \sigma_1|e_{m2}| \tag{43}$$

By substituting Equations (41)–(43) into Equation (40) and by utilizing $e_{w4} \geq |\varphi_w|$, then Equation (40) can be typified by:

$$\begin{aligned} dy_{m2}/dt &= -k_{m1}e_{m1}^2 - k_{m3}e_{m2}^2 + e_{m2}\varphi_w - e_{m2}(\mathbf{M} - \mathbf{M}^*)^T\mathbf{T} - e_{m2}\hat{e}_{w4}\mathrm{sgn}(e_{m2}) \\ &\quad + (\hat{e}_{w4} - e_{w4})\sigma_1^{-1}\sigma_1|e_{m2}| + \gamma_1^{-1}(\mathbf{M} - \mathbf{M}^*)^T\gamma_1 e_{m2}\mathbf{\Gamma} \\ &= -k_{m1}e_{m1}^2 - k_{m3}e_{m2}^2 + e_{m2}\varphi_w - e_{w4}|e_{m2}| \\ &\leq -k_{m1}e_{m1}^2 - k_{m3}e_{m2}^2 - |e_{m2}|(e_{w4} - |\varphi_w|) \\ &\leq -k_{m1}e_{m1}^2 - k_{m3}e_{m2}^2 \\ &\leq 0 \end{aligned} \tag{44}$$

Equation (44) portrays $dy_{m2}(t)/dt$ to be negative semi-definite, i.e., $y_{m2}(t) \le y_{m2}(0)$, meaning that e_{m1} and e_{m2} are bounded. By utilizing the Barbalat's lemma [29,30], it can be represented that $y_{m2}(t) \to 0$ at $t \to \infty$ by way of Equations (29), (30) and (44), i.e., e_{m1} and e_{m2} will converge to zero at $t \to \infty$. The stability of the backstepping control with three adaptive rules by using RRSPPNN with RGWO and recouped controller can be ensured and consequently the control block diagram is portrayed in Figure 5.

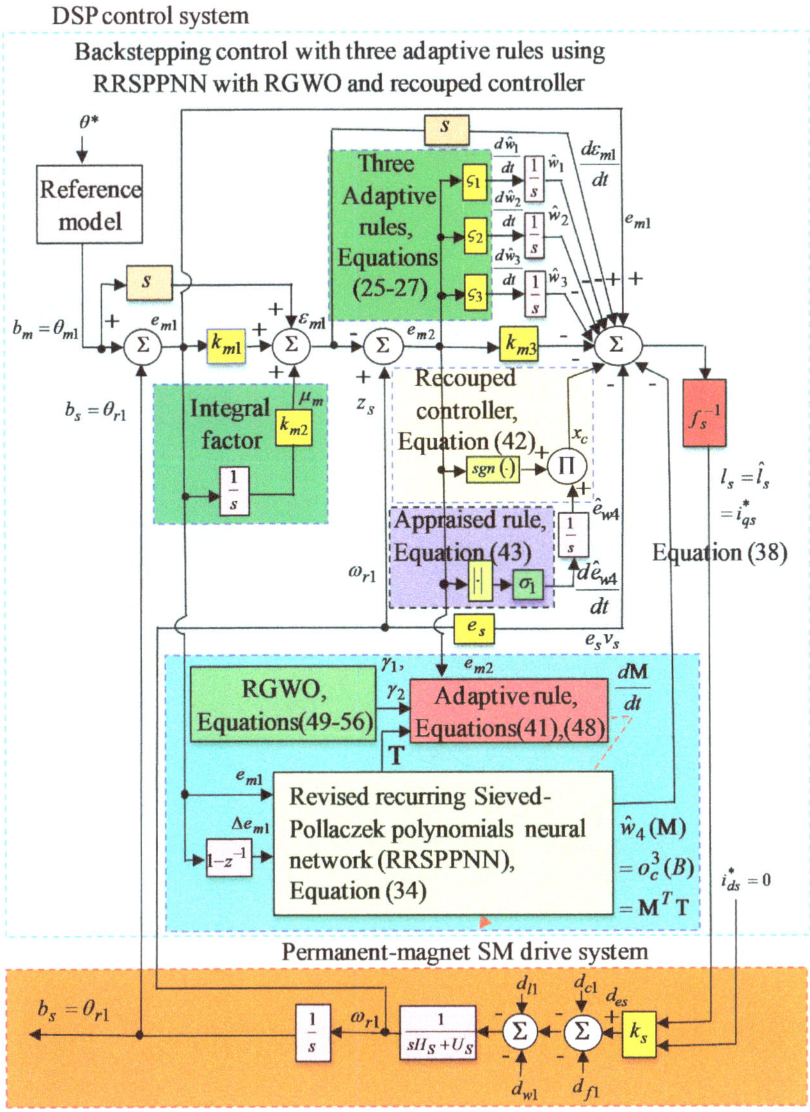

Figure 5. Block diagram by using backstepping control with three adaptive rules using RRSPPNN with RGWO and recouped controller.

3.2.3. Training of the RRSPPNN

By utilizing the Lyapunov stability and the gradient descent skill with the chain rule, a training skillfulness of parameters in the RRSPPNN can be derived. The RGWO with two adjusted factors is applied to look for two better learning rates in the RRSPPNN to acquire faster convergence. The connecting weight parametric presented in Equation (41) can be typified by:

$$db_{cb}^2/dt = \gamma_1\, e_{m2}\, o_b^2 \tag{45}$$

A goal function that explains the online training procedure of the RRSPPNN is typified by

$$L_2 = e_{m2}^2/2 \tag{46}$$

The adaptive learning rule of the connecting weight is typified by:

$$\frac{db_{cb}^2}{dt} = -\gamma_1 \frac{\partial L_2}{\partial b_{cb}^2} = -\gamma_1 \frac{\partial L_2}{\partial o_c^3}\frac{\partial o_c^3}{\partial ne_c^3}\frac{\partial ne_c^3}{\partial b_{cb}^2} = -\gamma_1 \frac{\partial L_2}{\partial o_c^3} o_b^2 \tag{47}$$

It is well-known that $\partial L_2/\partial o_c^3 = -e_{m2}$ by way of Equations (45) and (47). Hence, the adaptive learning rule of recurring weight b_{ac}^1 is typified by

$$\frac{db_{ac}^1}{dt} = -\gamma_2 \frac{\partial L_2}{\partial o_c^3}\frac{\partial o_c^3}{\partial o_b^2}\frac{\partial o_b^2}{\partial ne_b^2}\frac{\partial ne_b^2}{\partial o_a^1}\frac{\partial o_a^1}{\partial ne_a^1}\frac{\partial ne_a^1}{\partial b_{ac}^1} = \gamma_2\, e_{m2}\, b_{cb}^2 SP_b^{\lambda}(\cdot) q_a^1(B) o_c^3(B-1) \tag{48}$$

where γ_2 stands for the learning rate. To acquire better convergence, the RGWO is applied to look for two changeable learning rates in the RRSPPNN. Additionally, for improving convergence and looking for two perfect learning rates, the RGWO with two adjusted factors is proposed in this study.

3.2.4. Algorithm of Reformed Grey Wolf Optimization (RGWO)

In the RGWO, the optimization is conducted by α, β, and ρ. The RGWO algorithm can be typified by:

$$H(a_1+1) = [H_1(a_1) + H_2(a_1) + H_3(a_1)]/3 \tag{49}$$

where $H(a_1+1) = [\gamma_1\ \gamma_2]$ is a vector two learning rates, $H_1(a_1)$, $H_2(a_1)$, $H_3(a_1)$ are typified by:

$$H_1(a_1) = |\alpha(a_1) - R_1(a_1)\cdot[M_1(a_1)\cdot\alpha(a_1) - H(a_1)]| \tag{50}$$

$$H_2(a_1) = |\beta(a_1) - R_2(a_1)\cdot[M_2(a_1)\cdot\beta(a_1) - H(a_1)]| \tag{51}$$

$$H_3(a_1) = |\rho(a_1) - R_3(a_1)\cdot[M_3(a_1)\cdot\rho(a_1) - H(a_1)]| \tag{52}$$

where $\alpha(a_1)$, $\beta(a_1)$, $\rho(a_1)$ stand for the three vectors as the three best solutions. $R_1(a_1)$, $R_2(a_1)$, $R_3(a_1)$ and $M_1(a_1)$, $M_2(a_1)$, $M_3(a_1)$ can be typified by:

$$R_1(a_1) = R_2(a_1) = R_3(a_1) = 2c_1(a_1)\varphi_1 - d_1(a_1) \tag{53}$$

$$M_1(a_1) = M_2(a_1) = M_3(a_1) = 2\varphi_2 \tag{54}$$

where φ_1 and φ_2 stand for two random vectors. The updated numbers of two adjusted factors $c_1(a_1)$ and $d_1(a_1)$ control the tradeoff between exploration and exploitation. Two exponential-functional adjustable factors $c_1(a_1)$ and $d_1(a_1)$ stand for the updated values at iteration according to the following presentation by:

$$c_1(a_1) = 2e^{-a_1/a_{t1}} \tag{55}$$

$$d_1(a_1) = 2e^{-a_1/a_{t2}} \tag{56}$$

where a_1 stands for the iteration number; a_{t1} and a_{t2} stand for the total numbers of iteration allowed for the optimization. At last, $H(a_1+1) = [\gamma_1\ \gamma_2]$ stands for the best solution in connection with the learning rates $\gamma_i(t)$, $i = 1$, 2 of the two weights in the RRSPPNN. Hence, the better numbers could be optimized by utilizing RGWO with two adjusted factors that yield two changeable learning rates for two weights to look for two perfect values and to speed-up the convergence of two weights.

4. Test Results

A block diagram of the FOC permanent-magnet SM drive system utilizing the DSP controller is portrayed in Figure 1. A photo of the examination structure is portrayed in Figure 6. The sampling time of the control program in the examination is set as 2 ms.

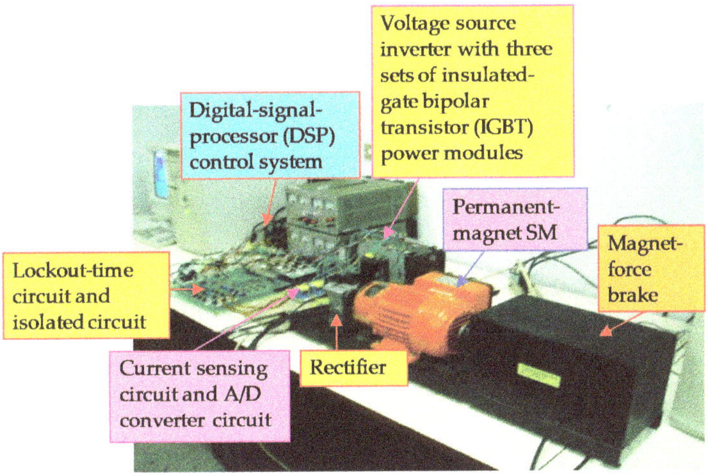

Figure 6. A photo picture of the examination structure.

A DSP controller involves 18 channels of input/output (I/O) ports with 6 channels of pulse-width-modulation (PWM) ports, 6 channels of analog-digital (A/D) converters, and 2 channel encoder connective ports. The coordinate transformation in the field-oriented control (FOC) is realized by DSP controller. The used control technologies in the real-time realization by utilizing the DSP controller are composed of the core program and the sub-core interrupt service routine (SCISR) in the DSP controller as portrayed in Figure 7. In the core program, parameters and I/O initialization are processed. The interrupt time for the SCISR is set. After permitting the interruption, the core program is used to monitor control data. The SCISR with 2 msec sampling time is used for reading the rotor position of the permanent-magnet SM drive system from encoder and three-phase currents by way of A/D converter circuit, calculating reference model and position error, executing lookup table and coordinate transformation, executing the backstepping control with three adaptive rules using RRSPPNN with RGWO and recouped controller, and outputting three-phase current mandates to swap sinusoidal pulse-width-modulation (SPWM) voltage source inverter with three-sets of insulated-gate bipolar transistor (IGBT) power modules by way of the lockout-time and isolated circuits. The SPWM voltage source inverter with three-sets of IGBT power modules is carried out by SPWM control with a switching frequency of 15 kHz. Additionally, the tested bandwidth of the position control loop and the tested bandwidth of the current control loop are about 90 and 900 Hz for the permanent-magnet SM drive system under the nominal event. The proposed controllers are realized by the DSP controller. The coordinate transformation in the FOC is realized by the DSP controller. The control goal is to control the rotor to rotate 6.28 rad cyclically. Then, when the mandate is a sinusoidal reference locus, the reference is set one.

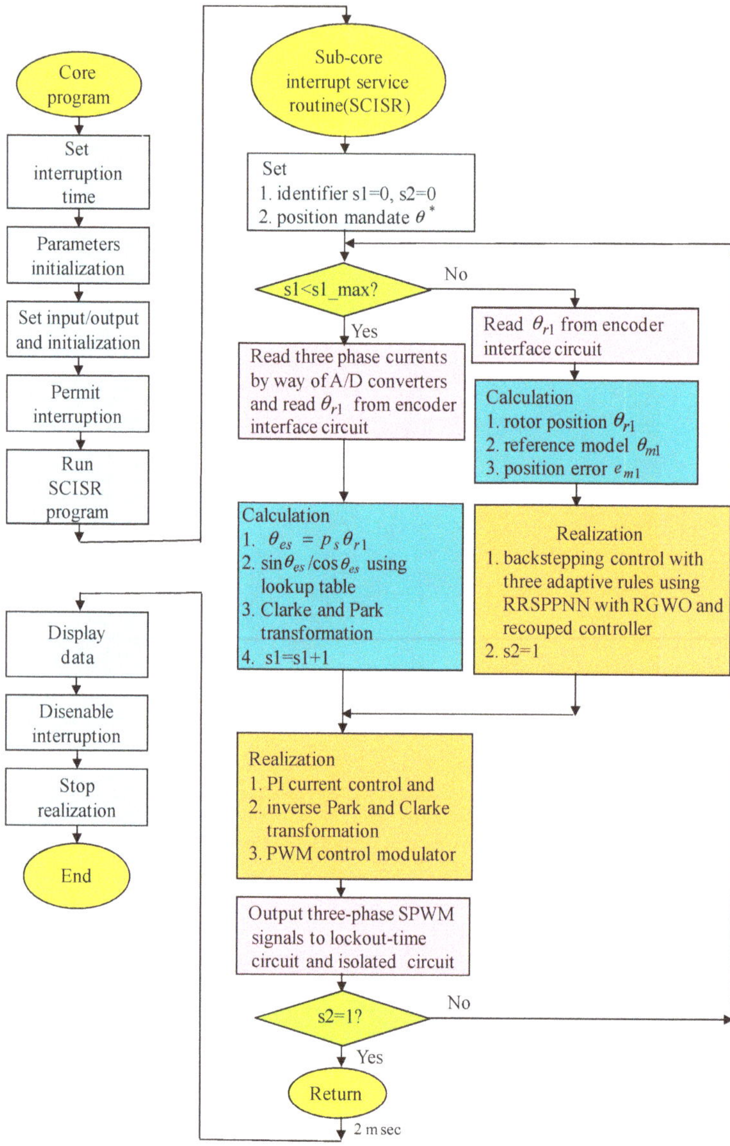

Figure 7. Flow chart of the realized program by utilizing DSP controller.

For a comparison of control performance with the four controllers, five events are provided in the experiment. The four controllers are the popular PI controller as the controller FC1, the backstepping control with three adaptive rules and a swapping function as the controller FC2, the modified recurrent Rogers–Szego polynomials neural network with DGWO [10] as the controller FC3, and the backstepping control with three adaptive rules using RRSPPNN with RGWO and recouped controller as the controller FC4. Five tested events are as follows. Event CQ1 is the nominal event at periodic step command from 0 rad to 6.28 rad. Event CQ2 is the cogging torque, the column friction torque, and the Stribeck effect torque, and the parameters variations event with 4 times the nominal value at periodic step command from 0 rad to 6.28 rad. Event CQ3 is the nominal event due to periodic sinusoidal command from

−6.28 rad to 6.28 rad. Event CQ4 is the cogging torque, the column friction torque and the Stribeck effect torque and the parameters variations event with 4 times the nominal value due to periodic sinusoidal command from −6.28 rad to 6.28 rad. Event CQ5 is the adding load torque disturbance $\tau_{lr} = 2.5$ Nm.

Two control gains of the popular PI controller as the controller FC1 are $k_{pp} = 4.5$, and $k_{ip} = k_{pp}/T_{ip} = 1.8$ by using the Kronecker method to construct a stability boundary in the k_{pp} and k_{ip} plane [32–34] on the tuning of the PI controller at Event CQ1 in the position trace so as to reach fine steady-state and transient-state control responses.

Some control gains of the backstepping control with three adaptive rules and a swapping function as the controller FC2 are $k_{m1} = 2.2$, $k_{m2} = 2.6$, $k_{m3} = 2.1$, $\varsigma_1 = 0.1$, $\varsigma_2 = 0.1$, $\varsigma_3 = 0.1$, $\beta_1 = 9.1$ according to heuristic knowledge [4] at Event CQ1 in the position trace so as to reach fine steady-state and transient-state control representation.

Some control gains of the modified recurrent Rogers–Szego polynomials neural network with DGWO [10] as the controller FC3 are given as $c_{x1} = 2.2$, $c_{x2} = 2.6$, $c_{x3} = 2.1$, $\eta_1 = 0.1$, $\eta_2 = 0.1$, $\eta_3 = 0.1$, $\gamma = 0.2$, $\tau = 0.1$ according to heuristic knowledge [4] at Event CQ1 in the position trace so as to reach fine steady-state and transient-state control responses. Moreover, numbers of neurons in the input layer, the hidden layer, and the output layer of the modified recurrent Rogers-Szego polynomials neural network are 2 neurons, 4 neurons and 1 neuron, respectively, so as to demonstrate the effectiveness of the controller adopting small neuron numbers. The method proposed by Lewis et al. [35] is used to initialize some parameters of the modified recurrent Rogers–Szego polynomials neural network. The adjustment process of these parameters involve a continuous reaction for the duration of the examination.

Some control gains of the backstepping control with three adaptive rules using RRSPPNN with RGWO and recouped controller as the controller FC4 are $k_{m1} = 2.2$, $k_{m2} = 2.6$, $k_{m3} = 2.1$, $\varsigma_1 = 0.1$, $\varsigma_2 = 0.1$, $\varsigma_3 = 0.1$, $\sigma_1 = 0.2$, $\delta = 0.1$ according to heuristic knowledge [4] at Event CQ1 in the position trace so as to reach fine steady-state and transient-state control responses. Moreover, the number of neurons in the first layer, the second layer, and the third layer of the RRSPPNN are 2 neurons, 4 neurons, and 1 neuron, respectively, so as to demonstrate the effectiveness of the controller by adopting small neuron numbers. The method proposed by Lewis et al. [35] is used to initialize some parameters of the RRSPPNN. The adjustment process of these parameters is keeping continusly reaction in the duration of the experiments.

All of the experiments obtained by utilizing the four controllers for controlling the permanent-magnet SM drive system at five events are as follows. Figure 8a–d are rotor position responses via experiments obtained by utilizing the controllers FC1, FC2, FC3, and FC4 at Event CQ1. Figure 9a–d are rotor speed responses by utilizing the controllers FC1, FC2, FC3 and FC4 at Event CQ1. Figure 10a–d are mandate control propulsion responses by utilizing the controllers FC1, FC2, FC3, and FC4 at Event CQ1. Figure 11a–d are rotor position responses by utilizing the controllers FC1, FC2, FC3, and FC4 at Event CQ2. Figure 12a–d are rotor speed responses by utilizing the controllers FC1, FC2, FC3, and FC4 at Event CQ2. Figure 13a–d are mandate control propulsion responses by utilizing the controllers FC1, FC2, FC3, and FC4 at Event CQ2. Figure 14a–d are rotor position responses by utilizing the controllers FC1, FC2, FC3, and FC4 at Event CQ3. Figure 15a–d are rotor speed responses by utilizing the controllers FC1, FC2, FC3, and FC4 at Event CQ3. Figure 16a–d are mandate control propulsion responses by utilizing the controllers FC1, FC2, FC3, and FC4 at Event CQ3. Figure 17a–d are rotor position responses by utilizing the controllers FC1, FC2, FC3, and FC4 at Event CQ4. Figure 18a–d are rotor speed responses by utilizing the controllers FC1, FC2, FC3, and FC4 at Event CQ4. Figure 19a–d are mandate control propulsion responses by utilizing the controllers FC1, FC2, FC3, and FC4 at Event CQ4. Figure 20a–d are measured rotor position responses by utilizing the controller FC1, FC2, FC3, and FC4 at Event CQ5.

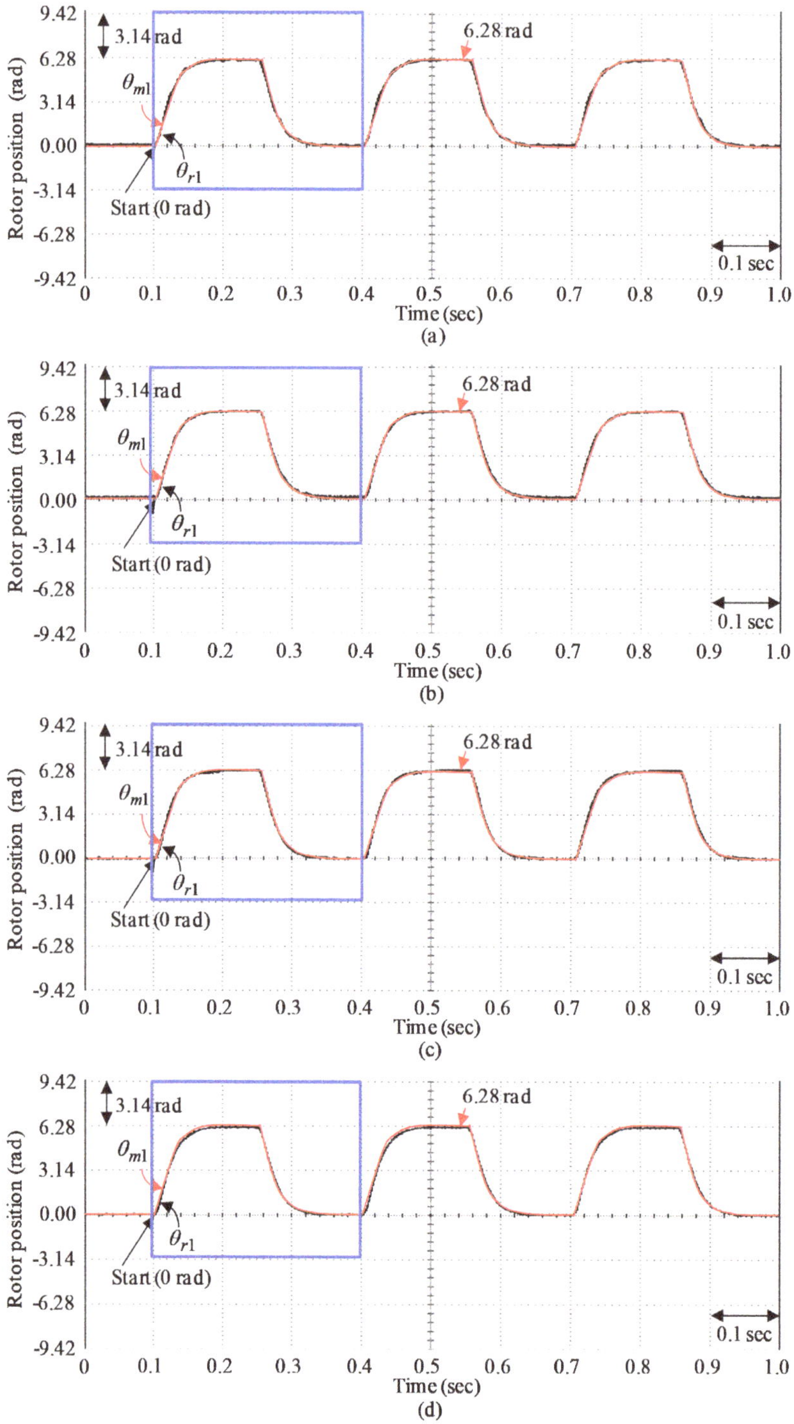

Figure 8. Rotor position responses via experiments at Event CQ1 by utilizing the controller: (**a**) FC1, (**b**) FC2, (**c**) FC3, (**d**) FC4.

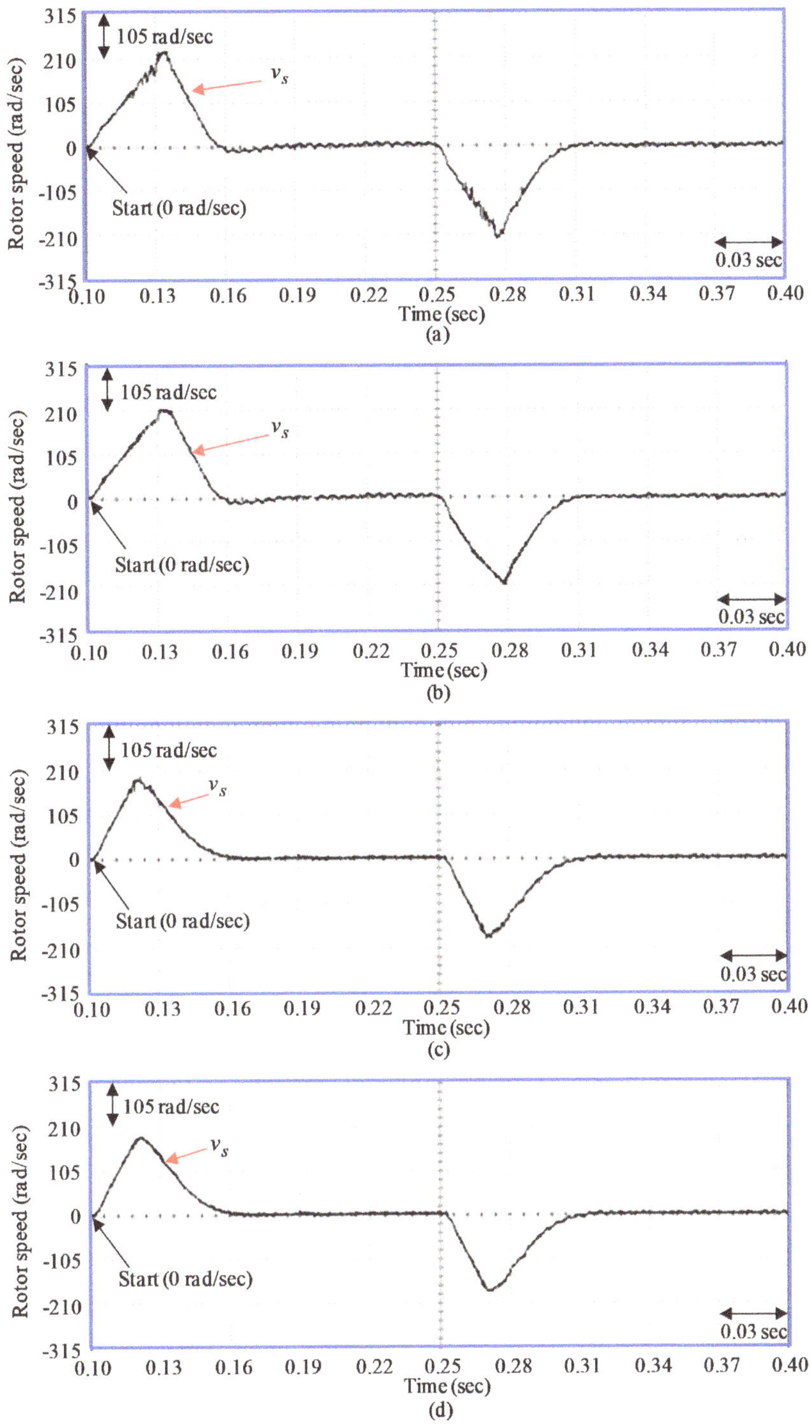

Figure 9. Rotor speed responses via experiments at Event CQ1 by utilizing the controller: (**a**) FC1, (**b**) FC2, (**c**) FC3, (**d**) FC4.

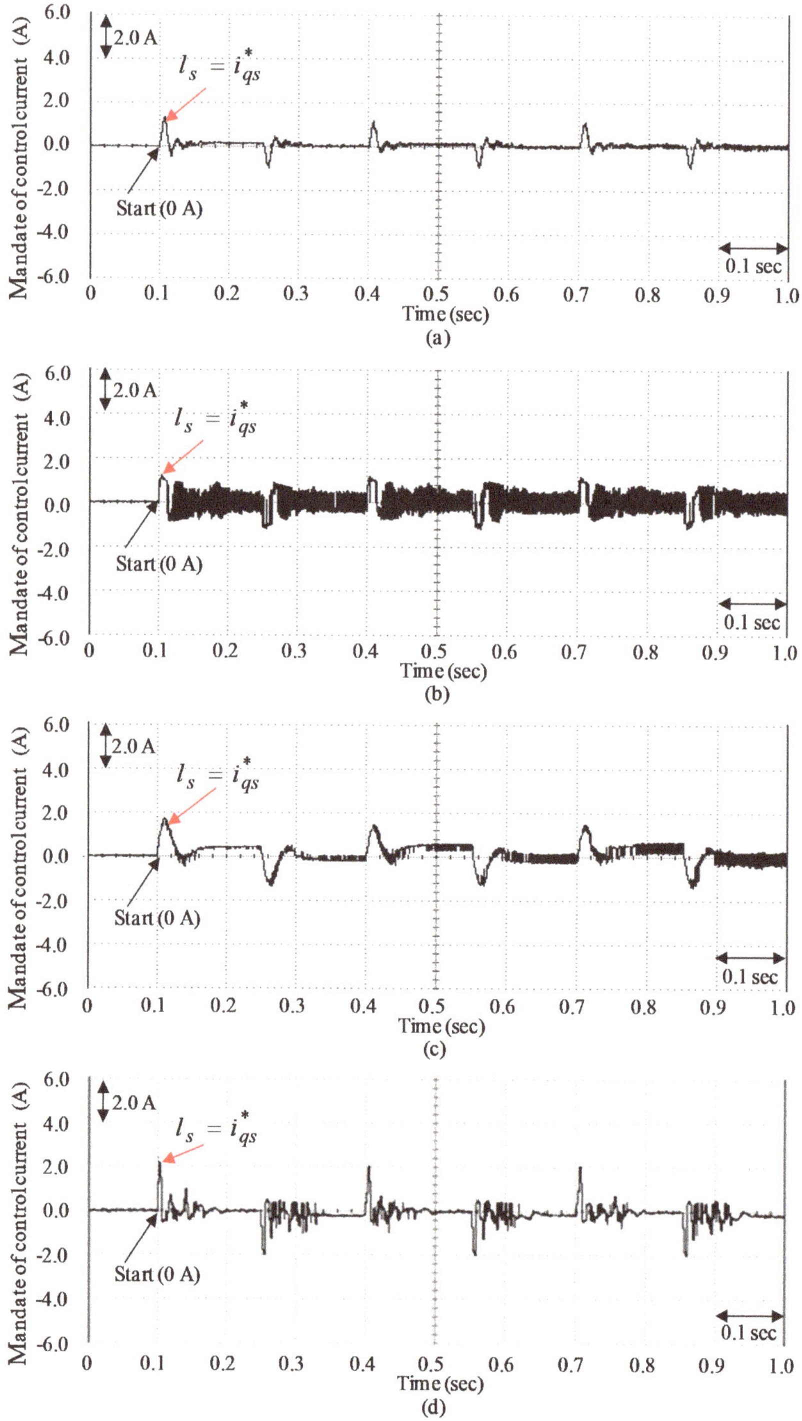

Figure 10. Control propulsion responses via experiments at Event CQ1 by utilizing the controller: (**a**) Figure 1, (**b**) FC2, (**c**) FC3, (**d**) FC4.

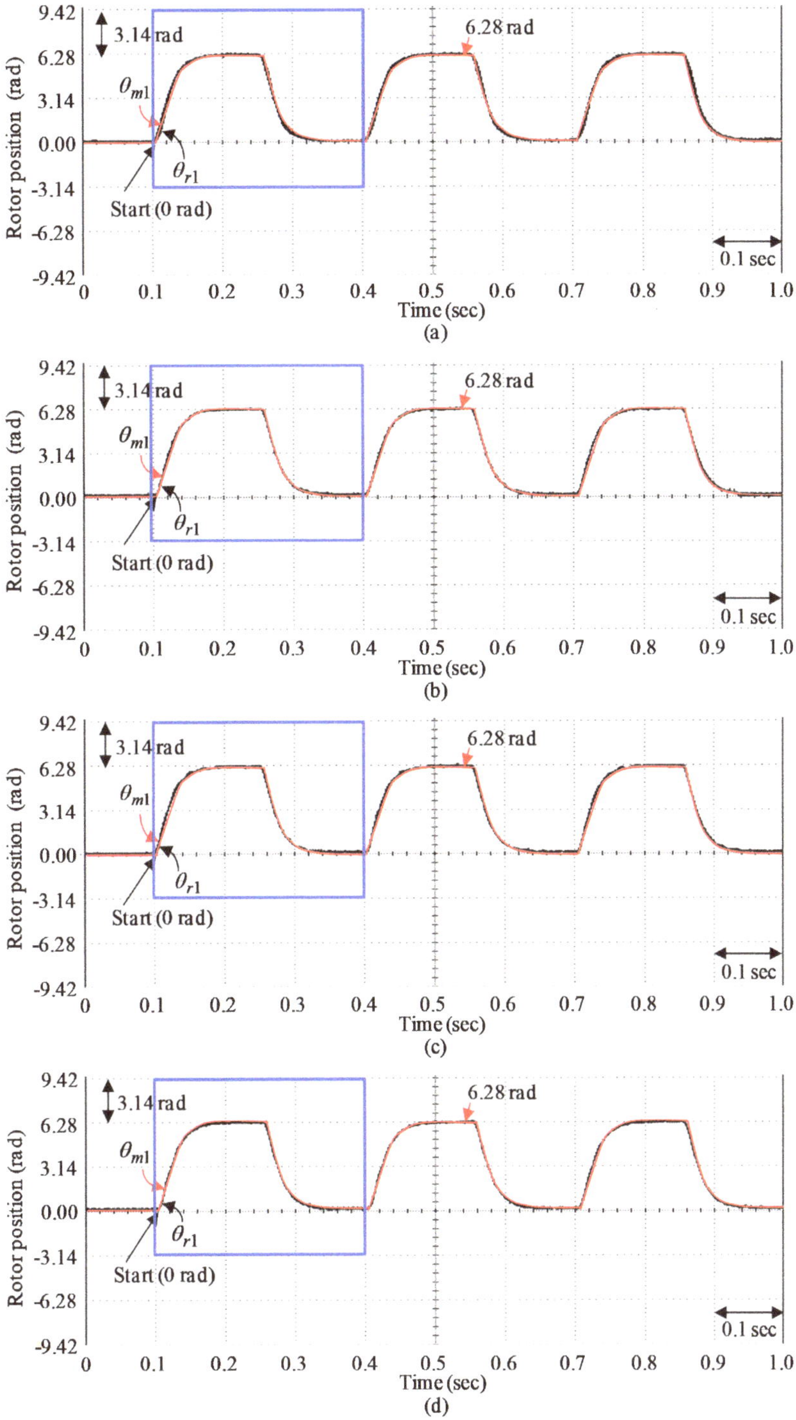

Figure 11. Rotor position responses via experiments at Event CQ2 by utilizing the controller: (**a**) FC1, (**b**) FC2, (**c**) FC3, (**d**) FC4.

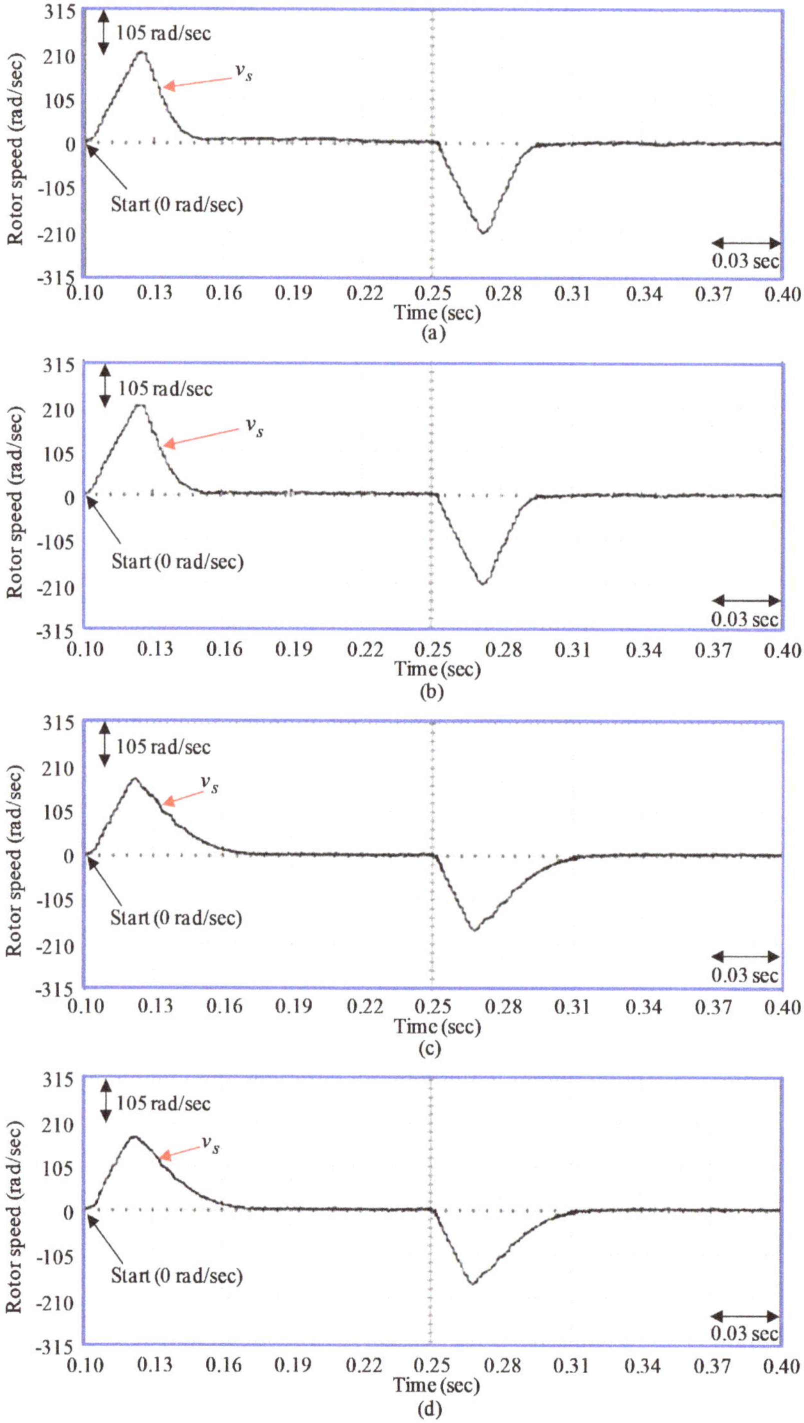

Figure 12. Rotor speed responses via experiments at Event CQ2 by utilizing the controller: (**a**) FC1, (**b**) FC2, (**c**) FC3, (**d**) FC4.

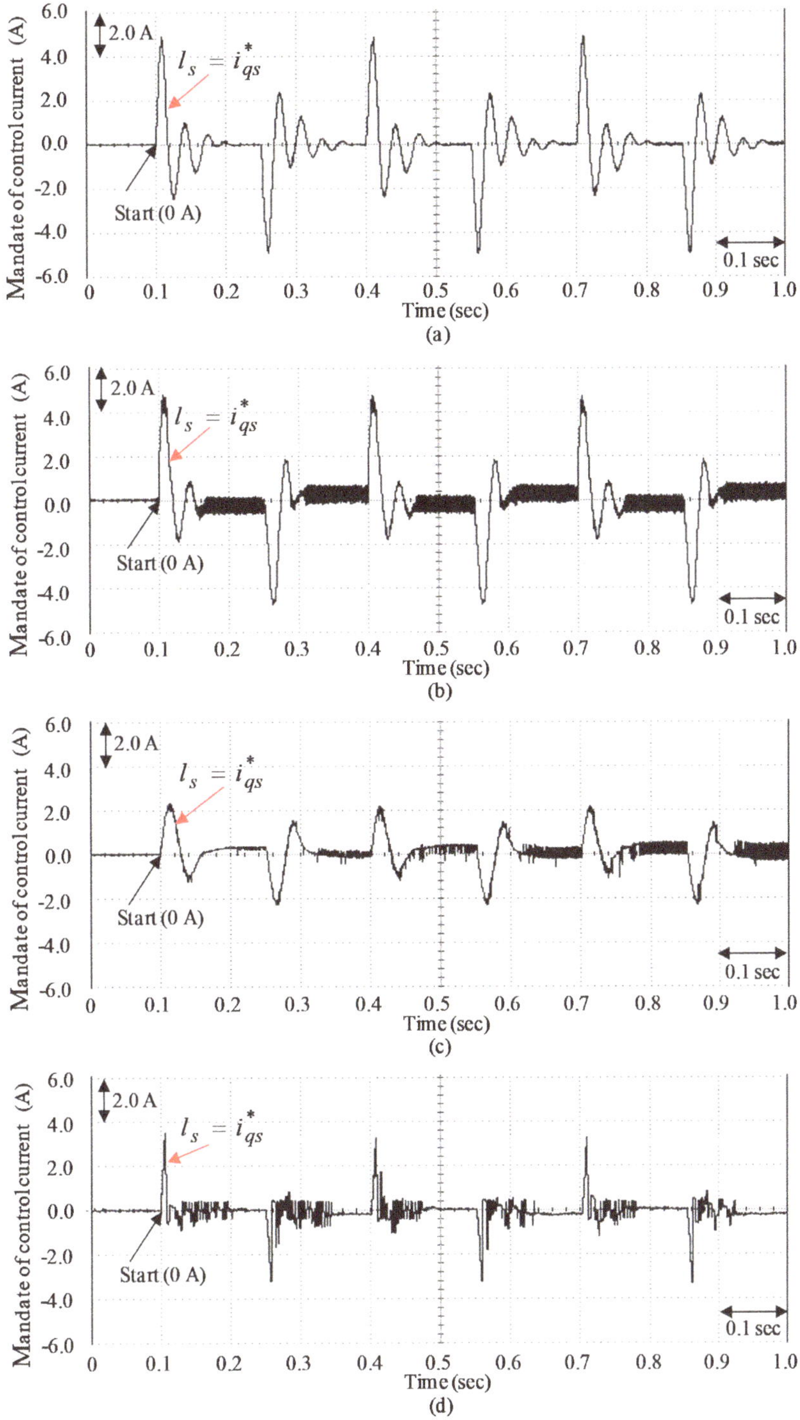

Figure 13. Control propulsion responses via experiments at Event CQ2 by utilizing the controller: (**a**) Figure 1, (**b**) FC2, (**c**) FC3, (**d**) FC4.

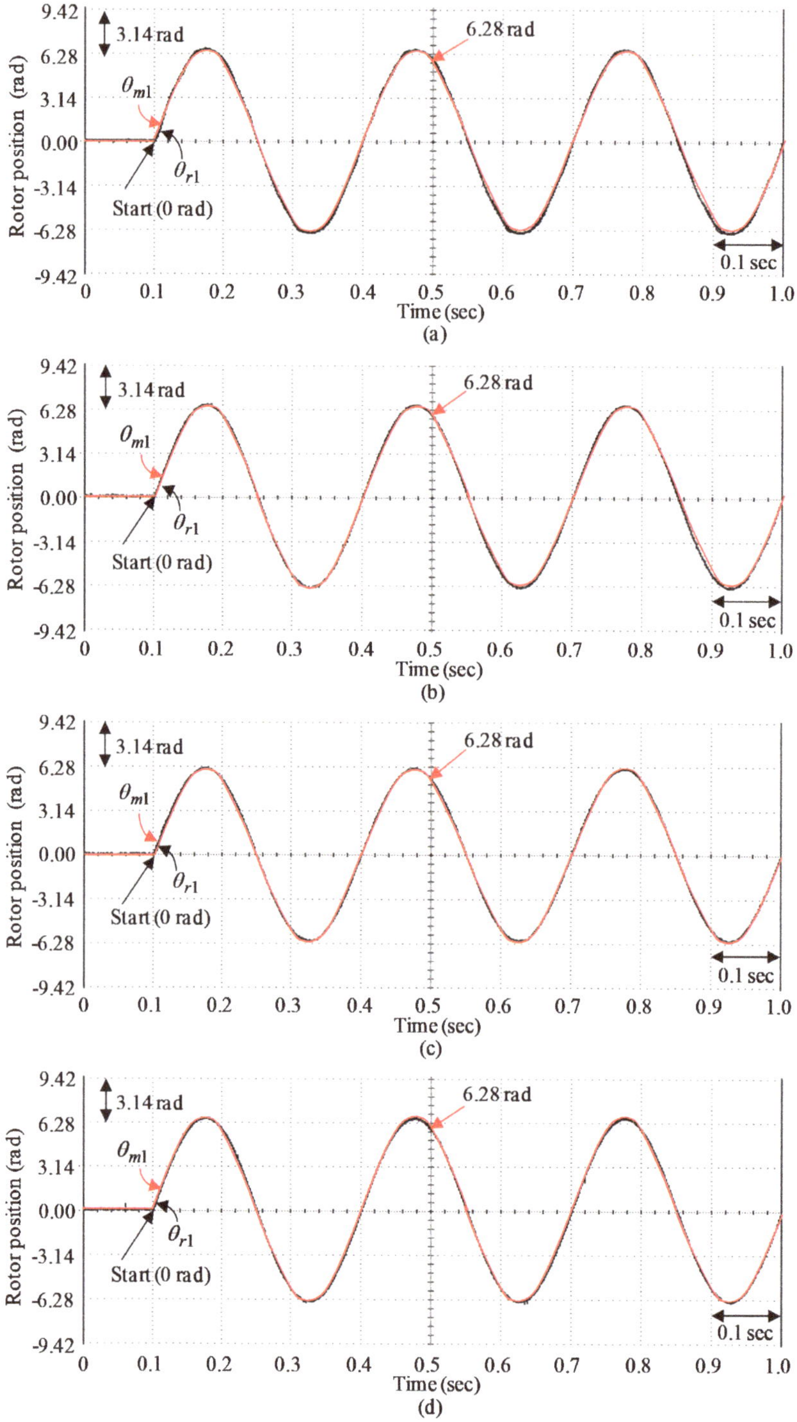

Figure 14. Rotor position responses via experiments at Event CQ3 by utilizing the controller: (**a**) FC1, (**b**) FC2, (**c**) FC3, (**d**) FC4.

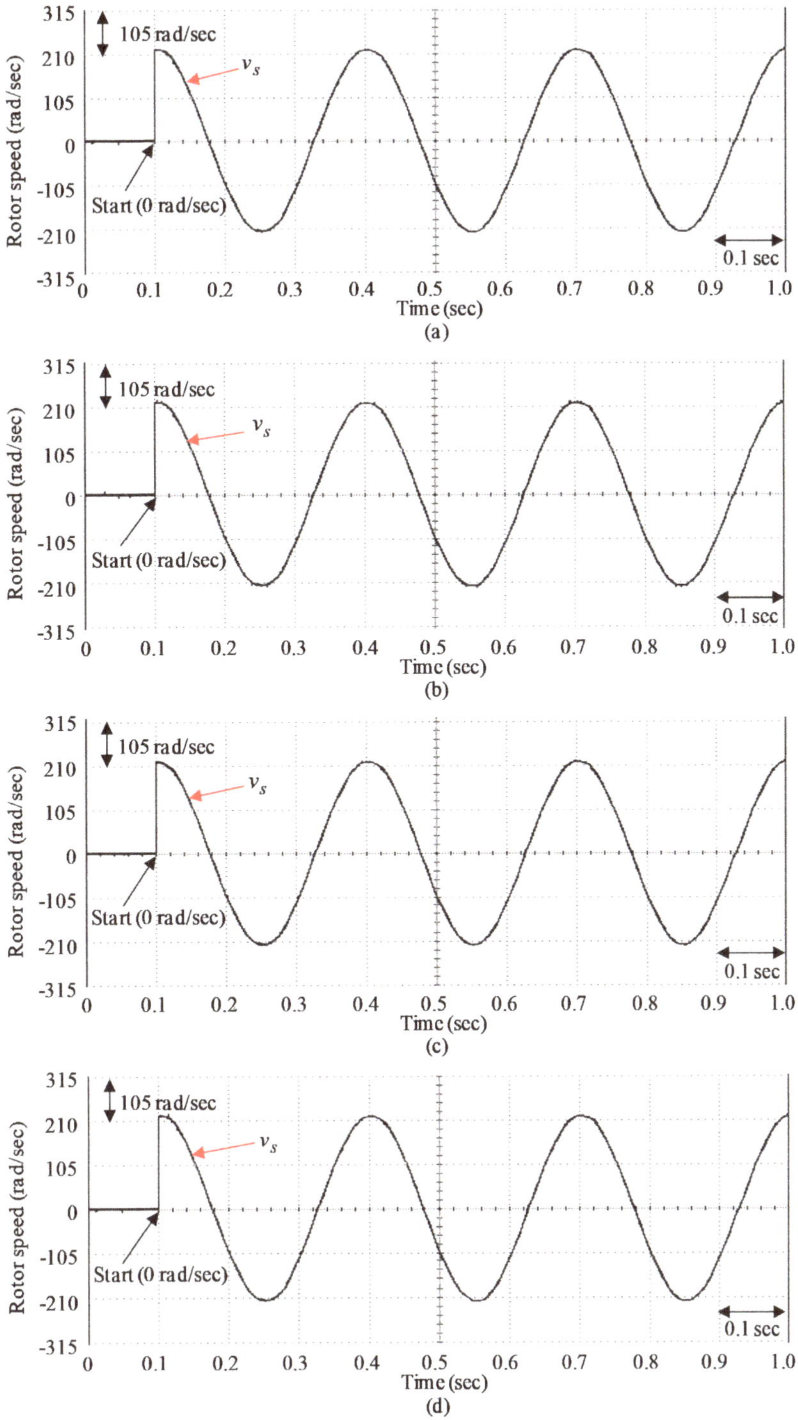

Figure 15. Rotor speed responses via experiments at Event CQ3 by utilizing the controller: (**a**) FC1, (**b**) FC2, (**c**) FC3, (**d**) FC4.

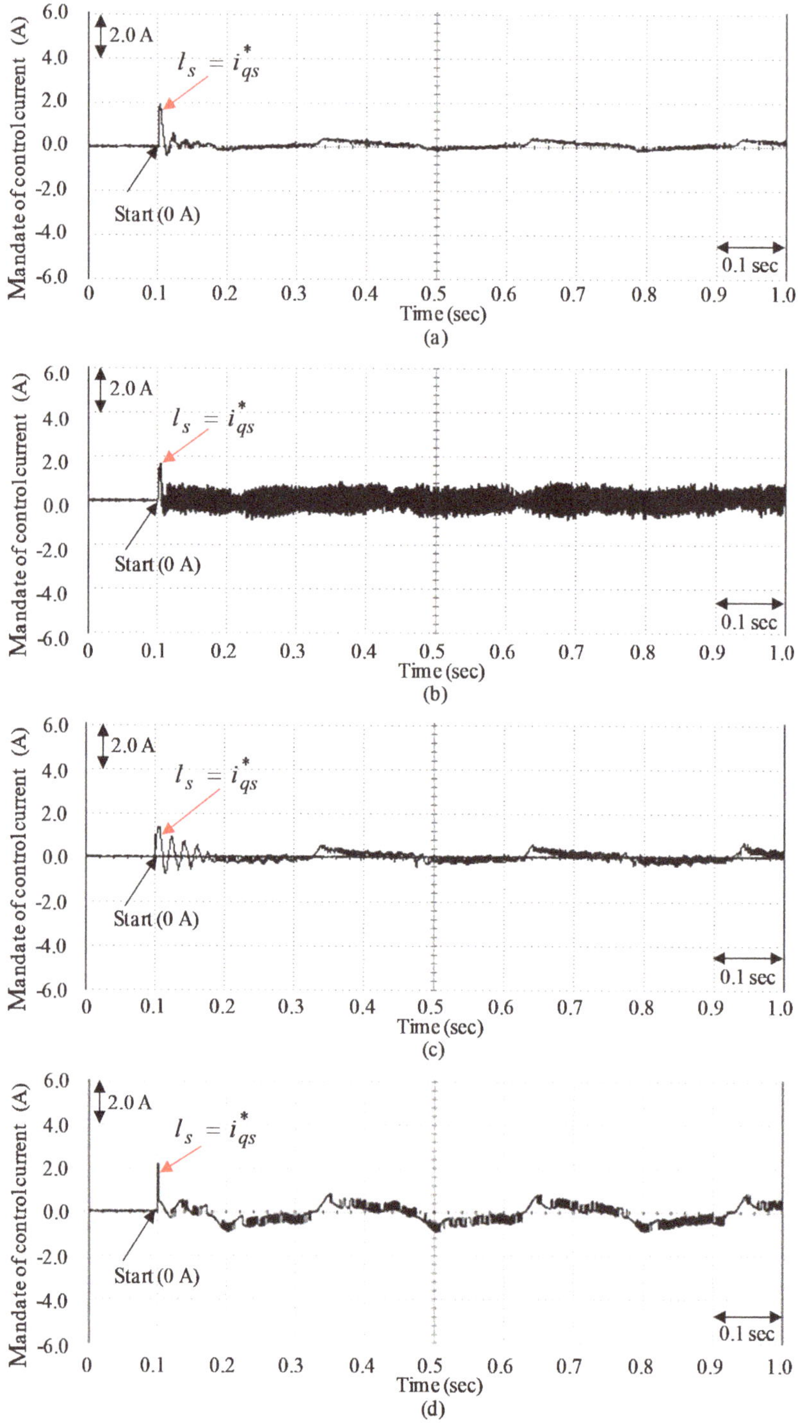

Figure 16. Control propulsion responses via experiments at Event CQ3 by utilizing the controller: (**a**) FC1, (**b**) FC2, (**c**) FC3, (**d**) FC4.

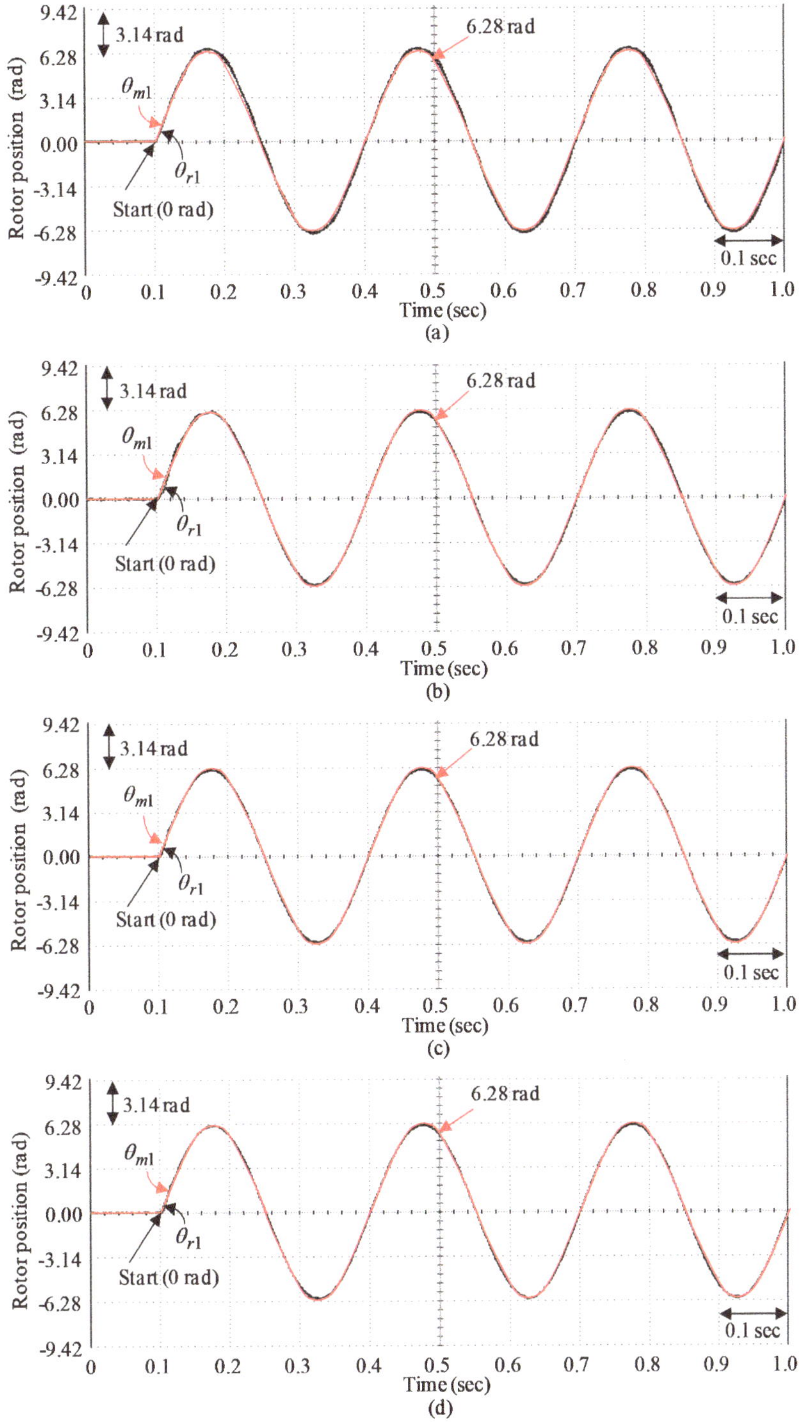

Figure 17. Rotor position responses via experiments at Event CQ4 by utilizing the controller: (**a**) FC1, (**b**) FC2, (**c**) FC3, (**d**) FC4.

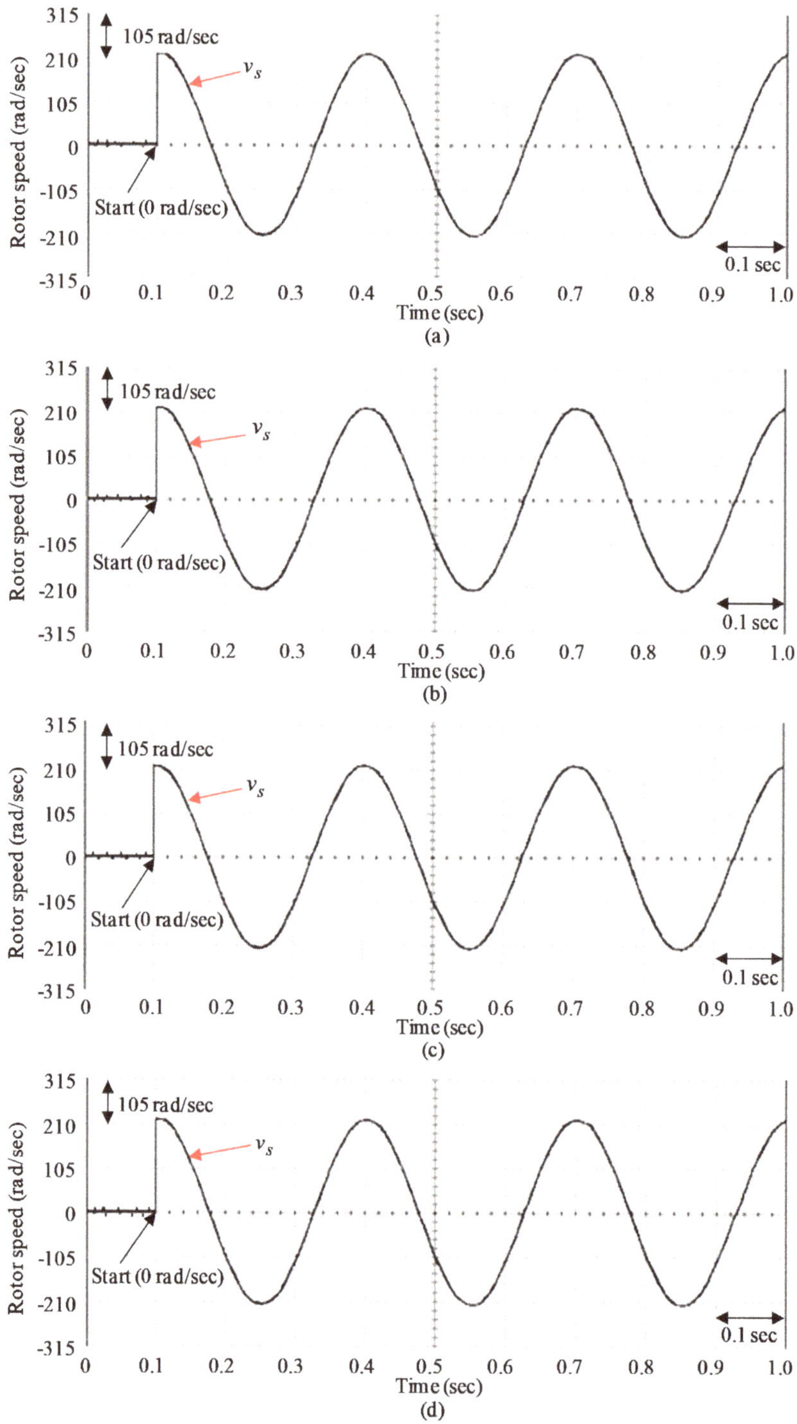

Figure 18. Rotor speed responses via experiments at Event CQ4 by utilizing the controller: (**a**) FC1, (**b**) FC2, (**c**) FC3, (**d**) FC4.

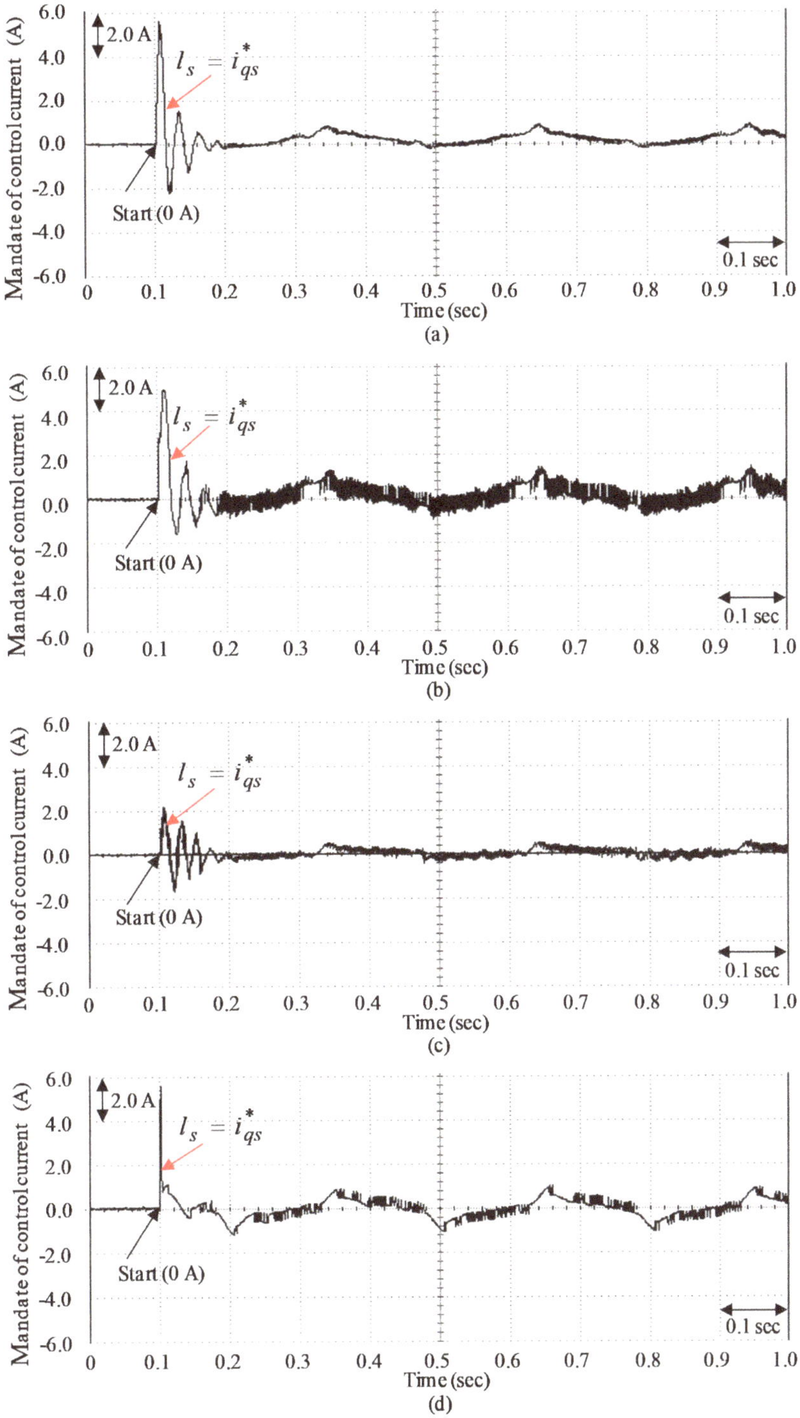

Figure 19. Test results of the control propulsion responses at Event CQ4 by utilizing the controller: (**a**) FC1, (**b**) FC2, (**c**) FC3, (**d**) FC4.

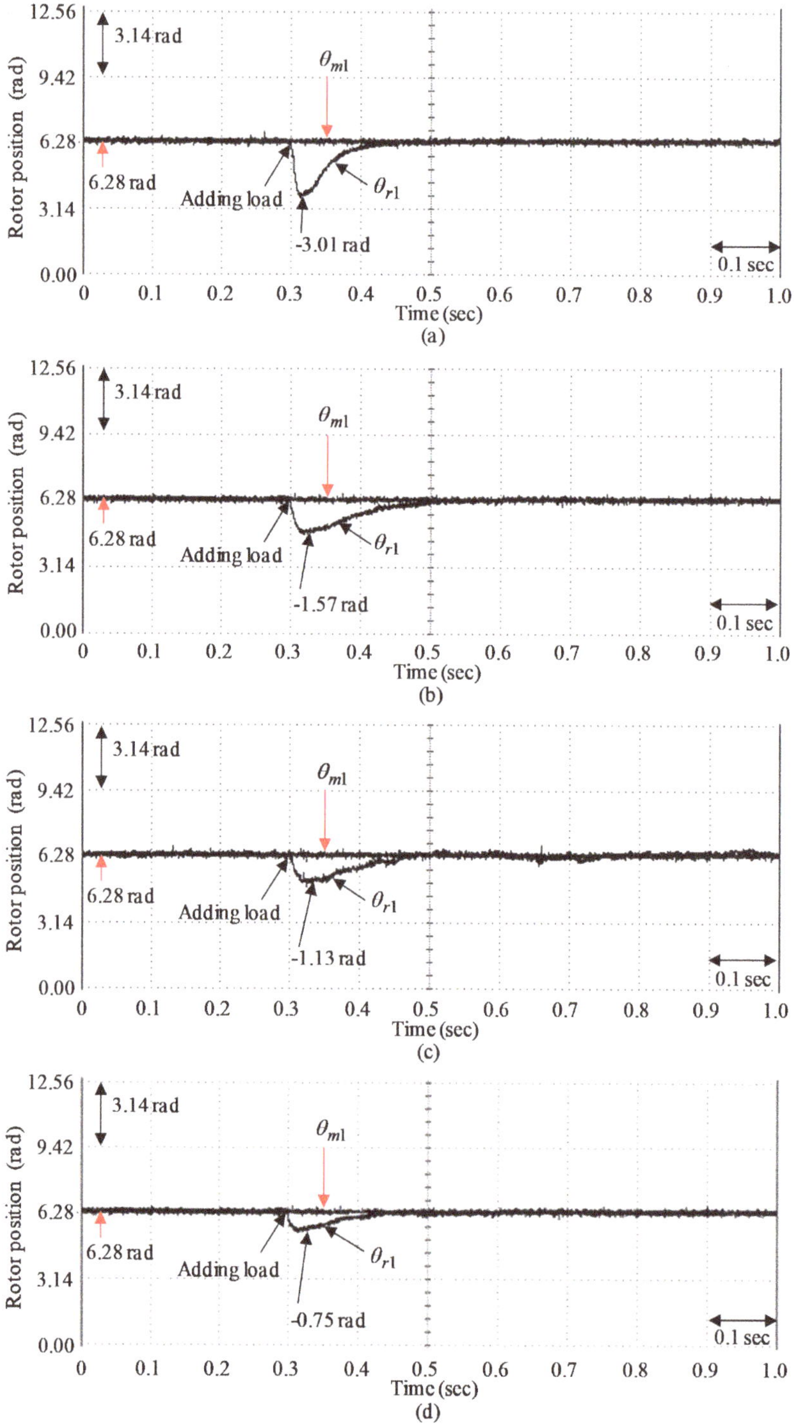

Figure 20. Regulated responses of the rotor positions at Event CQ5 by utilizing the controller: (**a**) FC1, (**b**) FC2, (**c**) FC3, (**d**) FC4.

Figures 8a and 14a obtained by utilizing the controller FC1 for controlling the permanent-magnet SM drive system at Event CQ1 and Event CQ3 are displayed as fine trace responses of the rotor positions. Figures 11a and 17a obtained by utilizing the controller FC1 for controlling the permanent-magnet SM drive system at Event CQ2 and Event CQ4 are displayed in sluggish trace responses of the rotor position owing to bigger nonlinear disturbance. Because of inappropriate tuning gains or the degenerate nonlinear effect, the linear controller has weak robustness under bigger nonlinear disturbance.

Figures 8b, 11b, 14b and 17b obtained by utilizing the controller FC2 for controlling the permanent-magnet SM drive system at Events CQ1, CQ2, CQ3, and CQ4 are displayed as good trace responses of the rotor positions. However, Figures 10b, 13b, 16b and 19b are displayed as serious vibration in the control propulsions by utilizing the swapping function with large upper bound at Events CQ1, CQ2, CQ3, and CQ4. It is a well-known fact that the control propulsions with serious vibration will wear the bearing mechanism and might excite unstable system dynamics.

Figures 8c, 11c, 14c and 17c obtained by utilizing the controller FC3 for controlling the permanent-magnet SM drive system at Events CQ1, CQ2, CQ3, and CQ4 are displayed as better trace responses of the rotor positions due to adaptive mechanism action. Figures 10c, 13c, 16c and 19c displayed a small vibration in the control propulsions at Events CQ1, CQ2, CQ3, and CQ4. Due to the on-line adaptive adjustment of the modified recurrent Rogers–Szego polynomials neural network [10], the magnitudes of vibration in the control propulsions at Events CQ1, CQ2, CQ3, and CQ4 displayed in Figures 10c, 13c, 16c and 19c have been slightly improved.

Figures 8d, 11d, 14d and 17d obtained by utilizing the controller FC4 for controlling the permanent-magnet SM drive system at Events CQ1, CQ2, CQ3, and CQ4 are displayed as best trace responses of the rotor positions due to on-line adaptive mechanism action. Figures 10d, 13d, 16d and 19d are displayed as smaller vibrations in the control propulsions at Events CQ1, CQ2, CQ3 and CQ4 due to on-line adaptive mechanism action of the RRSPPNN. Due to on-line adaptive adjustment of the RRSPPNN under bigger nonlinear disturbance the magnitudes of vibration in the control propulsions at Events CQ1, CQ2, CQ3, and CQ4 displayed in Figures 10d, 13d, 16d and 19d have been obviously improved.

Figure 20d obtained by utilizing the controller FC4 for controlling the permanent-magnet SM drive system at Event CQ5 under load regulation is better than the controller FC1, FC2, and FC3 displayed in Figure 20a–c.

5. Discussion and Analysis

Additionally, the control performances displayed in comparsion results by using the controllers FC1, FC2, FC3, and FC4 are listed in Table 1 in connection with five events with some test results. The 0.21, 0.19, 0.15, and 0.10 are the maximum errors of e_{m1} (rad) by utilizing the controllers FC1, FC2, FC3, and FC4 at Event CQ1, respectively. The 0.11, 0.09, 0.07, and 0.05 are the root-mean-square errors of e_{m1} (rad) by utilizing the controllers FC1, FC2, FC3, and FC4 at Event CQ1, respectively. The 0.56, 0.36, 0.28, and 0.19 are the maximum errors of e_{m1} (rad) by utilizing the controllers FC1, FC2, FC3, and FC4 at Event CQ2, respectively. The 0.27, 0.18, 0.13, and 0.09 are the root-mean-square errors of e_{m1} (rad) by utilizing the controllers FC1, FC2, FC3, and FC4 at Event CQ2, respectively. The 0.21, 0.18, 0.14, and 0.10 are the maximum errors of e_{m1} (rad) by utilizing the controllers FC1, FC2, FC3, and FC4 at Event CQ3, respectively. The 0.10, 0.09, 0.07, and 0.05 are the root-mean-square errors of errors of e_{m1} (rad) by utilizing the controllers FC1, FC2, FC3, and FC4 at Event CQ3, respectively. The 0.52, 0.37, 0.27, and 0.18 are the maximum errors of e_{m1} (rad) by utilizing the controllers FC1, FC2, FC3, and FC4 at Event CQ4, respectively. The 0.25, 0.18, 0.13, and 0.09 are the root-mean-square errors of e_{m1} (rad) by utilizing the controllers FC1, FC2, FC3, and FC4 at Event CQ4, respectively. The 3.01, 1.57, 1.13, and 0.75 are the maximum errors of e_{m1} (rad) by utilizing the controllers FC1, FC2, FC3, and FC4 at Event CQ5, respectively. The 1.50, 0.78, 0.56, and 0.32 are the root-mean-square errors of e_{m1} (rad) by utilizing the controllers FC1, FC2, FC3, and FC4 at Event CQ5, respectively. The controller FC4 has smaller trace error in comparison with the controllers FC1, FC2, and FC3. The controllers FC4 indeed yields the exellent control performance from Table 1.

Table 1. Performance comparison of four controllers.

Five Tested Events / Performance	Controller FC1				
	Event CQ1	Event CQ2	Event CQ3	Event CQ4	Event CQ5
Maximum error of e_{m1} (rad)	0.21	0.56	0.20	0.52	3.01
Root-mean-square error of e_{m1} (rad)	0.11	0.27	0.10	0.25	1.50
Five Tested Events / Performance	**Controller FC2**				
	Event CQ1	Event CQ2	Event CQ3	Event CQ4	Event CQ5
Maximum error of e_{m1} (rad)	0.19	0.36	0.18	0.37	1.57
Root-mean-square error of e_{m1} (rad)	0.09	0.18	0.09	0.18	0.78
Five Tested Events / Performance	**Controller FC3**				
	Event CQ1	Event CQ2	Event CQ3	Event CQ4	Event CQ5
Maximum error of e_{m1} (rad)	0.19	0.36	0.18	0.37	1.57
Maximum error of e_{m1} (rad)	0.15	0.28	0.14	0.27	1.13
Root-mean-square error of e_{m1} (rad)	0.07	0.13	0.07	0.13	0.56
Five Tested Events / Performance	**Controller FC4**				
	Event CQ1	Event CQ2	Event CQ3	Event CQ4	Event CQ5
Maximum error of e_{m1} (rad)	0.19	0.36	0.18	0.37	1.57
Maximum error of e_{m1} (rad)	0.10	0.19	0.10	0.18	0.75
Root-mean-square error of e_{m1} (rad)	0.05	0.09	0.05	0.09	0.32

Furthermore, control characteristic performance comparisons in the controllers FC1, FC2, FC3, and FC4 are listed in Table 2 for test results. In Table 2, various performances with regard to the control propulsion with vibration, the dynamic response, the ability of load regulation, the convergence speed, the position trace error, and the rejection ability of parameter disturbance in the controllers FC4 are superior to the controllers FC1, FC2, and FC3. Finally, the robust control performance of the controller FC4 demonstrates outstanding performance for controlling the permanent-magnet SM drive system in the trace of the periodic step and sinusoidal commands under the occurrence of parameter disturbance and load regulation due in large part to the on-line adaptive adjustment of the RRSPPNN.

Table 2. Control characteristic performance comparisons of controllers.

Four Controllers / Characteristic Performance	Controller FC1	Controller FC2	Controller FC3	Controller FC4
Vibration in control propulsion	Small	Middle	Smaller	Smallest
Dynamic response	Slow	Fast	Faster	Fastest
Ability of load regulation	Poor (maximum error as 3.01 (rad) with adding load at 6.28 (rad))	Good (maximum error as 1.57 (rad)with adding load at 6.28 (rad))	Better (maximum error as 1.13 (rad) with adding load at 6.28 (rad))	Best (maximum error as 0.09 (rad) with adding load at 6.28 (rad))

Table 2. *Cont.*

Four Controllers / Characteristic Performance	Controller FC1	Controller FC2	Controller FC3	Controller FC4
Convergence speed	Middle (traceerror response at 0.1 (rad) within 0.05 s)	Fast (traceerror respons at 0.1 (rad) within 0.04 s)	Faster (traceerror response at 0.1 (rad) within 0.03 s) (varied learning rate)	Fastest (traceerror response at 0.1 (rad) within 0.01 s) (varied learning rate)
Position traceerror	Middle	Small	Smaller	Smallest
Rejection ability for parameter disturbance	Poor	Good	Better	Best
Learning rate	None	None	Vary (optimal learning rate)	Vary (optimal learning rate)

6. Conclusions

The backstepping control with three adaptive rules and RRSPPNN with RGWO and recouped controller is used to determine the best values for parameters in neural network learning rules, and thereby robustness in learning control can be improved.

The main contribution of this study is as follows. Firstly, the field-oriented control has been smoothly applied to control the permanent-Magnet SM drive system to speed up the control response. Moreover, the backstepping controller with three adaptive rules and a swapping function has been smoothly derived to overcome influences under the external lumped force uncertainty disturbances in light of the Lyapunov function. Further, the backstepping control with three adaptive rules and RRSPPNN with RGWO and recouped controller to estimate the external lumped force uncertainty has been smoothly derived in the light of the Lyapunov function for diminishing the external lumped force uncertainty effect and improving the chattering phenomenon. The error recouped controller to recoup the smallest fabricated error of the error estimation law has been smoothly derived in light of the Lyapunov function. Two optimal learning rates of the RRSPPNN with two exponential-functional adjustable factors have been smoothly calculated by utilizing the RGWO algorithm to speed up the parameter's convergence.

Finally, some control performances regarding the chattering of control propulsion, the position response, the ability of load force adjustment, the position tracing error, and the refusal ability of unknown parameters disturbance by using the backstepping control with three adaptive rules and RRSPPNN with RGWO and recouped controller are more exceptional than the popular PI controller, the backstepping control with three adaptive rules and a swapping function, and the modified recurrent Rogers–Szego polynomials neural network with DGWO [10] from Tables 1 and 2.

Funding: This research received no external funding.

Acknowledgments: The author would like to acknowledge the financial support of the Ministry of Science and Technology of Taiwan under grant MOST 108-2221-E-239-011-MY2.

Conflicts of Interest: The author declares no conflict of interest.

List of Acronyms

SM	synchronous motor
RRSPPNN	revised recurring Sieved-Pollaczek polynomials neural network
RGWO	reformed grey wolf optimization
SRMs	switched reluctance motors
IMs	induction motors
CNC	computer numerical control
DGWO	decorated gray wolf optimization

SPPNN	feedforward Sieved-Pollaczek polynomials neural network
DSP	digital signal processor
FOC	field-oriented control
IGBT	insulated-gate bipolar transistor
PWM	pulse-width-modulation
SPWM	sinusoida pulse-width-modulation
SCISR	sub-core interrupt service routine
A/D	analog-digital
I/O	input/output
GWO	grey wolf optimization

References

1. Novotny, D.W.; Lipo, T.A. *Vector Control and Dynamics of AC Drives*; Oxford University Press: New York, NY, USA, 1996.
2. Leonhard, W. *Control of Electrical Drives*; Springer: Berlin, Germany, 1996.
3. Lin, C.H. Hybrid recurrent wavelet neural network control of PMSM servo-drive system for electric scooter. *Int. J. Autom. Controll.* **2014**, *12*, 177–187. [CrossRef]
4. Kanellakopoulos, I.; Kokotovic, P.V.; Morse, A.S. Systematic design of adaptive controller for feedback linearizable system. *IEEE Trans. Autom. Control* **1992**, *36*, 1241–1253. [CrossRef]
5. Fang, Y.; Fei, J.; Yang, Y. Adaptive backstepping design of a microgyroscope. *Micromachines* **2018**, *9*, 338. [CrossRef] [PubMed]
6. Zhang, S.; Cui, W.; Alsaadi, F.E. Adaptive backstepping control design for uncertain non-smooth strictfeedback nonlinear systems with time-varying delays. *Int. J. Control Autom. Syst.* **2019**, *17*, 2220–2233. [CrossRef]
7. Tran, T.T. Feedback linearization and backstepping: An equivalence in control design of strict-feedback form. *IMA J. Math. Control Inf.* **2019**. [CrossRef]
8. Guo, C.; Zhang, A.; Zhang, H.; Zhang, L. Adaptive backstepping control with online parameter estimator for a plug-and-play parallel converter system in a power switcher. *Energies* **2018**, *11*, 3528. [CrossRef]
9. Yang, C.; Yang, F.; Xu, D.; Huang, X.; Zhang, D. Adaptive command-filtered backstepping control for virtual synchronous generators. *Energies* **2019**, *12*, 2681. [CrossRef]
10. Chen, D.F.; Shih, Y.C.; Li, S.C.; Chen, C.T.; Ting, J.C. Permanent-magnet SLM drive system using AMRRSPNNB controller with DGWO. *Energies* **2020**, *13*, 2914.
11. Ko, E.; Park, J. Diesel mean value engine modeling based on thermodynamic cycle simulation using artificial neural network. *Energies* **2019**, *12*, 2823. [CrossRef]
12. Bagheri, H.; Behrang, M.; Assareh, E.; Izadi, M.; Sheremet, M.A. Free convection of hybrid nanofluids in a C-shaped chamber under variable heat flux and magnetic field: Simulation, sensitivity analysis, and artificial neural networks. *Energies* **2019**, *12*, 2807. [CrossRef]
13. Noureddine, B.; Djamel, B.; Vicente, F.B.; Fares, B.; Boualam, B.; Bachir, B. Maximum power point tracker based on fuzzy adaptive radial basis function neural network for PV-system. *Energies* **2019**, *12*, 2827.
14. Lee, D.; Kim, K. Recurrent neural network-based hourly prediction of photovoltaic power output using meteorological information. *Energies* **2019**, *12*, 215. [CrossRef]
15. Chen, Y.; Wang, Y.; Ma, J.; Jin, Q. BRIM: An accurate electricity spot price prediction scheme-based bidirectional recurrent neural network and integrated market. *Energies* **2019**, *12*, 2241. [CrossRef]
16. Li, G.; Wang, H.; Zhang, S.; Xin, J.; Liu, H. Recurrent neural networks based photovoltaic power forecasting approach. *Energies* **2019**, *12*, 2538. [CrossRef]
17. Han, L.; Jiao, X.; Zhang, Z. Recurrent neural network-based adaptive energy management control strategy of plug-in hybrid electric vehicles considering battery aging. *Energies* **2020**, *13*, 202. [CrossRef]
18. Lin, C.H. Comparative dynamic control for continuously variable transmission with nonlinear uncertainty using blend amend recurrent Gegenbauer-functional- expansions neural network. *Nonlinear Dyn.* **2017**, *87*, 1467–1493. [CrossRef]
19. Ting, J.C.; Chen, D.F. Novel mingled reformed recurrent Hermite polynomial neural network controller applied in continuously variable transmission system. *J. Mech. Sci. Technol.* **2018**, *32*, 4399–4412. [CrossRef]

20. Lin, C.H.; Chang, K.T. SCRIM drive system using adaptive backstepping control and mended recurrent Romanovski polynomials neural network with reformed particle swarm optimization. *Int. J. Adapt. Control Signal Process.* **2019**, *33*, 802–828. [CrossRef]
21. Waleed, A.S.; Allaway, W.R.; Askey, R. Sieved ultraspherical polynomials. *Trans. Am. Math. Soc.* **1984**, *284*, 39–55.
22. Emary, E.; Yamany, W.; Hassanien, A.E.; Snasel, V. Multi-objective gray-wolf optimization for attribute reduction. *Procedia Comput. Sci.* **2015**, *1*, 623–632. [CrossRef]
23. Mosavi, M.; Khishe, M.; Ghamgosar, A. Classification of sonar data set using neural network trained by gray wolf optimization. *Neural Netw. World* **2016**, *26*, 393–415. [CrossRef]
24. Khandelwal, A.; Bhargava, A.; Sharma, A.; Sharma, H. Modified grey wolf optimization algorithm for transmission network expansion planning problem. *Arab. J. Sci. Eng.* **2018**, *43*, 2899–2908. [CrossRef]
25. Mirjalili, S.; Mirjalili, S.M.; Lewis, A. Grey wolf optimizer. *Adv. Eng. Softw.* **2014**, *69*, 46–61. [CrossRef]
26. Sultana, U.; Khairuddin, A.B.; Mokhtar, A.S.; Zareen, N.; Sultana, B. Grey wolf optimizer based positionment and sizing of multiple distributed generation in the distribution system. *Energy* **2016**, *111*, 525–536. [CrossRef]
27. Parsian, A.; Ramezani, M.; Ghadimi, N. A hybrid neural network-gray wolf optimization algorithm for melanoma detection. *Biomed. Res.* **2017**, *28*, 3408–3411.
28. Duangjai, J.; Pongsak, P. Grey wolf algorithm with borda count for feature selection in classification. In Proceedings of the 3rd International Conference on Control and Robotics Engineering (ICCRE), Nagoya, Japan, 20–23 April 2018; pp. 238–242.
29. Astrom, K.J.; Wittenmark, B. *Adaptive Control*; Addison-Wesley: New York, NY, USA, 1995.
30. Slotine, J.J.E.; Li, W. *Applied Nonlinear Control*; Prentice-Hall: Englewood Cliffs, NJ, USA, 1991.
31. Ismail, M.H. On sieved orthogonal polynomials I: Symmetric Pollaczek analogues. *SIAM J. Math. Anal.* **1985**, *16*, 1093–1113. [CrossRef]
32. Astrom, K.J.; Hagglund, T. *PID Controller: Theory, Design, and Tuning*; Instrument Society of America: Research Triangle Park, NC, USA, 1995.
33. Hagglund, T.; Astrom, K.J. Revisiting the Ziegler-Nichols tuning rules for PI control. *Asian J. Control* **2002**, *4*, 364–380. [CrossRef]
34. Hagglund, T.; Astrom, K.J. Revisiting the Ziegler-Nichols tuning rules for PI control-part II: The frequency response method. *Asian J. Control* **2004**, *6*, 469–482. [CrossRef]
35. Lewis, F.L.; Campos, J.; Selmic, R. *Neuro-Fuzzy Control of Industrial Systems with Actuator Nonlinearities*; SIAM Frontiers in Applied Mathematics: Philadelphia, PA, USA, 2002.

energies

MDPI

Article

Design of High-Speed Permanent Magnet Motor Considering Rotor Radial Force and Motor Losses

Nai-Wen Liu [1], Kuo-Yuan Hung [1], Shih-Chin Yang [1,*], Feng-Chi Lee [2] and Chia-Jung Liu [2]

[1] Department of Mechanical Engineering, National Taiwan University, No. 1, Sec. 4, Roosevelt Road, Taipei 10617, Taiwan; d07522029@ntu.edu.tw (N.-W.L.); d07522030@ntu.edu.tw (K.-Y.H.)
[2] Department of Mechatronics Control, Industrial Technology Research Institute, 195, Sec. 4, Chung Hsing Road, Hsinchu 31057, Taiwan; lifengchi@itri.org.tw (F.-C.L.); CJLiu@itri.org.tw (C.-J.L.)
* Correspondence: scy99@ntu.edu.tw

Received: 30 September 2020; Accepted: 6 November 2020; Published: 10 November 2020

Abstract: Different from the design of conventional permanent magnet (PM) motors, high-speed motors are primarily limited by rotor unbalanced radial forces, rotor power losses, and rotor mechanical strength. This paper aimed to propose a suitable PM motor with consideration of these design issues. First, the rotor radial force is minimized based on the selection of stator tooth numbers and windings. By designing a stator with even slots, the rotor radial force can be canceled, leading to better rotor strength at high speed. Second, rotor power losses proportional to rotor frequency are increased as motor speed increases. A two-dimensional sensitivity analysis is used to improve these losses. In addition, the rotor sleeve loss can be minimized to less than 8.3% of the total losses using slotless windings. Third, the trapezoidal drive can cause more than a 33% magnet loss due to additional armature flux harmonics. This drive reflected loss is also mitigated with slotless windings. In this paper, six PM motors with different tooth numbers, stator cores, and winding layouts are compared. All the design methods are verified based on nonlinear finite element analysis (FEA).

Keywords: high-speed motor; permanent magnet motor; variable-frequency drive

1. Introduction

High-speed motors are widely implemented for fan, pump, and compressor applications where the operating speed is beyond 20~30 krpm. Compared to universal motors, high-speed PM motors use voltage source inverters instead of mechanical brushes for the AC voltage commutation [1,2]. Due to the progress on the pulse width modulation (PWM) control and high-density magnets, PM-type high-speed motors can further increase their power density by increasing the operating speed. Consequently, these high-density motors are compatible with moving applications under the size constraint [3].

Because of the high-speed rotation, the rotor radial force causes vibration, leading to additional mechanical losses and degradation of the bearing reliability. As reported in [4], the radial force is induced by the interaction between the rotor magnet flux and stator armature flux. An analytical model of radial force is derived in [5]. The radial force frequency is proportional to the rotor operating speed. Once the force frequency is close to the rotor mechanical natural frequency, the visible resonant vibration appears at high speed. Considering PM motors with surface-mounted magnets, the resonant vibration can also split magnets from the rotor surface. To improve the PM rotor structure, [6] proposed a rotor with a whole cylindrical magnet without an iron shaft to maintain the mechanical structure at high speed. In addition, [7] designed a rotor pole with several zigzag skew magnets to reduce the radial force. The cogging torque was reduced as well.

PM synchronous motors require armature windings to generate the electromagnetic torque. In general, three-phase armature windings can be fixed through stator teeth/slots design or direct attachment on the stator core. The slotless stator without iron teeth results in advantages at high speed [8,9]. In [10] with slotless stators, secondary armature flux harmonics are reduced due to the lack of a slotted effect during the rotation. In this case, rotor eddy current losses at high speed can be suppressed. However, the slotless stator requires a higher current density to maintain the same torque output. It causes the thermal issue on stator windings. In [11] three different slotless windings are compared among armature flux harmonics, current density, and rotor eddy current loss. The slotless windings with a diamond shape cause additional armature flux harmonics, which is not suited for high-speed operation. In [12] different steel thicknesses are compared for the slotless stator. At high speed, the thinner steel is recommended to reduce the axial eddy current losses reflected by the skin effect. However, slotless windings result in higher copper losses due to the lower armature flux density. Recently, [13] proposed slotless windings through a printed circuit board. The low winding resistance is the primary advantage. Based on these studies, the slotless windings might be suitable for high-speed motors; however, the high current density and low torque density are two concerns.

Considering the PM rotor structure at high speed, an additional rotor sleeve is required to fix magnets on the rotor surface. However as mentioned in [14] the rotor sleeve also degrades the energy conversion performance at high speed. In general, the rotor sleeve can be formulated by copper, steel, or carbon-fiber. For the copper sleeve, additional eddy current loss is induced once the armature flux flows across the sleeve. Besides, for the steel sleeve, leakage flux instead of eddy current is the result since the steel is an additional magnetic material on the rotor [14]. In [15] the rotor sleeve is realized by carbon-fiber. In this material, no eddy current and leakage flux is induced. However, carbon-fiber shows poor thermal conduction. This sleeve-covered rotor requires additional temperature consideration to dissipate the heat outside of the rotor. It is observed that the sleeve design is the tradeoff between the rotor strength and motor efficiency.

Six-step trapezoidal brushless DC drives are widely implemented for high-speed motors due to the simple controller implementation with negligible delay issues [16–18]. However, because only the two-phase conduction is applied, considerable third-order and fifth-order current harmonics are induced. In this case, relatively high eddy current loss occurs on the rotor, leading to the rotor thermal issue at high speed [19]. On the other hand, sinusoidal-based vector control uses space vector pulse width modulation (PWM) to manipulate the sinusoidal current. Without secondary current harmonics, the eddy current loss in the rotor can be minimized. However, as reported in [20], the complicated signal process, including the voltage delay compensation and inductance cross-coupling decoupling, is the major challenge. For trapezoidal six-step commutation, a PM motor with the minor influence of non-sinusoidal current harmonics is desired to minimize the motor rotor loss.

This paper designed a suitable high-speed PM motor for the fan system. The desired power is 400 W at the speed 80 krpm. Different from the design of conventional motors, high-speed PM motors are primarily limited by rotor unbalanced radial forces, rotor losses, and rotor mechanical strength. These three design issues are improved through several design considerations. First, the rotor radial force is minimized based on the selection of tooth numbers and windings. Second, rotor losses are minimized through the motor geometric design using a two-dimensional sensitivity analysis. In order to maintain the rotor structure at high speed, the cooper sleeve is typically added on magnets. However, the rotor sleeve-reflected loss dependent on rotor frequency is significant at high speed. This sleeve loss can be minimized with slotless windings. In this paper, six PM motors with different tooth numbers, stator core, and winding layouts are compared among rotor radial forces, rotor losses, and sleeve effects. All the motor design methods are verified based on nonlinear FEA.

2. Reference Case Study

This section defines the targeted specification of the high-speed PM motor. The main characteristics are listed in Table 1.

Table 1. Test PM machine characteristics.

Characteristics	Values
Rotor poles	2-pole
Rated torque	45.4 mNm
Rated current	10 A
Rated speed	80 krpm (1.33 kHz elec. frequency)
DC bus voltage	24 V
Stator outer radius	58 mm
Rotor outer radius	12.5 mm

In this paper, the reference is based on a 400 W 80 krpm commercial motor reported by Nidec [21]. This motor is primarily designed for the high-speed air blower. The stator and rotor outer diameter are respectively 58 and 12.5 mm. The DC bus voltage is 24 V for the use of battery DC power.

2.1. Reference Motor

Figure 1 shows a photograph of the (a) stator and (b) rotor for the reference motor. For the stator, three coils are designed with concentrated windings. By contrast, the rotor contains one pole pair magnet where the copper sleeve is included on the rotor to fix the magnet. In this paper, these geometries are fixed to compare the performance difference among six PM motor topologies.

(a) (b)

Figure 1. Photograph of reference motor specification: (**a**) stator and (**b**) rotor.

2.2. Six Different Motor Topologies

In order to design a suitable motor for high-speed operation, six PM motors with different tooth numbers, stator core, and winding layouts are compared among rotor radial forces, rotor losses, and sleeve effects. As seen from the reference motor in Figure 1, a 2-pole rotor is designed to minimize the ratio of mechanical speed to electrical frequency. This 3-coil-2-pole motor design is commonly applied for high-speed applications. However, asymmetric 3-coil windings might cause unbalanced radial force on the rotor due to the interaction between radial and tangential magnet flux. To improve the influence of radial force, six motors with different stator designs are compared in this paper.

Table 2 lists the main characteristics of these six different motors. For example, the motor identifier, the 3-coil-2-pole motor with double winding across one slot, is referred to as the 3Slot-1CS. The corresponding FEA models are shown in Figure 2. It is noted that the winding topologies are different, leading to the different torque density under the same ampere turns.

Table 2. The main characteristics of six different motor topologies.

Motor Identifier	3Slot-1CS	6Slot-2CS	6Slot-3CS
Rated power (W)	400	400	400
Rated speed (krpm)	80	80	80
Pole	2	2	2
Coil	3	6	6
Coil span	1	2	3
Winding factor	0.866	0.866	1.000
Motor Identifier	**3Sless-1CS**	**6Sless-2CS**	**6Sless-3CS**
Rated power (W)	400	400	400
Rated speed (krpm)	80	80	80
Pole	2	2	2
Coil	3	6	6
Coil span	1	2	3
Winding factor	0.866	0.866	1.000

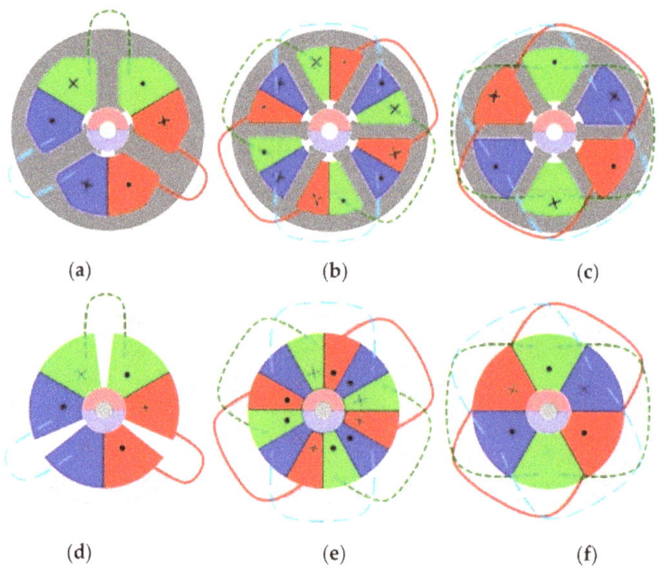

Figure 2. The corresponding FEA (Finite Element Analysis) models of six different analyzed PM motors: (**a**) 3Slot-1CS, (**b**) 6Slot-2CS, (**c**) 6Slot-3CS, (**d**) 3Sless-1CS, (**e**) 6Sless-2CS, and (**f**) 6Sless-3CS.

In general, the torque density is dependent on the magnitude of the winding factor, K_{WF}. As reported in [22] K_{WF} can be evaluated by:

$$K_{WF} = \sin\left(\frac{\xi}{2}\right) \cdot \frac{\sin\left(N_S \cdot \frac{\theta_S}{2}\right)}{N_S \cdot \sin\left(\frac{\theta_S}{2}\right)} \tag{1}$$

where ξ is the coil pitch, N_s is the number of coils per phase per pole, and θ_s is the electrical angle between two coils.

In general, K_{WF} is equal to unity when the coil pitch is equal to the pole pitch. Taking the FEA model in Figure 2c,f, the coil pitch is equal to the pole pitch with 180 deg. At this time, K_{WF} is equal to unity. Under this effect, the relatively highest torque density is expected.

3. Rotor Unbalanced Force Analysis

This section analyzes the rotor unbalanced force for these six motors based on FEA. To investigate the performance of the motor, including the nonlinear properties and the effect of the complex geometry, the electromagnetic FEA is a popular method. On this basis, FEA is a numerical method to solve the physical system modeled by complicated differential equations. For AC motors, Maxwell's equations are the main target equations to be solved. FEA solves the magnetic vector potential within a well-defined 2-D or 3-D region. The magnetic flux density and magnetic energy can then be derived. To analyze the mechanical properties of the motion system, e.g., electromagnetic torque, the virtual-work method is applied.

Figure 2a illustrates the original stator with only 3 coils. The corresponding characteristics are listed by 3Slot-1CS in Table 2. However, the design of 3Slot-1CS results in several disadvantages. First, the asymmetric stator tooth arrangement and windings layout causes an unbalanced radial force. Besides, 3Slot-1CS contains a poor winding factor with only 0.866 as calculated by (1). These issues are investigated as follows.

3.1. Rotor Magnetic Force

The unbalanced magnetic force is one key issue in high-speed motors. In general, the magnetic force results in rotor eccentricity, acoustic noise, vibration, and mechanical loss. These side effects all increase as the speed increases, limiting the high-speed region. The magnetic force consists of radial force F_{rad} and tangential force F_{tan}. As reported from [23,24], the Maxwell stress tensor equation can be used to calculate the magnetic force in the rotor. It is shown to be:

$$F_{rad} = \frac{L_s}{2\mu_0} \oint_{l_c} (B_n^2 - B_t^2)\, dl \tag{2}$$

$$F_{tan} = \frac{L_S}{\mu_0} \oint_{l_c} rB_nB_t\, dl \tag{3}$$

where L_s is the stack length, μ_0 is the magnetic permeability of air, l_c is the integration contour, B_n is the radial magnetic flux density, and B_t is the tangential flux density. It is important to note that F_{tan} contributes to the electromagnetic torque while F_{rad} results in the unbalanced rotor vibration and hence acoustic noises.

Based on FEA, Figure 3 illustrates the magnetic radial force F_{rad} versus the rotor position. Regarding the rotor eccentricity simulation, Figure 4a illustrates the ideal FEA motor simulation without rotor eccentricity. In this case, both the stator and rotor have the same center location. At no load without the armature flux, F_{rad} contains the spatial information with respect to $3\theta_e$. The peak values occur when the rotor positions are respectively at 0, 120, and 240 deg. It is observed that the peak radial force occurs when the rotor magnet is aligned with the stator tooth. By contrast at full load, the location of the peak value slightly moves away from three positions with the increased magnitude.

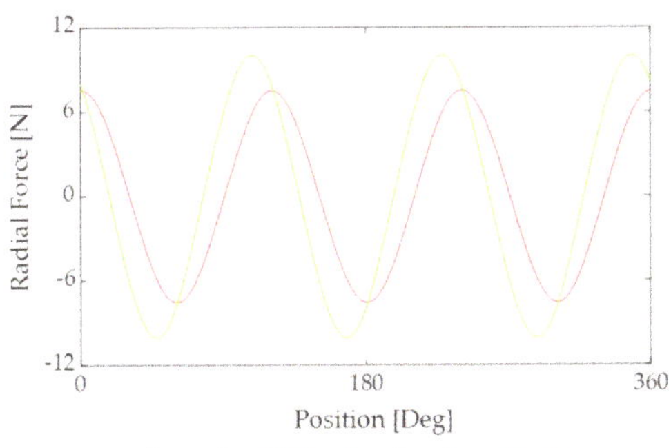

Figure 3. Comparison of unbalanced radial force for the 3Slot-1CS motor topology.

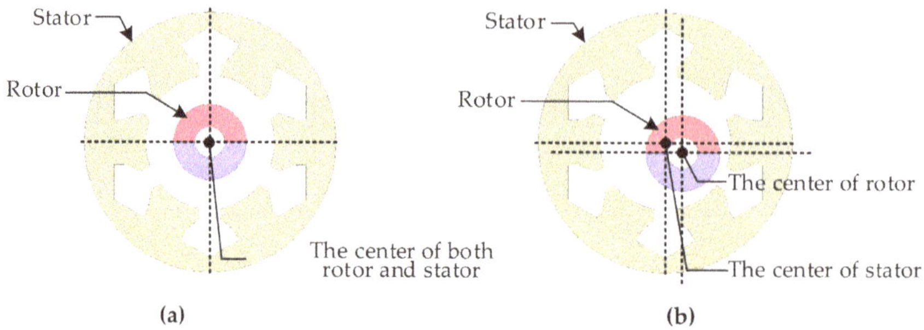

Figure 4. Illustrations of rotor eccentricity: (**a**) no eccentricity and (**b**) the eccentricity from the stator center.

Table 3 summarizes the peak radial force among the six analyzed PM motors in Figure 2. On this basis, the radial magnetic force is a result of the interaction between rotor magnets and stator teeth. Considering the slotted stator, 3Slot-1CS contains the worst radial magnetic force because of the asymmetric stator tooth with the odd coils of 3. By designing the symmetric stator with even coils, this radial force disappears due to the force cancellation between two teeth with same-phase windings. By contrast, for the slotless stator, the radial force should be zero because of the lack of an iron tooth for the interaction with the magnets. However, a small amount of radial force still appears, resulting from the influence of armature flux as load increases, e.g., 3Sless-1CS. Similar to slotted stators, this force is cancelled with symmetric stator coils.

Table 3. Radial force of six analyzed motors under an ideal situation.

Motor Identifier	**3Slot-1CS**	**6Slot-2CS**	**6Slot-3CS**
No Load (N)	7.55	≈0.00	≈0.00
Rated Load (N)	10.08	≈0.00	≈0.00
Motor identifier	**3Sless-1CS**	**6Sless-2CS**	**6Sless-3CS**
No Load (N)	≈0.00	≈0.00	≈0.00
Rated Load (N)	0.94	≈0.00	≈0.00

It is noted that the radial force analysis in Table 3 is calculated under the assumption of ideal rotor fabrication. For actual motors, the assembly tolerance due to the rotor installation should be considered. Figure 4b demonstrates the rotor eccentricity. In this case, the rotor center deviates from the original stator center. Under this effect, Table 4 investigates the peak radial force with manufacturing error. In this simulation, 10% rotor eccentricity is analyzed considering the motor fabrication tolerance. With additional rotor eccentricity, the radial force increases even for motors with symmetric stator. However, compared to motors with an asymmetric stator, the peak radial force magnitude is still sufficiently lower than the magnitude with a symmetric stator. Based on this simulation, it is concluded that the unbalanced radial force can be cancelled using the symmetric stator with even coil numbers, e.g., 6 coils. More importantly, the high-speed motors for the slotless stator without the iron tooth significantly reduce the radial force compared to slotted stator motors.

Table 4. Radial force of six analyzed motors under 10% rotor eccentricity.

Motor Identifier	3Slot-1CS	6Slot-2CS	6Slot-3CS
No Load (N)	7.56	0.58	0.56
Rated Load (N)	10.09	0.69	0.57
Motor identifier	**3Sless-1CS**	**6Sless-2CS**	**6Sless-3CS**
No Load (N)	0.00	0.01	0.01
Rated Load (N)	3.52	0.01	0.01

3.2. Coil Span Design

This part explains the coil span design to maximize the torque output. Although both the 6Slot-2CS and 6Slot-3CS stator achieve the same cancellation performance on radial force, the torque output can be different. As mentioned in (1), the torque output is proportional to the corresponding winding factor. For the 6Slot-2CS with the coil span of 2 in Figure 2b, the relatively low winding factor of 0.866 is the result. By contrast, for 6Slot-3CS, the coil pitch is equal to the pole pitch. Thus, the unity winding factor is the result.

Figure 5 compares the rated torque output versus the rotor position among three different slotted stator motors: 3Slot-1CS, 6Slot-2CS, and 6Slot-3CS. Regarding the mechanical load simulation, the FEA-based transient electromagnetic analysis is applied. The desirable current waveforms are fed into the motor model. In this case, the condition of the electromagnetic field is equivalent to the specific load condition. In FEA, the rated load means to supply the rated current.

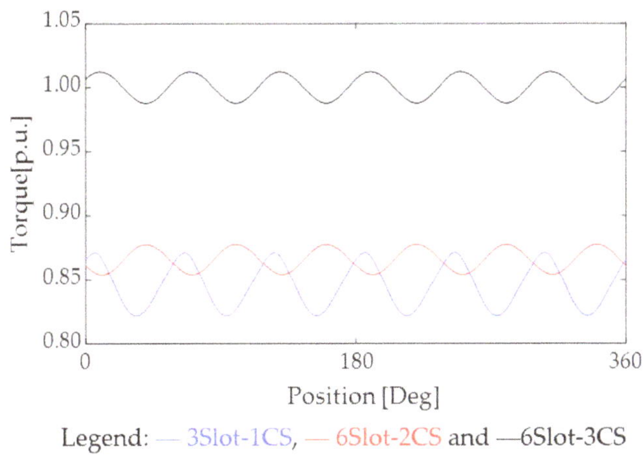

Figure 5. Torque output comparison among three analyzed slotted stators: 3Slot-1CS, 6Slot-2CS, and 6Slot-3CS (the same ampere turns).

In this simulation, the per unit value with respect to the average torque of 3Slot-1CS is defined to easily compare the torque difference. The 3Slot-1CS with concentrated windings in Figure 2a has the lowest average torque output. Although 6Slot-2CS and 3Slot-1CS contain the same winding factor of 0.866, 6Slot-2CS with distributed windings results in a nearly sinusoidal armature flux distribution, reducing the torque ripple. More importantly, 6Slot-3CS provides the highest torque of ~15% more torque under the same ampere turns. It is concluded that 6Slot-3CS with the unity winding factor and even slot numbers is the suitable motor candidate, balancing the radial force and torque at high speed.

3.3. Slotless Windings

The stator topology can be designed based on slotted stators in Figure 2a–c or slotless stators in Figure 2d–f. However, the armature flux distribution can be different. This part compares the armature flux between these two stator cores.

For example, the 3-coil stator can be realized by Figure 2a 3Slot-1CS and Figure 2d 3Sless-1CS. For the slotted core, the iron teeth are added in the stator for the arrangement of coil windings. Since the iron teeth are a magnetic material, the equivalent air gap can be reduced. By contrast, for the slotless core, the windings are directly attached on the surface of the stator backiron. Under this effect, the equivalent air gap is increased compared to the slotted core.

Although the tooth design reduces the air gap, secondary flux harmonics are induced because of the discontinuous air gap distribution. Figure 6 analyzes the air gap flux density versus the rotor position between (a) 3Slot-1CS and (b) 3Sless-1CS. In this simulation, the magnet flux density at no load is analyzed between two different stators. Compared to the slotted core, the slotless core has the lower flux density magnitude due to the larger equivalent air gap. However, a nearly sinusoidal flux distribution is observed, leading to the advantage of rotor magnet loss reduction at high speed. In order to balance the torque output and motor loss, the geometric optimal design will be applied in the next section to find the best-suited high-speed PM motor.

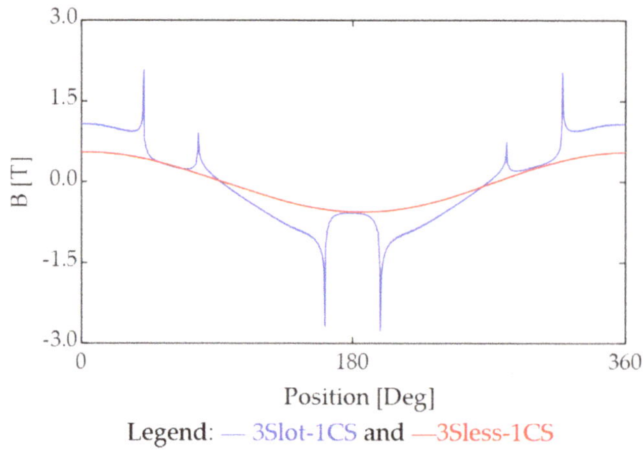

Figure 6. Comparison of magnet flux density distribution along the air gap: 3Slot-1CS and 6Sless-2CS (no load).

4. Geometric Design for Loss Minimization

Although high-speed motors achieve high power density, the considerable high loss density is also a result, leading to the rotor thermal issue. This section aims to minimize the motor rotor loss based on the geometric design. In general PM motor losses consist of (a) copper loss, (b) iron loss, and (c) magnet eddy current loss. In this section, the property of these losses is explained. The stator design is then performed to minimize the motor loss balancing the torque output.

Considering firstly slotted stator motors, key stator parameters are selected as the tooth depth t_d and tooth width t_w [25] Figure 7a defines both t_d and t_w in a slotted stator. It is noted that one additional parameter, slot width s_w, is shown in this figure. However, s_w changes are dependent on t_w. Under this effect, the motor copper loss and iron loss can be minimized in slotted motors by the design of t_d and t_w. By contrast, for slotless motors without an iron tooth, the parameters of slot span s_{spn} and slot depth s_d are selected to determine the slotless stator geometry. Similar to slotted stator motors, the balance design of copper loss and iron loss can be applied on slotless motors by adjusting s_{spn} and

s_d. Figure 7b illustrates both s_{spn} and s_d in a slotless motor. In the following, the property of copper loss, iron loss, and magnet loss will be investigated. FEA is then applied to verify the performance comparison between different stator cores.

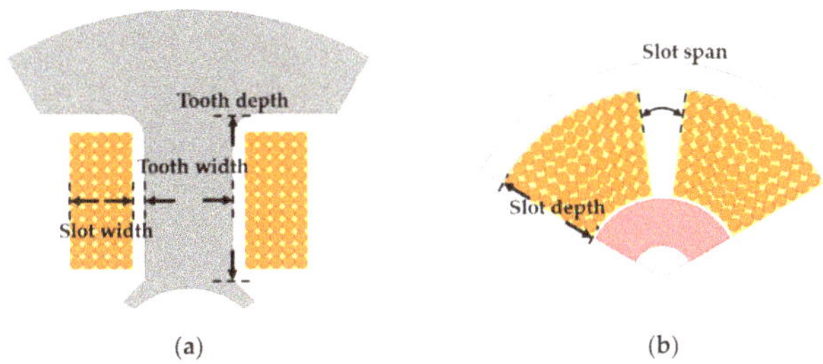

Figure 7. Illustration of key stator geometric parameters for the (**a**) slotted stator and (**b**) slotless stator PM motor.

4.1. Copper Loss

The copper loss is induced by the winding resistances. The copper loss P_{Cu} can be estimated by:

$$P_{Cu} = (N_{coil} \cdot I)^2 \cdot \rho_{Cu} \cdot \frac{\ell_{slot} + \ell_{end}}{K_{ratio} \cdot A_{end}} \tag{4}$$

where N_{coil} is the number of turns per coil, I is the phase current, ρ_{Cu} is the resistivity of the copper conductor, l_{slot} is the slot axial end, l_{end} is the average length of end winding, K_{ratio} is the slot ratio, and finally A_{end} is the slot cross-section. It is noteworthy that the copper loss causes additional heat when phase current passes through coil conductors. For slotless stators, more copper loss is expected because more windings are required to produce the same torque output. However, the low rotor iron loss is the design tradeoff.

4.2. Iron and Magnet Loss

Different from copper loss, the iron loss and magnet loss are both induced by the variation of the magnetic field in the iron core. In general, the iron loss can appear in both the stator and rotor. However, for the surface PM motors analyzed in Figure 2, the rotor iron loss can be negligible due to the large air gap, including the magnet height.

The iron loss can be subdivided into hysteresis loss and eddy current loss. The hysteresis loss appears when the ferromagnetic material is repeatedly magnetized with the AC magnetic field. It causes mutual friction between two internal magnetic domains, resulting in the energy loss [25]. On the other hand, the eddy current loss is caused by the local circulating current due to the armature reflected magnetic field. Considering both hysteresis loss P_{hys} and eddy current loss P_{eddy}, the total iron loss P_{iron} can be shown to be:

$$P_{iron} = P_{hys} + P_{eddy} = K_{hys} B_{peak}{}^2 f + \frac{\pi^2 d^2 \sigma}{6}\left(B_{peak} f\right)^2 + 8.67 K_{eddy}\left(B_{peak} f\right)^{1.5} \tag{5}$$

where K_{hys} is the hysteresis loss coefficient, B_{peak} is the peak flux density, f is the rotor operating frequency, d is the strip lamination thickness, σ is the strip lamination conductivity, and K_{mag} is the anomalous eddy-current loss coefficient.

It is noteworthy that in (5), only the iron eddy current loss is considered in P_{eddy}. In addition to iron, the eddy current also appears in magnets, e.g., magnet loss P_{mag}. For standard PM motors, P_{mag} can be negligible at low speed. However, for high-speed motors, the magnet loss P_{mag} must be taken into account to estimate the overall loss [26]. P_{mag} is mainly caused by air gap armature flux harmonics, which is formulated by:

$$P_{mag} = \int \rho_{Fe} \cdot J^2 dV \quad (6)$$

where ρ_{Fe} is the electric resistivity of the conduction body and J is the armature current density.

4.3. Stator Geometric Design

The stator design of the high-speed motor is performed in this part. Considering firstly the slotted stator motor in Figure 2a–c, key stator geometric features are selected by the tooth depth t_d and tooth width t_w [27]. For the geometric design, the copper loss P_{Cu} and iron loss P_{iron} are both minimized by adjusting t_d and t_w. For example, P_{iron} decreases as t_w increases due to more flux across the area. However, P_{Cu} might increase as t_w increases because of less winding space. By contrast, for the slotless motors in Figure 2d–f, the winding area instead of teeth area determines the magnitude of loss P_{Cu} and P_{iron}. The windings area can be designed through the parameters of slot span s_{spn} and slot depth s_d.

In this paper, the two-dimensional parameter sensitivity analysis is used to design the stator geometry. Taking the example of 3Slot-1CS in Figure 8a, the total loss versus both t_d and t_w is evaluated. It is found that the minimum loss is achieved at 13.83W when t_d = 9.5 mm and t_w = 9 mm. By contrast, for 6Slot-3CS in Figure 8b, the minimum loss is 14.55 W after the geometric design.

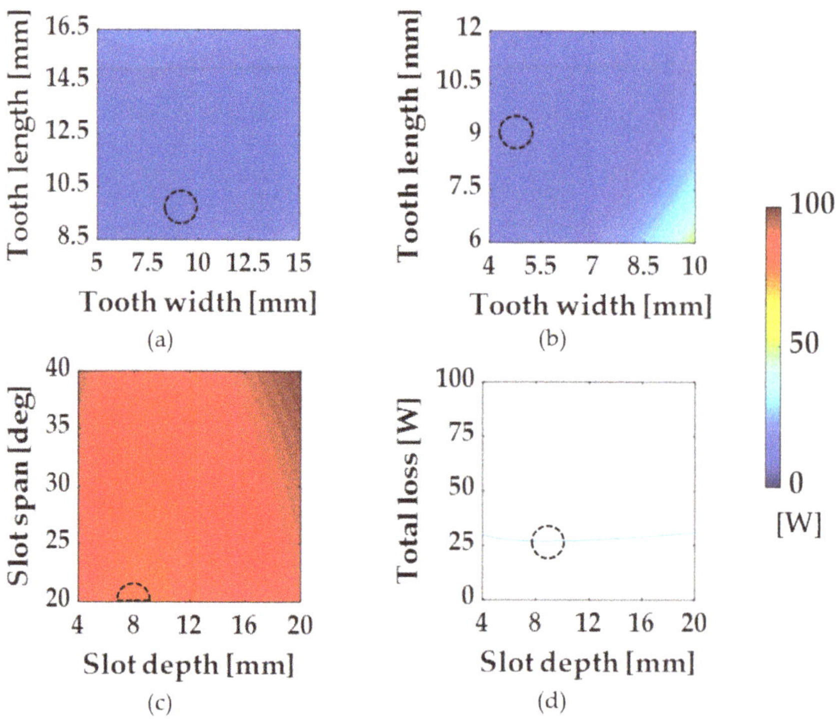

Figure 8. Stator geometric design for the total loss minimization: (**a**) 3Slot-1CS, (**b**) 6Slot-3CS, (**c**) 3Sless-1CS, and (**d**) 6Sless-3CS (no sleeve, 1 p.u. torque and rated speed 80 krpm).

Similarly, Figure 8c,d compares the total loss respectively for 3Sless-1CS and 6Sless-3CS. The minimum loss is resultant, respectively, at 78.88 and 26.90 W. It is noteworthy that the design of 3Sless-1CS considers both s_{spn} and s_d because the slot span in Figure 7b can be adjusted with concentrated windings. By contrast, only sd is adjusted for 6Sless-3CS with distributed windings because s_{spm} obviously needs to be kept at the maximum value for the optimization of space utilization and efficiency.

Table 5 lists the different loss components among the six PM motors analyzed in Figure 2. In this comparison, the torque output is maintained the same to evaluate their loss distributions. The rotor speed is selected at the 80 krpm rate speed. Because of the larger equivalent air gap, the slotless stators result in the higher copper loss than compared to in slotted stators to maintain the same torque. However, for six coils, the copper loss visibly decreased on 6Sless-2CS and 6Sless-3CS. Although the total loss is still higher than 6Slot-2CS and 6Slot-3CS, the negligible iron and magnet loss is the primarily advantage. This property is suited as the topology of the high-speed motor to improve the rotor thermal issue and magnet demagnetization.

Table 5. Different loss components of six different motors without a sleeve.

Motor Identifier	3Slot-1CS	6Slot-2CS	6Slot-3CS
Tooth Depth (mm)	9.5	9	9
Tooth Width (mm)	9	5	5
Copper Loss (W)	4.20	5.07	5.43
Iron Loss (W)	6.65	7.75	7.73
Magnet Loss (W)	2.98	1.39	1.39
Total Loss (W)	13.83	14.21	14.55
Motor Identifier	**3Sless-1CS**	**6Sless-2CS**	**6Sless-3CS**
Slot Span (deg)	20	60	120
Slot depth (mm)	8	9	9
Copper Loss (W)	77.10	26.88	25.85
Iron Loss (W)	1.00	0.97	1.04
Magnet Loss (W)	0.78	0.01	0.01
Total Loss (W)	78.88	27.86	26.90

4.4. Rotor Sleeve Effect

In Section 4.3, the normal PM rotor without sleeve was analyzed. However, at high speed, the rotor sleeve is required to maintain the magnets on the rotor surface. Under this effect, the influence of the rotor sleeve is investigated in this part. In general, the rotor sleeve is realized by a thin layer of non-magnetic material, e.g., copper, iron or stainless steel. For motors with hundreds of watt output, the copper sleeve is preferred due to the better ductility and machinability. Unfortunately, the copper sleeve causes additional eddy loss, increasing the rotor loss as well as the rotor temperature.

Similar to Figure 8, two-dimensional parameter sensitivity analysis is used to determine the stator geometry of the six analyzed PM motors with a rotor copper sleeve. Figure 9 shows the resulting geometric design with respect to the total loss. Considering the slotted stators in Figure 9a,b, the total loss significantly increases compared to the same slotted stators without a copper sleeve. Table 6 summarizes the different motor loss components for PM motors with a copper sleeve. In this table, the sleeve and magnet loss are combined together since the rotor thermal dissipation is directly dependent on these losses. It is shown that the sleeve-reflected loss is dominant for high-speed slotted PM motors, especially for 3Slot-1CS.

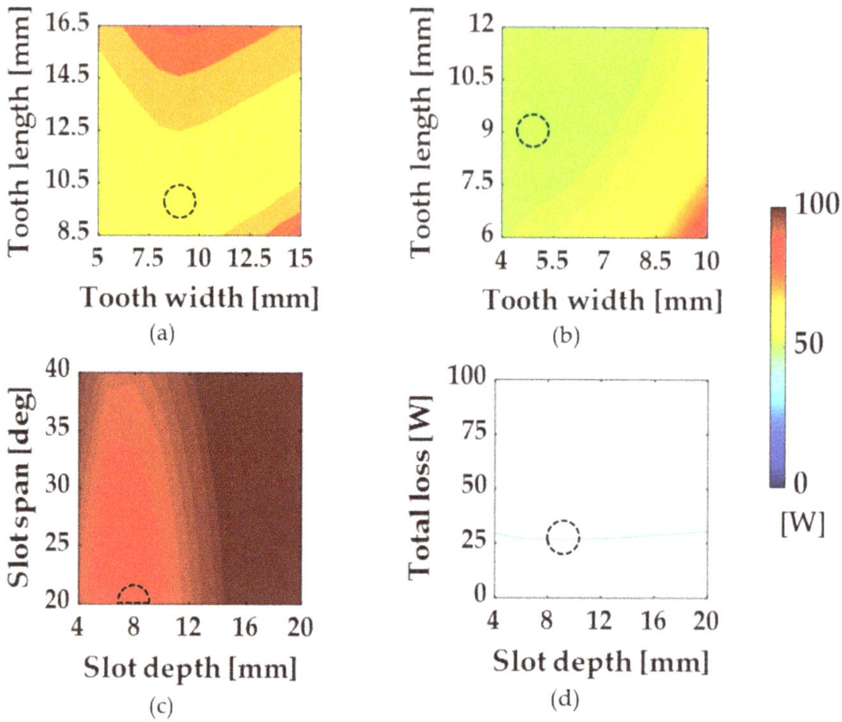

Figure 9. Stator geometric design for the total loss minimization: (**a**) 3Slot-1CS, (**b**) 6Slot-3CS, (**c**) 3Sless-1CS, and (**d**) 6Sless-3CS (copper sleeve, 1 p.u. torque and rated speed 80 krpm).

Table 6. Different loss components of six different motors with sleeve.

Motor Identifier	3Slot-1CS	6Slot-2CS	6Slot-3CS
Tooth Depth (mm)	9.5	9	9
Tooth Width (mm)	9	5	5
Copper Loss (W)	7.23	8.39	6.29
Iron Loss (W)	6.61	7.72	7.71
Magnet Loss (W)	50.78	36.86	36.74
Total Loss (W)	64.62	52.97	50.74
Motor Identifier	**3Sless-1CS**	**6Sless-2CS**	**6Sless-3CS**
Slot Span (deg)	20	60	120
Slot depth (mm)	8	9	9
Copper Loss (W)	79.58	27.99	26.88
Iron Loss (W)	1.00	0.98	0.96
Magnet Loss (W)	7.32	0.12	0.01
Total Loss (W)	87.9	29.09	27.85

Figure 10 analyzes the distribution of the armature flux density among these six PM motors. Because of the nearly trapezoidal wave for 3Slot-1CS with concentrated windings in Figure 10a, the total loss greatly increases from no sleeve 13.83W to sleeve 64.69 W. By using the distributed windings, this sleeve loss is slightly decreased due to the nearly sinusoidal armature flux distribution, e.g., 3Slot-2CS in Figure 10b and 3Slot-3CS in (c). However, the total loss is still significantly higher than the no-sleeve slotted stator motors with coils.

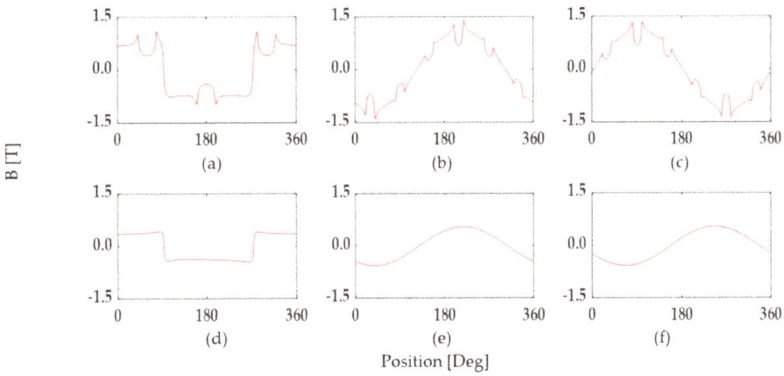

Figure 10. Distribution of armature flux density among six different PM motors: (**a**) 3Slot-1CS, (**b**) 3Slot-2CS, (**c**) 3Slot-3CS, (**d**) 6Sless-1CS, (**e**) 6Sless-2CS, and (**f**) 6Sless-3CS (rated speed).

Figure 9c,d analyzes the total loss for the slotless motors 3Sless-1CS and 6Sless-3CS with respect to the change of s_{spn} and s_d. For 3Sless-1CS, a certain amount of sleeve loss is observed due to the concentrated windings, leading to the trapezoidal armature flux in Figure 10d. By contrast, for 6Sless-3CS with distributed windings, the total loss is similar to that on the no-sleeve motor in Figure 8d. Table 6 also lists the different loss components among the three slotless PM motors. Compared to slotted motors with a copper sleeve, slotless motors with distributed windings achieve both the lowest total loss and sleeve loss. As a result, it is concluded that 6Sless-3CS with distributed windings and a unity winding factor is best suited as the PM motor topology for high-speed operation.

5. Motor Loss Caused by Trapezoidal Commutation

Prior stator design results considered the assumption of three-phase ideal sinusoidal current inputs. However, at high speed, six-step trapezoidal commutation is preferred due to the implementation challenge on the sinusoidal space-vector pulse width modulation (SVPWM) [19]. In addition, the six-step drive has better inverter efficiency because of lower switches compared to SVPWM [19]. Figure 11 explains the six-step trapezoidal drive system analyzed in this section. In general, three Hall sensors are required to obtain the rotor position. In Figure 11a, the relative location between the motor and Hall sensors is shown. These three Hall sensors can be attached respectively at 0, 120, and 240 deg with respect to the motor stator. Besides, (b) demonstrates the corresponding signal flowchart for the six-step trapezoidal commutation. Three-phase voltages $V_A/V_B/V_C$ are manipulated based on the desired torque command through six-step commutation.

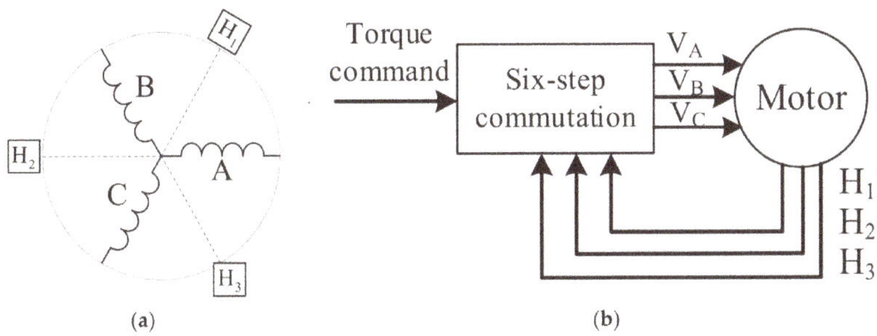

Figure 11. Illustration of six-step trapezoidal drive system: (**a**) the relative location between motor and Hall sensors, and (**b**) the corresponding six-step commutation signal process.

In this section, FEA is used to investigate the motor loss caused by the six-step drive. For the test condition, the trapezoidal wave is used as three-phase current inputs. The full PWM duty is assumed considering the operation at rated speed. Table 7 shows same loss distributions for six PM motors considering the trapezoidal current input. It is found that the magnet and sleeve loss all increase among six motors. For slotted stator motors, the magnet and sleeve loss in the rotor greatly increase compared to Table 5 with sinusoidal current inputs. By contrast, for slotless motors, magnet and sleeve loss increase as well. The rotor loss is approximately one third lower than the same loss on slotted motors. However, compared to slotless motors with sinusoidal currents, the advantage of negligible rotor loss disappears, resulting in the rotor temperature consideration. Based on this comparison, it is concluded that slotless stator motors with distributed windings must be designed as the high-speed PM motor considering the drive of six-step commutation. More importantly, the SVPWM sinusoidal drive is strongly recommended for high-speed motors by minimizing the rotor magnet and sleeve loss.

Table 7. Different loss components of six different motors.

Motor Identifier	**3Slot-1CS**	**6Slot-2CS**	**6Slot-3CS**
Tooth Depth (mm)	9.5	9	9
Tooth Width (mm)	9	5	5
Copper Loss (W)	10.81	6.07	7.63
Iron Loss (W)	16.50	10.08	9.59
Magnet Loss (W)	84.02	48.91	46.92
Total Loss (W)	111.33	65.06	64.14
Motor Identifier	**3Sless-1CS**	**6Sless-2CS**	**6Sless-3CS**
Slot Span (deg)	20	60	120
Slot depth (mm)	8	9	9
Copper Loss (W)	56.86	35.67	31.65
Iron Loss (W)	1.91	1.38	1.41
Magnet Loss (W)	37.28	16.46	17.13
Total Loss (W)	96.05	53.51	50.19

6. Conclusions

This paper proposed a systematic design for high-speed PM motors with consideration of the radial force, rotor loss, rotor sleeve, and trapezoidal commutation. First, the unbalanced rotor radial force is canceled using an even number of stator coils. Considering the rotor manufacturing error, the proposed even coils also reduce the radial force below 0.01N considering 10% rotor eccentricity. Second, the rotor loss at high speed is minimized based on the two-dimensional sensitivity analysis. Specifically, the sleeve reflected eddy current loss is less than 8.3% of the total loss with the slotless stator. It greatly reduces the demagnetization risk at high speed. Third, the trapezoidal commutation drive induces additional 33~65% magnet losses at high speed. These losses are also improved once slotless windings are selected. Considering all high-speed design issues, it is concluded that the 6Sless-3CS topology in Figure 2f with a slotless stator, even coils, and full-pitch distributed windings is the best-suited PM motor, balancing the torque output and motor loss.

Author Contributions: N.-W.L. and S.-C.Y. developed main idea. N.-W.L. and K.-Y.H. implemented and verified the motor design theory through FEA software. F.-C.L. and C.-J.L. contributed analysis tools. All authors have read and agreed to the published version of the manuscript.

Funding: The authors gratefully acknowledge the financial and equipment support from the Ministry of Science and Technology (MOST) and National Chung-Shan Institute of Science and Technology (NCSIST), Taiwan, R.O.C. under Grant 109-3116-F-002-005 -CC1 and 09HT512029.

Conflicts of Interest: The authors declare no conflict of interest.

References

1. Bianchi, N.; Bolognani, S.; Luise, F. Analysis and Design of a PM Brushless Motor for High-Speed Operations. *IEEE Trans. Energy Convers.* **2005**, *20*, 629–637. [CrossRef]
2. Cros, J.; Viarouge, P.; Chalifour, Y.; Figueroa, J. A New Structure of Universal Motor Using Soft Magnetic Composites. *IEEE Trans. Ind. Appl.* **2004**, *40*, 550–557. [CrossRef]
3. Tenconi, A.; Vaschetto, S.; Vigliani, A. Electrical Machines for High-Speed Applications: Design Considerations and Tradeoffs. *IEEE Trans. Ind. Electron.* **2013**, *61*, 3022–3029. [CrossRef]
4. Krotsch, J.; Piepenbreier, B. Radial Forces in External Rotor Permanent Magnet Synchronous Motors with Non-Overlapping Windings. *IEEE Trans. Ind. Electron.* **2011**, *59*, 2267–2276. [CrossRef]
5. Park, S.; Kim, W.; Kim, S.-I. A Numerical Prediction Model for Vibration and Noise of Axial Flux Motors. *IEEE Trans. Ind. Electron.* **2014**, *61*, 5757–5762. [CrossRef]
6. Xu, S.; Liu, X.; Le, Y. Electromagnetic Design of a High-Speed Solid Cylindrical Permanent-Magnet Motor Equipped with Active Magnetic Bearings. *IEEE Trans. Magn.* **2017**, *53*, 1–15. [CrossRef]
7. Wang, S.; Hong, J.; Sun, Y.; Cao, H. Effect Comparison of Zigzag Skew PM Pole and Straight Skew Slot for Vibration Mitigation of PM Brush DC Motors. *IEEE Trans. Ind. Electron.* **2020**, *67*, 4752–4761. [CrossRef]
8. Wallmark, O.; Kjellqvist, P.; Meier, F. Analysis of Axial Leakage in High-Speed Slotless PM Motors for Industrial Hand Tools. *IEEE Trans. Ind. Appl.* **2009**, *45*, 1815–1820. [CrossRef]
9. Looser, A.; Baumgartner, T.; Kolar, J.W.; Zwyssig, C. Analysis and Measurement of Three-Dimensional Torque and Forces for Slotless Permanent-Magnet Motors. *IEEE Trans. Ind. Appl.* **2012**, *48*, 1258–1266. [CrossRef]
10. Bianchi, N.; Bolognani, S.; Luise, F. Potentials and Limits of High Speed PM Motors. *IEEE Trans. Ind. Appl.* **2004**, *40*, 40–1570. [CrossRef]
11. Jumayev, S.; Boynov, K.O.; Paulides, J.J.H.; Lomonova, E.; Pyrhonen, J. Slotless PM Machines With Skewed Winding Shapes: 3-D Electromagnetic Semianalytical Model. *IEEE Trans. Magn.* **2016**, *52*, 1–12. [CrossRef]
12. Millinger, J.; Wallmark, O.; Soulard, J. High-Frequency Characterization of Losses in Fully Assembled Stators of Slotless PM Motors. *IEEE Trans. Ind. Appl.* **2018**, *54*, 2265–2275. [CrossRef]
13. Neethu, S.; Nikam, S.P.; Pal, S.; Wankhede, A.K.; Fernandes, B.G. Performance Comparison Between PCB-Stator and Laminated-Core-Stator-Based Designs of Axial Flux Permanent Magnet Motors for High-Speed Low-Power Applications. *IEEE Trans. Ind. Electron.* **2019**, *67*, 5269–5277.
14. Jun, H.-W.; Lee, J.; Lee, H.-W.; Kim, W.-H. Study on the Optimal Rotor Retaining Sleeve Structure for the Reduction of Eddy-Current Loss in High-Speed SPMSM. *IEEE Trans. Magn.* **2015**, *51*, 1–4.
15. Huang, Z.; Fang, J.; Liu, X.; Han, B. Loss Calculation and Thermal Analysis of Rotors supported by Active Magnetic Bearings for High-speed Permanent Magnet Electrical Machines. *IEEE Trans. Ind. Electron.* **2015**, *63*, 2027–2035. [CrossRef]
16. Kim, T.-H.; Ehsani, M. Sensorless Control of the BLDC Motors From Near-Zero to High Speeds. *IEEE Trans. Power Electron.* **2004**, *19*, 1635–1645. [CrossRef]
17. Chen, S.; Liu, G.; Zheng, S. Sensorless Control of BLDCM Drive for a High-Speed Maglev Blower Using Low-Pass Filter. *IEEE Trans. Power Electron.* **2016**, *32*, 8845–8856. [CrossRef]
18. Liu, G.; Cui, C.; Wang, K.; Han, B.; Zheng, S. Sensorless Control for High-Speed Brushless DC Motor Based on the Line-to-Line Back EMF. *IEEE Trans. Power Electron.* **2014**, *31*, 4669–4683. [CrossRef]
19. Schwager, L.; Tüysüz, A.; Zwyssig, C.; Kolar, J.W. Modeling and Comparison of Machine and Converter Losses for PWM and PAM in High-Speed Drives. *IEEE Trans. Ind. Appl.* **2014**, *50*, 995–1006. [CrossRef]
20. Huang, C.-L.; Chen, G.-R.; Yang, S.-C.; Hsu, Y.-L. Comparison of High Speed Permanent Magnet Machine Sensorless Drive using Trapezoidal BLDC and Sinusoidal FOC under Insufficient PWM Frequency. In Proceedings of the IEEE Energy Conversion Congress and Exposition, Baltimore, MD, USA, 29 September–3 October 2019; pp. 321–325.
21. Shiozawa, K.; Takaki, H. Motor, Blower, and Vacuum Cleaner. U.S. Patent US20190082917A1, 20 December 2019.
22. Krishnan, R. *Permanent Magnet Synchronous and Brushless DC Motor Drives*; CRC Press: Boca Raton, FL, USA, 2009; ISBN 978-0-8247538-4-9.
23. Islam, R.; Husain, I. Analytical Model for Predicting Noise and Vibration in Permanent-Magnet Synchronous Motors. *IEEE Trans. Ind. Appl.* **2010**, *46*, 2346–2354. [CrossRef]
24. Gieras, J.F.; Wang, C.; Lai, J.C. *Noise of Polyphase Electric Motors*; CRC Press: Boca Raton, FL, USA, 2006.

25. Fiorillo, F.; Novikov, A. An improved approach to power losses in magnetic laminations under nonsinusoidal induction waveform. *IEEE Trans. Magn.* **1990**, *26*, 2904–2910. [CrossRef]
26. Tuysuz, A.; Zwyssig, C.; Kolar, J.W. A Novel Motor Topology for High-Speed Micro-Machining Applications. *IEEE Trans. Ind. Electron.* **2013**, *61*, 2960–2968. [CrossRef]
27. Wang, X.; Zhou, S.; Wu, L.; Zhao, M.; Hu, C. Iron Loss and Thermal Analysis of High Speed PM motor Using Soft Magnetic Composite Material. In Proceedings of the International Conference on Electrical Machines and Systems, Harbin, China, 11–14 August 2019; pp. 321–325.

energies MDPI

Article

Combined Optimal Torque Feedforward and Modal Current Feedback Control for Low Inductance PM Motors

Roland Kasper * and Dmytro Golovakha

Chair of Mechatronics, Otto von Guericke University, 39106 Magdeburg, Germany; dmytro.golovakha@ovgu.de
* Correspondence: roland.kasper@ovgu.de

Received: 7 October 2020; Accepted: 22 November 2020; Published: 25 November 2020

Abstract: Small sized electric motors providing high specific torque and power are required for many mobile applications. Air gap windings technology allows to create innovative lightweight and high-power electric motors that show low phase inductances. Low inductance leads to a small motor time constant, which enables fast current and torque control, but requires a high switching frequency and short sampling time to keep current ripples and losses in an acceptable range. This paper proposes an optimal torque feedforward control method, minimizing either torque ripples or motor losses, combined with a very robust and computation-efficient modal current feedback control. Compared to well-known control methods based on the Clarke-Park Transformations, the proposed strategy reduces torque ripples and motor losses significantly and offers a very fast implementation on standard microcontrollers with high robustness, e.g., against measurement errors of rotor angle. To verify the accuracy of the proposed control method, an experimental setup was used including a wheel hub motor built with a slotless air gap winding of low inductance, a standard microcontroller and GaN (Gallium Nitride) Power Devices allowing for high PWM switching frequencies. The proposed control method was validated first by correlation of simulation and experimental results and second by comparison to conventional field-oriented control.

Keywords: optimal control; modal current control; feedforward torque control; feedback current control; torque ripples and loss minimization; low inductance permanent magnet motor

1. Introduction

Manufacturers of electric vehicles or electric planes require electric motors with small size, lightweight construction and maximum efficiency, providing at the same time high torque and power. For low-power applications (<200 W), electrical motor weight can be reduced by air gap winding designs offered, e.g., by Faulhaber and Maxxon [1]. A new slotless air gap winding design for a 15-inch wheel-hub motor providing high power of 40 kW at a speed of 1500 rpm and a very low weight of 20 kg is presented in References [2–4]. Wheel hub motors take a new step in technology of electric vehicles as they allow to create completely new designs of mobile applications without a lot of the mechanics used today and a lot more room for passengers and baggage. High specific power and torque densities are the main indicators of the quality of wheel hub motors. This way, the very lightweight wheel hub motors based on air gap windings [2–4] are most promising.

Usually, motors are based on slotted windings, which significantly increase the motor weight and reduce specific torque and power density. Slotless air gap winding design reduces the weight of electric motors, as it avoids a big part of stator back-iron needed for conventional slotted machines. Additionally, a very high torque can be generated without any torque ripple induced by slotted geometry. However, low number of turns and low iron volume in air gap winding designs lead to very low phase inductances of only several μH, which makes control a challenge [4,5]. For this reason, a high pulse width modulation (PWM) switching frequency of about 100 kHz is needed to keep current ripples, and accordingly torque ripples and associated losses, in a demanded range. In addition, the motor control must be adapted to low phase inductance and high PWM frequency for microcontrollers with a very small sampling step size.

Depending on the phase current waveform, conventional control methods for permanent magnet (PM) motors can be divided into two groups: based on the block commutation [5–9] and based on the Clarke-Park Transformations [10–16]. The first one is usually used for low-cost and low-power applications because of low manufacturing costs and simple design. On the other hand, this leads to a high torque ripple waveform caused by non-ideal commutation currents [5–9]. This disadvantage is even more significant for PM motors with low inductance due to the inverse relationship between phase inductances and current ripples. Trigonometric Clarke-Park Transformations transfer phase currents into a torque generating Q-part and a field generating D-part (DQ control), dependent on rotor position [10–16]. This gives the opportunity for smoother phase commutations, which decreases current and torque ripple but assumes pure sinusoidal waveforms for B-field (magnetic flux density) and phase currents. For high sampling frequencies, calculation of transformations is a high burden for the microcontroller. If the B-field consists of significantly higher harmonics, they should be compensated by adding additionally higher harmonic parts to the phase currents. This method may be unacceptable for high-frequency control, and only iterative approaches are known [10–14].

This article proposes a combined optimal torque feedforward and modal current feedback control method for PM motors with very low phase inductance. Previously, in References [17,18], a high-frequency Combined Optimal Torque and Modal Current Control (OTMIC) was proposed for low inductance PM motors focused on the minimization of motor losses. In this article, this method for loss minimization is described as mathematically complete. Further, it is extended to torque ripple minimization to give a powerful optimization method flexibly adaptable to different goals. Limitations of References [17,18] given by growing size of symbolic expressions will be avoided by a new matrix formulation of the algorithm that allows for adaptive real-time applications. It is possible to switch online between both optimization strategies to fulfill changing demands depending on certain operation points (e.g., low speed → minimal torque ripple, high speed → minimal loss). In case of motors based on air gap winding designs, motor B-field is nearly independent of speed, torque and phase currents. This conclusion was used to explicitly take into consideration all harmonics of B-field for torque control. In an off-line optimization step, optimal current values for a set of electric angles are stored in a table for on-line usage under hard real-time conditions. By means of a very robust modal current control approach, which works independently on rotor angle, the 3-phase current control problem can be reduced to a 2-phase modal one, based on a linear model of the PM machine with constant coefficients. It becomes evident that the non-controllable state (one of the 3-phase currents) can be eliminated without any pre-conditions, e.g., on B-field waveform, and all statements on stability are global using this approach. Additionally, modal transformation matrices are constant, especially independent of electrical angle and any physical parameter. This reduces computation effort significantly compared, e.g., to Park-Clarke Transformation, where transformation matrices change with the angle. Errors of angle measurement or estimation have only a very limited effect on current control. Thus, independence of motor parameters increases the robustness of the transformation due to any physical changes, further allowing for a very fast software implementation on a low-cost microcontroller with an acceptable sampling step size. Optimization as well as control problem can be solved symbolically, allowing for very fast and adaptive implementations.

In the second section of this article, a new approach of PM motor control based on the mathematical model of slotless air gap winding PM motor with very low phase inductance is presented. The proposed method combines modal feedback control to guarantee the desired stability, dynamics and robustness margins with optimal feedforward control to minimize important motor features like ohmic losses or torque ripple. The last section shows the experimental and simulation results of the proposed method and comparison of the proposed control with conventional Field-Oriented Control (FOC) based on DQ

Clarke-Park Transformations (DQ control). A PM wheel hub motor with slotless air gap windings and very small inductances (near 1.5 μH per phase) was used as a test object.

2. Design of the Proposed Combined Optimal Feedforward Torque and Modal Current Feedback Control for Low Inductance PM Motors

2.1. Operation Principle of PM Motor with Very Low Phase Inductance

The dynamic model of a PM motor with slotless air gap winding is obtained considering the following assumptions [3]:

1. Rotor consists of surface-mounted permanent magnets.
2. No cogging torque due to slotless design.
3. Harmonic magnetic field density generated by permanent magnets.
4. B-field harmonics and inductance are constant, e.g., independent from speed and phase currents due to very small current-induced magnetic fields of the windings.
5. Stator has symmetric 3-phase star-connected windings commutated by a 3-phase inverter (B6-bridge).

A model is shown in Figure 1, where R—phase resistance, L, M—phase self and mutual inductance, $\underline{e} = \begin{bmatrix} e_a & e_b & e_c \end{bmatrix}^T$—phase back-EMFs (back electromotive force), $\underline{u} = \begin{bmatrix} u_a & u_b & u_c \end{bmatrix}^T$, $\underline{i} = \begin{bmatrix} i_a & i_b & i_c \end{bmatrix}^T$—phase voltages and currents, u_S—voltage at star point and u_{DC}, i_{DC}—input direct voltage, current to PM motor inverter [2–4].

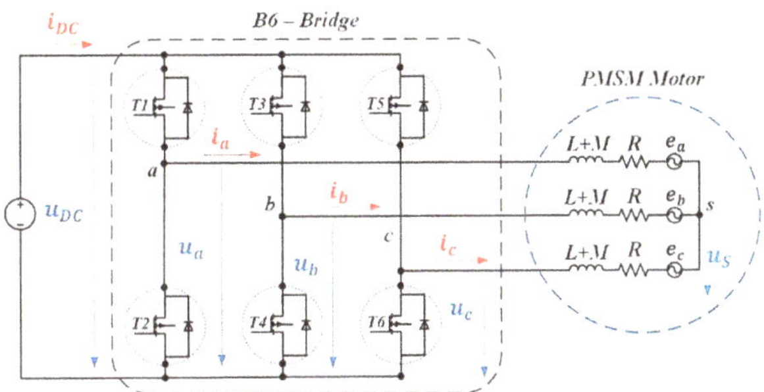

Figure 1. B6-bridge-fed 3-phase permanent magnet (PM) drive connections.

Considering the Y-connection, Kirchhoff's first Law

$$i_a + i_b + i_c = 0 \tag{1}$$

and Kirchhoff's second Law

$$\begin{bmatrix} u_a \\ u_b \\ u_c \end{bmatrix} = \begin{bmatrix} R & 0 & 0 \\ 0 & R & 0 \\ 0 & 0 & R \end{bmatrix} \begin{bmatrix} i_a \\ i_b \\ i_c \end{bmatrix} + \begin{bmatrix} L & -M & -M \\ -M & L & -M \\ -M & -M & L \end{bmatrix} \begin{bmatrix} di_a/dt \\ di_b/dt \\ di_c/dt \end{bmatrix} + \begin{bmatrix} e_a \\ e_b \\ e_c \end{bmatrix} + \begin{bmatrix} 1 \\ 1 \\ 1 \end{bmatrix} u_S \tag{2}$$

or more compact

$$\underline{u} = \underline{R}\,\underline{i} + \underline{L}\,d\underline{i}/dt + \underline{e} + \underline{u}_S \tag{3}$$

state the mathematical model of a 3-phase PM motor. The 3-phase back-EMF values

$$\underline{e} = \omega\, k_M\, \underline{B}(\varphi), \tag{4}$$

depend on ω—the mechanical angular velocity of the rotor, k_M—a motor geometric constant and $\underline{B} = \begin{bmatrix} B_a(\varphi) & B_b(\varphi) & B_c(\varphi) \end{bmatrix}^T$—the vector of averaged B-fields acting on air gap winding phases [3], which itself depends on φ—the electrical angle. Obviously, $B_a(\varphi) = B(\varphi)$, $B_b(\varphi) = B\left(\varphi - \frac{2\pi}{3}\right)$ and $B_c(\varphi) = B\left(\varphi - \frac{4\pi}{3}\right)$ are given by the harmonic magnetic field density generated by permanent magnets, calculated by:

$$B(\varphi) = \sum_k b_k \sin(k\,\varphi). \tag{5}$$

as a function of electric angle and odd harmonic coefficients, b_k. Either FE analysis of the underlying magnetic circuit or line-fitting of measured back-EMF values give the B-field harmonic coefficients with very good accuracy [2–4].

Electrical torque is generated by Lorentz force [2–4]:

$$T_E = k_M \begin{bmatrix} B_a & B_b & B_c \end{bmatrix} \begin{bmatrix} i_a \\ i_b \\ i_c \end{bmatrix} = k_M \underline{B}^T \underline{i} \tag{6}$$

acting on the airgap winding. Loss torque:

$$T_L = d\,\omega + c \tag{7}$$

collects linear mechanical friction together with back-iron and airgap winding eddy losses into the linear friction coefficient d and mechanical coulomb friction together with back-iron hysteresis losses into the constant friction coefficient c. It will be noted that loss torque T_L and resulting loss power depends only on speed and is independent of motor current. Electrical torque T_E reduced by loss torque T_L gives motor output torque:

$$T_M = T_E - T_L. \tag{8}$$

Equations (1)–(8) describe the motor model in terms of back-EMF, output torque and phase currents based on information of angular position φ, angular velocity ω, B-field harmonics and motor parameters k_M, d and c. Torque Equation (6) shows that the adaptation of phase current waveforms taking into account the B-field harmonics is the key to motor torque control [2–4].

Phase currents $i(\varphi)$, like B-field, can also be defined as harmonic functions of electrical angle φ:

$$i(\varphi) = \sum_k a_k \sin(k\,\varphi). \tag{9}$$

Cosine terms are omitted. If included, they would vanish during optimization. From (1), we get additional constraints:

$$a_{3k} = 0 \tag{10}$$

for current coefficients. Average electrical torque:

$$T_A = \frac{1}{2\pi} \int_0^{2\pi} T_E\, d\varphi = \frac{3}{2} k_M \sum_k a_k b_k \tag{11}$$

can then be expressed by Fourier coefficients of phase current and B-Field. Using Equations (5) and (8)–(11), in this paper will be proposed a new method to generate optimal phase currents to generate a given torque T_{Ref} for an arbitrary harmonic B-field, while minimizing either the motor losses or alternatively while minimizing the output torque ripples.

2.2. Optimal Control for Minimal Motor Losses

Generation of a given torque T_{Ref} with maximum efficiency or, equivalently expressed, with minimum losses P_L is a typical design requirement of an electrical motor. Averaged total motor losses consist of ohmic and non-ohmic losses [3]:

$$\begin{aligned} P_L &= \tfrac{1}{2\pi} \int_0^{2\pi} \left(R\left(i_a^2 + i_b^2 + i_c^2\right) + d\,\omega^2 + c\,\omega \right) d\varphi \\ P_L &= \tfrac{3R}{2} \sum_k a_k^2 + d \cdot \overline{\omega}^{\,2} + c \cdot \overline{\omega}, \end{aligned} \tag{12}$$

where non-ohmic losses depend only on the average angular speed in a pole pair segment:

$$\overline{\omega} = \int_0^{2\pi} \omega d\varphi \quad \text{and} \quad \overline{\omega}^{\,2} = \int_0^{2\pi} \omega^2 d\varphi \tag{13}$$

and ohmic losses can be expressed by Fourier coefficients of phase currents according to (9). Given demands lead to the mathematical optimization problem for minimization of motor losses P_L while generating a demanded torque T_{Ref}:

$$Min = P_L \quad \text{with constraint} \quad T_A = T_{Ref} \tag{14}$$

for unknown parameters a_k of the phase current harmonics based on the given harmonics b_k of the B-field. The solution of this quadratic optimization problem can be determined analytically, e.g., by combining cost function and constraints in a Lagrange function, calculating derivatives and solving first-order conditions for unknown coefficients:

$$a_m = \frac{2}{3} \frac{T_{Ref}}{k_M} \frac{b_m}{\overline{b}^2} \tag{15}$$

with the sum of squares of selected harmonics of B-field:

$$\overline{b}^2 = \sum_m b_m^2 \tag{16}$$

and the selection indices:

$$m = 1, 5, 7, 11, 13, 17, 19, 23 \ldots \tag{17}$$

Remaining parameters a_k are zero. Selection indices show that some odd harmonics of phase current {3, 9, 15, ... } will not contribute to torque. The optimal current waveforms can be calculated from (15) for each electrical angle taking into account the harmonics of B-field, motor constant k_M and given reference torque T_{Ref}. This approach achieves that each demanded torque can be generated with minimum motor losses for arbitrary B-field and regardless of motor speed. Air gap winding PM motor design allows to have nearly ideally constant B-field harmonics with minor sensitivity to temperature changes. If necessary, e.g., for application to other motor types, the effect of temperature on B-field and phase resistances can be easily incorporated into the model and adapted online.

2.3. Optimal Control for Minimal Torque Ripples

Generation of a given torque T_{Ref} with minimum torque ripples is another typical design requirement of an electrical motor, especially if vibrations and noise are critical. Torque ripple is defined as mean square deviation between electrical torque T_E and average torque T_A:

$$T_R^2 = \frac{1}{2\pi} \int_0^{2\pi} \left(T_E - T_A\right)^2 d\varphi \tag{18}$$

In this case, optimal current can be determined solving the optimization problem:

$$Min = T_R^2 \quad \text{with constraint} \quad T_A = T_{Ref} \tag{19}$$

Inserting current definition (9) into the integral (18), the optimization problem stated in (19) again can be solved analytically by formulating Lagrange function, calculating derivatives and solving the first-order conditions. Besides this straightforward path, there exists another approach, which leads to more simple expressions and allows further insight into the solution. Starting from (5), (6) harmonics

$$t_k = \frac{1}{\pi} \int_0^{2\pi} T_E \cos(6\, k\, \varphi) d\varphi \tag{20}$$

of electrical torque T_E can be determined and collected in vector

$$\underline{t} = \underline{A}\, \underline{a} \tag{21}$$

with matrix

$$\underline{A} = \frac{2}{3} k_M \begin{bmatrix} 2b_1 & 2b_5 & 2b_7 & 2b_{11} & 2b_{13} & 2b_{17} & 2b_{19} & 2b_{23} \\ b_7 - b_5 & b_{11} - b_1 & b_{13} + b_1 & b_{17} + b_5 & b_{19} + b_7 & b_{23} + b_{11} & b_{13} & b_{17} \\ b_{13} - b_{11} & b_{17} - b_7 & b_{19} - b_5 & b_{23} - b_1 & b_1 & b_5 & b_7 & b_{11} \\ b_{19} - b_{17} & b_{23} - b_{13} & -b_{11} & -b_7 & -b_5 & -b_1 & b_1 & b_5 \\ -b_{23} & -b_{19} & -b_{17} & -b_{13} & -b_{11} & -b_7 & -b_5 & -b_1 \\ 0 & 0 & -b_{23} & -b_{19} & -b_{17} & -b_{13} & -b_{11} & -b_7 \\ 0 & 0 & 0 & 0 & -b_{23} & -b_{19} & -b_{17} & -b_{13} \\ 0 & 0 & 0 & 0 & 0 & 0 & -b_{23} & -b_{19} \end{bmatrix} \tag{22}$$

and vector of current harmonics

$$\underline{a} = [a_1, a_5, a_7, a_{11}, a_{13}, a_{17}, a_{19}, a_{23}] \tag{23}$$

for an example with 23 harmonics. Only average value at zero and multiples of the sixth harmonic appear. All other coefficients are zero. Matrix $\underline{A}$ is square and has full rank. Defining the torque harmonics

$$\underline{t} = \left[2\, T_{Ref}, 0, 0, 0, 0, 0, 0, 0\right] \tag{24}$$

in a way that the zero coefficient is given by the demanded torque T_{Ref} and all other torque harmonics vanish, delivers the solution for the above-mentioned optimization problem by solving the linear system of equations defined by (21). For practical applications, a much smaller number of harmonics than given in (23) will be sufficient. The larger number presented here helps to demonstrate the structure of the matrix. It is important to state that also the torque ripples minimization leads to an analytical solution for optimal currents that is well suited for online and adaptive applications.

2.4. Modal Current Control

So far, we have two feedforward strategies to implement either loss minimal or torque ripples minimal operation of the motor. Both strategies rely on current control, which is able to deal with higher current harmonics, which means also with higher frequencies. A modal current control method, exploiting all structural benefits of given control problem and allowing for very high-speed control, will be presented in more detail in this section. The complete model of the PM motor with air gap winding is defined by (1)–(3). Node Equation (1) shows that only 2 of the 3 phase currents are independent. Taking this dependency into account, (3) can be written as

$$\underline{G}(\underline{u} - \underline{e}) = \underline{R}\, \underline{i} + \underline{L} d\underline{i}/dt \tag{25}$$

with

$$\underline{G} = \frac{1}{3}\begin{bmatrix} 2 & -1 & -1 \\ -1 & 2 & -1 \\ -1 & -1 & 2 \end{bmatrix}. \quad (26)$$

For simplicity, back-EMF $\underline{e}$ can be neglected and later added to input voltage $\underline{u}$ as a feedforward compensation or simply treated as a disturbance for the controller. This coupled system for the 3 phase currents $\underline{i}$ can be completely decoupled with a modal transformation

$$\begin{bmatrix} J_\Sigma \\ J_1 \\ J_2 \end{bmatrix} = \underbrace{\frac{1}{3}\begin{bmatrix} 1 & 1 & 1 \\ -1 & -1 & 2 \\ -1 & 2 & -1 \end{bmatrix}}_{\underline{W}} \begin{bmatrix} i_a \\ i_b \\ i_c \end{bmatrix} \quad (27)$$

applied to system (25) in a way that modal inductance matrix

$$\underline{\Lambda} = \underline{W}\,\underline{L}\,\underline{W}^{-1} = \begin{bmatrix} L-2M & 0 & 0 \\ 0 & L+M & 0 \\ 0 & 0 & L+M \end{bmatrix} \quad (28)$$

gets diagonal. Phase currents $\underline{i}$ are transformed to modal currents $\underline{J} = \begin{bmatrix} J_\Sigma & J_1 & J_2 \end{bmatrix}^T$. Considering (1) and the first row of the transformation matrix defined in (27), modal current J_Σ will be zero. Thus, only 2 out of 3 modal currents associated with the modal inductances $L+M$ are relevant. The decoupled modal system

$$\frac{1}{3}\begin{bmatrix} 0 & 0 & 0 \\ -1 & -1 & 2 \\ -1 & 2 & -1 \end{bmatrix}\begin{bmatrix} u_a \\ u_b \\ u_c \end{bmatrix} = \begin{bmatrix} R & 0 & 0 \\ 0 & R & 0 \\ 0 & 0 & R \end{bmatrix}\begin{bmatrix} J_\Sigma \\ J_1 \\ J_2 \end{bmatrix} + \begin{bmatrix} L-2M & 0 & 0 \\ 0 & L+M & 0 \\ 0 & 0 & L+M \end{bmatrix}\begin{bmatrix} dJ_\Sigma/dt \\ dJ_1/dt \\ dJ_2/dt \end{bmatrix} \quad (29)$$

shows that modal current J_Σ is not controllable as the associated row of the modal input matrix is completely zero. The remaining 2 modal currents, J_1 and J_2, come from 2 identical first-order systems with time constant $\frac{L+M}{R}$. Eliminating non-controllable current J_Σ gives the reduced controllable system

$$\underbrace{\frac{1}{3}\begin{bmatrix} -1 & -1 & 2 \\ -1 & 2 & -1 \end{bmatrix}}_{\underline{\Gamma}}\begin{bmatrix} u_a \\ u_b \\ u_c \end{bmatrix} = \begin{bmatrix} R & 0 \\ 0 & R \end{bmatrix}\begin{bmatrix} J_1 \\ J_2 \end{bmatrix} + \begin{bmatrix} L+M & 0 \\ 0 & L+M \end{bmatrix}\begin{bmatrix} dJ_1/dt \\ dJ_2/dt \end{bmatrix} \quad (30)$$

Thus, the 3-phase current control problem is reduced to the control of two completely identical but decoupled first-order systems, associated with modal currents J_1 and J_2. The redundant total current J_Σ displays any errors in current control loop and can be used as an indicator of the quality of the system online. Modal currents J_1 and J_2 are not connected and can be controlled completely independent by modal input voltages

$$\underline{V} = \begin{bmatrix} V_1 \\ V_2 \end{bmatrix} = \underline{\Gamma} \cdot \underline{u}, \quad (31)$$

as shown in Figure 2. This simple separation based on the modal transformation given in (27) does not use any trigonometric calculations based on the rotor angle as in conventional Clarke-Park Transformations, is independent of motor parameters and is well defined for any shape of B-field.

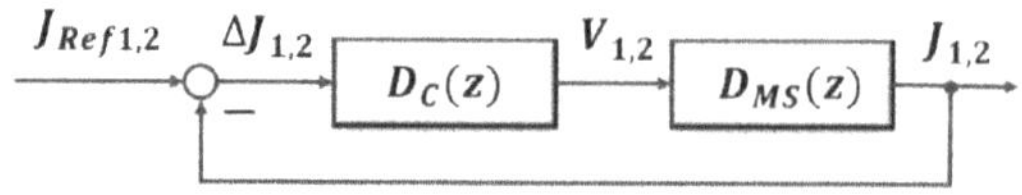

Figure 2. Modal control loop for modal currents $J_{1,2}$.

According to (30). motor transfer function is given by

$$G_M(s) = \frac{1/R}{1 + \frac{L+M}{R}s}. \tag{32}$$

For high-speed control of low inductance PM motors with high number of poles, current sensor dynamics must be included in the control loop with a first-order sensor model

$$G_S(s) = \frac{1}{1 + T_S s} \tag{33}$$

and sensor time constant T_S. As a microcontroller-driven B6-bridge generates PWM pulses at frequency $f_{PWM} = 1/\Delta t$ to control motor input voltage and ADC (Analog-to-Digital Converter) samples current sensor signals with the same rate, time continuous transfer function

$$G_{MS}(s) = G_M(s)G_S(s) \tag{34}$$

has to be discretized with that frequency to get the discrete transfer function

$$D_{MS}(z) = \frac{(\beta - 1 + (1-\alpha)\delta)z + ((\alpha - 1)\delta - \alpha)\beta + \alpha}{R(\delta - 1)(z - \alpha)(z - \beta)} \tag{35}$$

with sampling time Δt, discrete eigenvalues of motor $\alpha = e^{-\frac{R\Delta t}{L+M}}$, sensor $\beta = e^{-\frac{\Delta t}{T_S}}$ and motor/sensor speed ratio $\delta = \frac{L+M}{R\,T_S}$. Very small motor time constant due to low phase inductance allows to achieve very fast current and electrical torque response with requested time constant T_{Req}, which expands the possibilities of design for more complex control systems like dynamic stability control or autonomous driving. To get a controlled system of first order

$$D(z) = \frac{1 - z_R}{z - z_R} \tag{36}$$

with requested eigenvalues $z_R = e^{-\frac{\Delta t}{T_{Req}}}$, it is necessary to use a discrete PID (Proportional-Integral-Derivative) controller with filtered derivative action:

$$D_C(z) = K_P + \frac{K_I}{z-1} + \frac{K_D}{N_D + \frac{1}{z-1}}. \tag{37}$$

Control parameters can be determined by comparison of coefficients of discrete closed loop

$$D(z) = \frac{D_C(z)D_{MS}(z)}{1 + D_C(z)D_{MS}(z)} \tag{38}$$

and reference transfer function given in (36), as

$$\begin{aligned}
K_P &= \frac{(1-\alpha+\delta(\beta-1))R(z_R-1)}{(\delta-1)(\beta-1)(\alpha-1)};\\
K_I &= R(z_R-1);\\
K_D &= \frac{R\delta(z_R-1)(\beta-\alpha)^2}{(\delta-1)^2(\beta-1)^2(\alpha-1)^2};\\
N_D &= \frac{\beta-1+\delta(1-\alpha)}{(\delta-1)(\beta-1)(\alpha-1)}.
\end{aligned} \tag{39}$$

A saturation of output and I-part of the PID controller guarantees safe operation in the case of limited supply voltages. Outputs from PID controllers $D_C(z)$ are two modal control voltages, V_1 and V_2. According to (31), V_1 and V_2 must be transformed back to three control voltages $\underline{u}$ by means of a Moore Penrose pseudo-inverse transformation matrix $\underline{\Gamma}^+$

$$\begin{bmatrix} u_a \\ u_b \\ u_c \end{bmatrix} = \underbrace{\begin{bmatrix} -1 & -1 \\ 0 & 1 \\ 1 & 0 \end{bmatrix}}_{\underline{\Gamma}^+} \begin{bmatrix} V_1 \\ V_2 \end{bmatrix}. \tag{40}$$

Modal current control ensures system stability regardless of fluctuations in system parameters caused by changes of motor temperature and other factors. Moreover, it can be implemented for PM synchronous motors with any waveform of B-field. However, it should be noted, that the motor model should satisfy the mathematical model described with Equations (1)–(3), which can be very good when performed for PM motors with air gap windings and also other PM motors.

2.5. Control Diagram of OTMIC

The proposed control diagram of Combined Optimal Torque and Modal Current Control (OTMIC) is shown in Figure 3. The control algorithm uses a torque reference T_{Ref} and measured electrical angle φ to estimate optimal phase current waveforms for motor loss or torque ripples minimization, $\underline{i}_{Ref} = \begin{bmatrix} i_{aRef}(\varphi) & i_{bRef}(\varphi) & i_{cRef}(\varphi) \end{bmatrix}^T$, based on one of the optimization methods described in Sections 3.1 and 3.2. The optimal currents $\underline{i}_{Ref}$ are used as a reference for measured phase currents $\underline{i}$. Current errors $\Delta\underline{i}$ as differences between reference and measured currents are fed to the Modal Current Control described in Section 2.4 and used to generate the control voltages $\underline{u}$, which set PM motor in motion with the optimal torque waveform via a SVPWM (Space Vector Pulse Width Modulation) and a B6-bridge. Architecture of OTMIC control allows to simplify the microcontroller torque/current control loop by using constant matrices with simple coefficients for modal transformation.

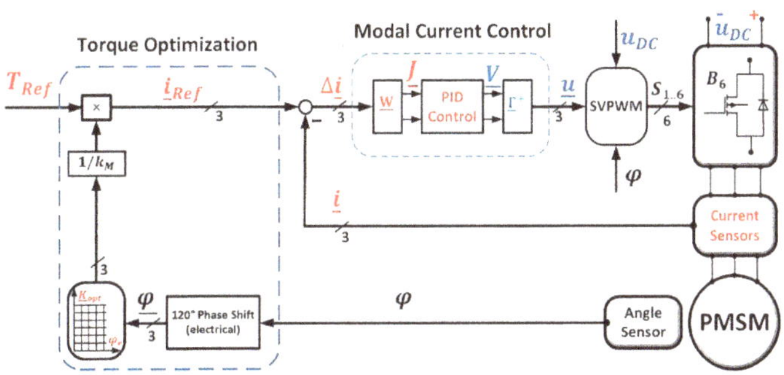

Figure 3. Proposed Combined Optimal Torque and Modal Current Control (OTMIC) diagram of PM motor drive system.

In addition, optimal reference currents according to (9) and (15)–(17) for minimal Loss Control or from (9) and (21)–(24) for minimal Torque Ripple Control can be calculated off-line for a set of angles and stored in a table. These simplifications allow to implement the control loop with very small step size of $\Delta t < 10\ \mu s$ on a common low-cost microcontroller.

3. Units Simulation and Experimental Results

3.1. Motivation and Experimental Setup

A wheel hub motor for a hub-less scooter, shown in Figure 4 on the experimental setup (left above) and in the dissembled state (right) for better demonstration of the air gab winding, will be used for verification of OTMIC control method. The motor has a very low total weight of 2.7 kg, a high nominal power of 4.8 kW and a nominal torque of 80 Nm. More motor data is given in Table 1, where B odd Harmonics are 1st, 3rd, 5th and 7th harmonic numbers. The experimental platform shown in Figure 4 was used to verify the proposed OTMIC method with either minimized motor losses or minimized torque ripples for low inductance PM motors.

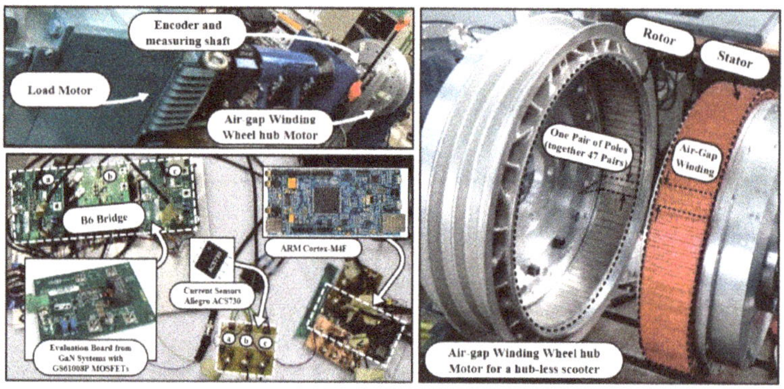

Figure 4. Experimental setup for OTMIC verification [18].

Table 1. Air gab winding wheel-hub motor parameters [17,18].

Symbol	Description	Value	UOM
	Motor Parameters		
u_{DC}	DC voltage	48	V
$L_s + M_s$	phase inductance	1.5	μH
R_s	phase resistance	0.026	Ω
t_M	motor time constant	58	μs
p	number of poles	94	–
k_M	motor constant	0.304	Nm/(TA)
$\underline{b}_k$	B odd harmonics	[1.15 0.2 0.06 0.01] T	
c	constant of coulomb friction and hysteresis loss	0.0832	Nm
d	constant of mechanical friction and eddy loss	0.0008	$Nm{\cdot}s/rad$
	Control Parameters		
K_P	Proportional term	0.166	V/A
K_I	Integral term	0.026	V/A
K_D	Derivative term	0.013	V/A
N_D	PID filter term	0.9854	–
T_S	sensor time constant	1	μs
Δt	sampling time	10	μs
f_{PWM}	switching frequency	100	kHz
T_{Req}	requested time constant	20	μs

Due to elimination of stator iron, motor inductance is very low (near 1.5 µH per phase). Thus, motor needs an adapted control with high switching frequency of about 100 kHz to keep current ripples and total losses in an acceptable range. Control parameters calculated according to (39) are also given in Table 1.

To drive the motor, six GS61008P MOSFETs (metal–oxide–silicon transistors) from company GaN Systems, able to switch up to 90 A at a rate up to 300 kHz, were connected as a B6-bridge. Due to limited MOSFETs cooling of experimental boards, phase current was restricted to 20 A for the experiments. Hall-effect-based sensors from Allegro ACS730 were used for phase current measurement with a very small sensor time constant of 1 µs. Rotor angle is delivered by an encoder using 5 k increments. 100 kHz PWM switching frequency to power the B6-bridge requires a very small step size in the microcontroller (lower than 10 µs). To achieve this requirement, an ARM Cortex-M4F has been used running at 200 MHz. It allows to implement the control loop shown in Figure 3 in 8.3 µs. In comparison to the proposed method, a conventional DQ control needs a minimum of 9.4 µs, which underlines the effectiveness of the proposed method. This comparison can be seen in Figure 5.

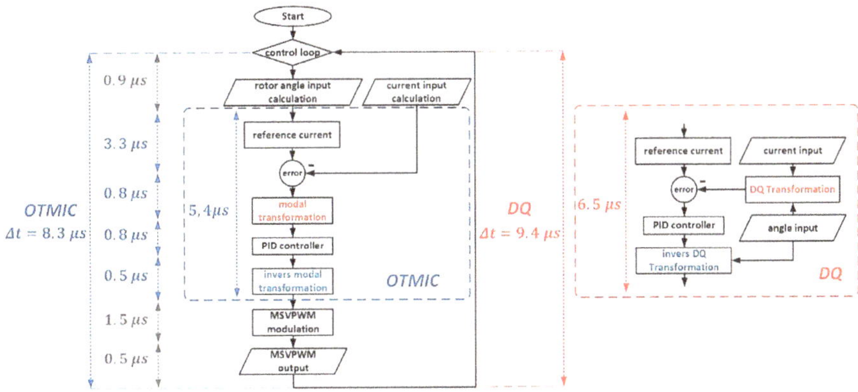

Figure 5. Step response of simulated and measured torque for OTMIC and FOC (DQ).

This means that OTMIC control can be calculated faster and in addition, gives the optimal torque in contrast to FOC (DQ) based on the Clarke-Parke Transformations.

3.2. Verification of the Proposed Method

The simulation results presented in this study have been performed through Matlab/Simulink environment based on the mathematical model of the PM motor described in Section 2 and the control model presented in Section 3. Conventional DQ field-oriented vector control used as a reference to the proposed OTMIC was also implemented in the experiment to control the wheel hub motor, with the parameters in Table 1.

The motor dynamic behavior can be verified by means of step response of electrical torque T_E. As shown in Figure 6, measured and simulated torque of proposed OTMIC and conventional DQ methods were carried out according to zero velocity, mechanically fixed rotor and 5 Nm reference torque T_{Ref} (green). Because torque sensor bandwidth was too low for this experiment, electrical torque generated in air gap winding (grey) was calculated using (6) based on the values of measured currents and the electric angle with a given motor geometric constant k_M, and compared with the simulation for OTMIC (blue) and DQ (red) control methods. As can be seen form Figure 6, the difference between both methods is minimal due to the very low motor time constant. The exciting fast torque step response within 35 µs underlines the very high dynamics of this type of motor. The slight difference between simulation and measurement is due to parasitic inductances and capacitances in power lines and supply. This can be easily considered in the motor model but was not done in this experiment to show the robustness of the control.

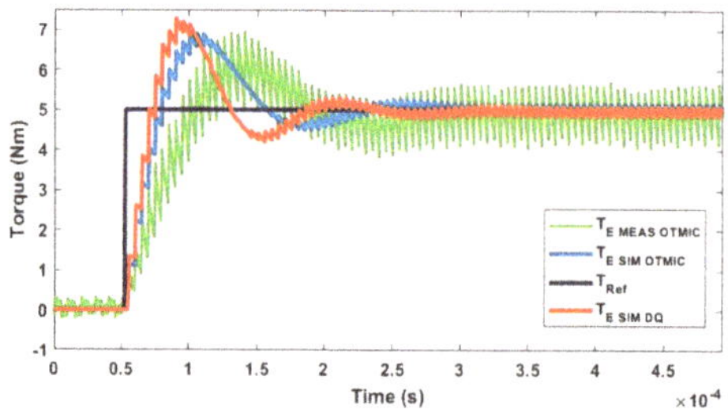

Figure 6. Step response of simulated and measured torque for OTMIC and DQ.

For a more detailed demonstration of the basic principles of the proposed control algorithm, Figure 7 shows the results of modeling the ideal optimal current waveforms to minimize the torque ripples (OTMIC TRO) and losses (OTMIC LO) for two waveforms of B-Field (Case A with odd harmonic coefficients b_k from Table 1 and Case B with b_k from Reference [19], which result from a change in the air gap between stator and rotor). For a more detailed analysis, the torque RMS (root mean square) and efficiency value are also presented for a sinusoidal current of conventional DQ control based on the Clarke-Park Transformations. Figure 7 shows the simulation results of an ideal system without any errors in the phase current or rotor position measurements. As can be seen, the optimal phase current waveforms are able to the minimize torque ripples or maximize motor efficiency based on the Equations (6) and (12). The advantages of the OTMIC TRO are seen very well in Figure 7. In an ideal system, OTMIC TRO eliminants any torque fluctuations for any waveform of the B-field. OTMIC with loss optimization (LO) changes the phase current waveform and improves loss rates and efficiency of the system with increased torque ripples. The advantages of the OTMIC LO in comparison with the standard DQ method are visible only in the presence of significant parasitic harmonics in the form of a B-field. Simply said, the ideal sinusoidal B-field leads to the same torque ripples and motor loss for every method.

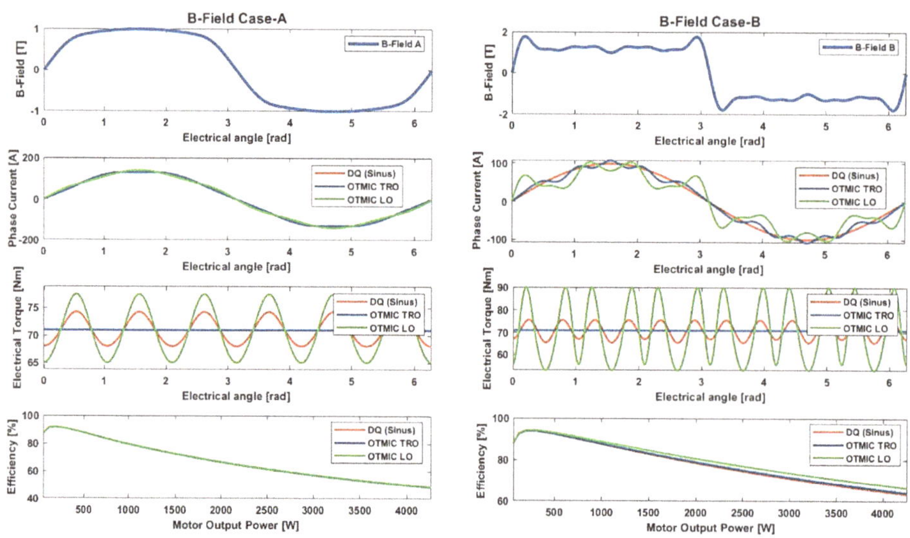

Figure 7. Working principles of the proposed OTMIC control algorithm for two B-Field waveforms.

Simulated and measured phase currents and torque results during normal operation of proposed OTMIC control with minimization of torque ripple (TRO Figure 8a—simulation, Figure 9a—measurement) and losses (LO Figures 8b and 9b) as well as a conventional vector DQ control (DQ Figures 8c and 9c) correspond to the references: $\omega = 8$ rad/s and $T_{Ref} = 10$ Nm. The behavior of phase currents and torque for both OTMIC optimizations given in Figure 9a,b has a slight asymmetry caused by speed fluctuations of load machine and errors in angle measurements, which could not be completely avoided in experimental setup. The current and torque harmonic distribution for the aforementioned control approaches are shown in Figure 8d for simulation and in Figure 9d for measured results. The amplitude of each harmonic component is normalized relative to that of a fundamental one. The FFT (Fast Fourier transform) analyses of simulation currents and torques underline the main idea of the proposed method. As shown in Figure 8d (left), the phase current waveforms of OTMIC have 5th and 7th harmonics expected to conventional control. On the other hand, the torque waveform of OTMIC with torque ripple optimization TRO (blue) is deprived of such harmonics, which leads to the very small torque fluctuations as opposed to DQ control, as shown in Figure 8d (right).

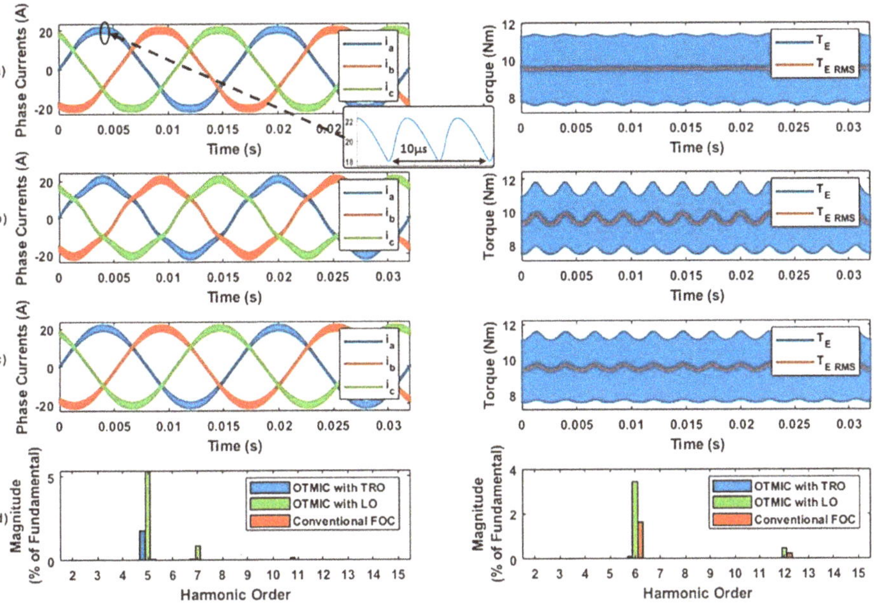

Figure 8. Simulation results of stator current and torque of: (**a**) OTMIC with torque ripples optimization, (**b**) OTMIC with losses optimization, (**c**) conventional DQ control and (**d**) harmonic spectrum of phase current (left) and torque (right) for every method [18].

The harmonic spectrum of measured phase current with OTMIC control shown in Figure 9d (left) has parasitical 2nd, 4th and 6th harmonics due to inaccuracies in experimental setup, mainly caused by interactions of control system and load machine, but they are less than 1% of that of the fundamental one. The necessary 5th and 7th harmonics have a slight deviation of about 0.5% of the fundamental one compared to simulation results. The current waveform of conventional DQ control should be quite similar to a sinusoidal, as shown in the simulation results in Figure 8c and harmonic analyses in Figure 8d (left). However, the harmonic spectrum of measured phase currents by the conventional method consist of high 3rd and 4th harmonics due to the high sensitivity of the standard method based on the trigonometric transformations to errors in the control system, especially errors in angle measurement.

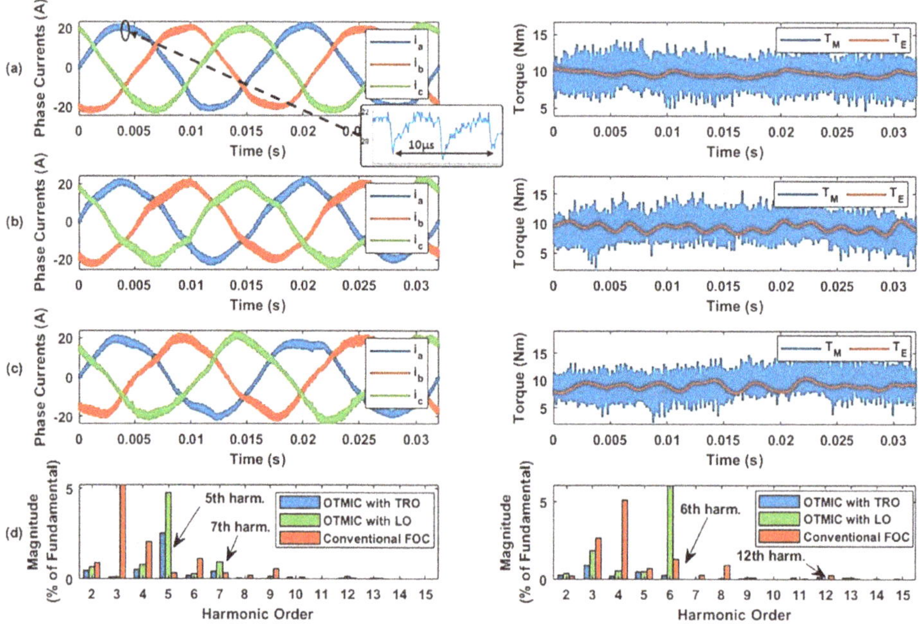

Figure 9. Experimental results of stator current and torque of: (**a**) OTMIC with torque ripples optimization, (**b**) OTMIC with losses optimization, (**c**) conventional DQ control and (**d**) harmonic spectrum of phase current (left) and torque (right) for every method [18].

The measured torque harmonic distribution shown in Figure 9d (right) consists of a large number of harmonics due to speed errors in rotating shaft and additional torque produced by load machine. However, the 6th and 12th harmonics retain their tendency and can be compared with the simulation. In addition, the FFT analyses of measured torques in Figure 9d (right) shows that the proposed OTMIC control with torque ripple minimization is able to compensate additional harmonics in B-field. Summarized, the measured results show very good correspondence to theoretical calculations. As opposed to OTMIC, the Clarke-Park Transformations need a very accurate rotor angle measurement, which can be seen in the larger asymmetries given in Figure 9c. Thus, the proposed OTMIC method is less sensitive to speed fluctuations and angle errors due to the absence of trigonometric transformations of phase currents based on the electric angle.

Figure 10 compares the measured values of motor efficiency for OTMIC control with LO and TRO and conventional DQ control. Motor efficiency was measured with given constant velocity of 30 rad/s, variable torque from 1 to 10 Nm and output power from 30 to 300 W. Results show that OTMIC control with loss minimization increases the system efficiency by 0.7% in compassion to the conventional method, while delivering the same torque. However, for a correct analysis of the proposed control in case of a motor loss and efficiency, any errors in the measurement system should be avoided and motor torque and current should be increased to the rated values, which could not be fully realized in our experimental setup. More detailed analyses of the system efficiency should be investigated in the future work.

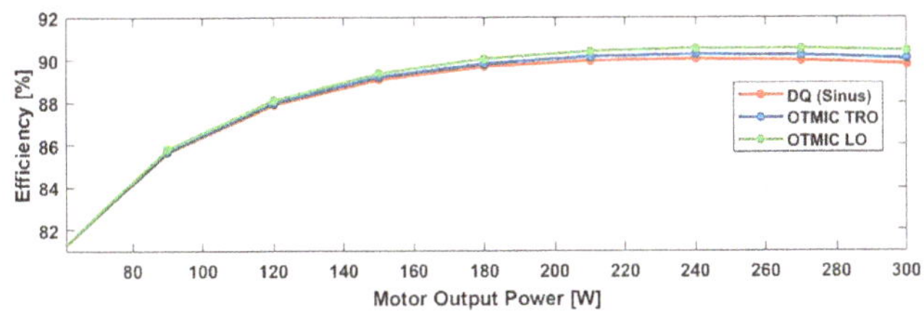

Figure 10. Motor efficiency for OTMIC and DQ control.

Simulation and experimental results show that the OTMIC control system is less sensitive to errors in the electrical angle and phase currents measurements compared to FOC. In addition, the proposed method worked very robustly independent of deviations of motor parameters like phase resistance due to the temperature and the speed of fluctuations due to the load motor.

4. Conclusions

This paper proposed a Combined Optimal Torque feedforward and Modal Current feedback Control architecture (OTMIC) for PM motors with very low inductance and higher harmonic B-field, which minimizes motor losses or torque ripples/motor noise by means of offline calculation of optimal current waveform. Optimal three-phase currents are used as reference values to the measured currents. The optimization algorithm considers all relevant coefficients of B-field harmonic distribution to calculate the appropriate current waveform off-line and employs an extremely fast table-lookup for real-time implementation in the microcontroller. Modal current feedback control provides stability, high system dynamics without deviation of a given reference and high robustness against variation of motor parameters or errors in rotor angle measurement.

Compared to classical field-oriented control methods based on the cumbersome and complex Clarke-Park Transformations, the proposed control is simpler for implementation on the microcontroller and allows to minimize either torque ripples or motor losses. As a result, the proposed OTMIC control can reach its full potential in high-frequency systems, where simplicity of the control algorithm and microcontroller calculation time matters. Due to the transformations used in the conventional DQ method, the phase currents' waveforms have to be nearly ideally sinusoidal. This means that conventional methods to achieve the specified requirements of motor torque or total losses demand the special concepts and additional real-time calculations to adapt the phase current waveform to the non-sinusoidal waveform of the magnetic field. It complicates the control system and increases the calculation time in the microcontroller. Moreover, in real high-frequency control systems with errors in current and angle measurements, this problem can be more significant. Experimental results show that the Clarke-Park Transformations for electric drives with a very small motor time constant and high number of poles introduces significant current deviations from the values given. In this case, the proposed OTMIC control method used for PM synchronous motors with a very low phase inductance and tested experimentally shows a very good match between simulated and measured waveforms of phase current and torque. The proposed control can also be applied to other electric motors such as brushless DC (BLDC), asynchronous or PM motors, with any waveform of B-field waveform.

Author Contributions: Conceptualization, R.K. and D.G.; methodology, R.K.; software, R.K. and D.G.; validation, D.G.; formal analysis, R.K. and D.G.; writing—original draft preparation, R.K. and D.G.; writing—review and editing, R.K. and D.G.; funding acquisition, R.K. All authors have read and agreed to the published version of the manuscript.

Funding: This work was supported in part by the European Commission under Grant ERDF (European Regional Development Fund) and the County of Saxony-Anhalt.

Conflicts of Interest: The authors declare no conflict of interest.

References

1. Ragot, P.; Markovic, M.; Perriard, Y. Analytical Determination of the Phase Inductances of a Brushless DC Motor with Faulhaber Winding. *IEEE Trans. Ind. Appl.* **2010**, *46*, 1360–1366. [CrossRef]
2. Borchardt, N.; Heinemann, W.; Kasper, R. Design of a wheel hub motor with air gap winding and simultaneous utilization of all magnetic poles. In Proceedings of the 2012 IEEE International Electric Vehicle Conference, Greenville, SC, USA, 4–8 March 2012; pp. 1–7.
3. Borchardt, N.; Kasper, R. Nonlinear design optimization of electric machines by using parametric Fourier coefficients of air gap flux density. In Proceedings of the IEEE International Conference on Advanced Intelligent Mechatronics (AIM), Banff, AB, Canada, 12–15 July 2016; pp. 645–650.
4. Kasper, R.; Borchardt, N. Boosting Power Density of Electric Machines by Combining Two Different Winding Types. In Proceedings of the 7th IFAC Symposium on Mechatronic Systems, Leicestershire, UK, 5–8 September 2016; pp. 322–329.
5. Tan, B.; Hua, Z.; Zhang, L.; Fang, C. A New Approach of Minimizing Commutation Torque Ripple for BLDCM. *Energies* **2017**, *10*, 173. [CrossRef]
6. Li, X.; Jiang, G.; Chen, W.; Shi, T.; Zhang, G.; Geng, Q. Commutation Torque Ripple Suppression Strategy of Brushless DC Motor Considering Back Electromotive Force Variation. *Energies* **2019**, *12*, 1932. [CrossRef]
7. Li, H.; Zheng, S.; Ren, H. Self-Correction of Commutation Point for High-Speed Sensorless BLDC Motor with Low Inductance and Nonideal Back EMF. *IEEE Trans. Power Electron.* **2017**, *32*, 642–651. [CrossRef]
8. Li, X.; Xia, C.; Cao, Y.; Chen, W.; Shi, T. Commutation Torque Ripple Reduction Strategy of Z-Source Inverter Fed Brushless DC Motor. *IEEE Trans. Power Electron.* **2016**, *31*, 7677–7690. [CrossRef]
9. Yoon, K.-Y.; Baek, S.-W. Robust Design Optimization with Penalty Function for Electric Oil Pumps with BLDC Motors. *Energies* **2019**, *12*, 153. [CrossRef]
10. Song, Q.; Li, Y.; Jia, C. A Novel Direct Torque Control Method Based on Asymmetric Boundary Layer Sliding Mode Control for PMSM. *Energies* **2018**, *11*, 657. [CrossRef]
11. Qian, J.; Ji, C.; Pan, N.; Wu, J. Improved Sliding Mode Control for Permanent Magnet Synchronous Motor Speed Regulation System. *Appl. Sci.* **2018**, *8*, 2491. [CrossRef]
12. Mohd Zaihidee, F.; Mekhilef, S.; Mubin, M. Robust Speed Control of PMSM Using Sliding Mode Control (SMC)—A Review. *Energies* **2019**, *12*, 1669. [CrossRef]
13. Wang, Q.; Yu, H.; Wang, M.; Qi, X. A Novel Adaptive Neuro-Control Approach for Permanent Magnet Synchronous Motor Speed Control. *Energies* **2018**, *11*, 2355. [CrossRef]
14. Liu, X.; Yu, H.; Yu, J.; Zhao, L. Combined Speed and Current Terminal Sliding Mode Control With Nonlinear Disturbance Observer for PMSM Drive. *IEEE Access* **2018**, *6*, 29594–29601. [CrossRef]
15. Zeb, K.; Din, W.U.; Khan, M.A.; Khan, A.; Younas, U.; Busarello, T.D.C.; Kim, H.J. Dynamic Simulations of Adaptive Design Approaches to Control the Speed of an Induction Machine Considering Parameter Uncertainties and External Perturbations. *Energies* **2018**, *11*, 2339. [CrossRef]
16. Kang, L.; Cheng, J.; Hu, B.; Luo, X.; Zhang, J. A Simplified Optimal-Switching-Sequence MPC with Finite-Control-Set Moving Horizon Optimization for Grid-Connected Inverter. *Electronics* **2019**, *8*, 457. [CrossRef]
17. Kasper, R.; Golovakha, D.; Süberkrüb, F. Combined Optimal Torque and Modal Current Control for Low Inductance PM Motor. In Proceedings of the IEEE International Conference on Mechatronics (ICM), Ilmenau, Germany, 18–20 March 2019; pp. 491–497.
18. Golovakha, D. Combined Optimal Torque and Modal Current Control for Low Inductance PM Motor. Ph.D. Thesis, Otto von Guericke University, Magdeburg, Germany, 2020.
19. Borchardt, N.; Kasper, R. Analytical magnetic circuit design optimization of electrical machines with air gap winding using a Halbach array. In Proceedings of the International Electric Machines and Drives Conference (IEMDC), Miami, FL, USA, 21–24 May 2017; pp. 1–7.

energies

MDPI

Article

Fault Diagnosis and Tolerant Control of Three-Level Neutral-Point Clamped Inverters in Motor Drives

Kuei-Hsiang Chao * and Chen-Hou Ke

Department of Electrical Engineering, National Chin-Yi University of Technology, Taichung 41170, Taiwan; e00402125@gmail.com

* Correspondence: chaokh@ncut.edu.tw; Tel.: +886-4-2392-4505 (ext. 7272); Fax: +886-4-2392-2156

Received: 3 October 2020; Accepted: 26 November 2020; Published: 29 November 2020

Abstract: This paper presents an extension theory-based assessment method to perform fault diagnosis for inverters in motor driving systems. First, a three-level neutral-point clamped (NPC) inverter is created using the PSIM software package to simulate faults for any power transistor in the NPC-type inverter. Fast Fourier transformation is used to transform the line current signals in the time domain into a spectrum in the frequency domain for analysis of the corresponding spectrum of features of the inverter for faults with different power transistors. Then, the relationships between fault types and specific spectra are established as characteristics for the extension assessment method, which is then used to create a smart fault diagnosis system for inverters. Fault-tolerant control (FTC) is used here when the rated output of a faulty inverter is decreased in order to maintain balanced output in three phases by changing the framework of the transistor connection. This is performed to reinforce the reliability of the inverter. Finally, by the simulation and experimental results, the feasibility of the proposed smart fault diagnosis system is confirmed. The proposed fault diagnosis method is advantageous due to its minimal use of data and lack of a learning process, which thereby reduces the fault diagnosis time and makes the method easily used in practice. The proposed fault-tolerant control strategy allows both online and smooth switching in the wiring structure of the inverter.

Keywords: extension theory; smart fault diagnosis; three-level neutral-point clamped inverters; line current spectrum feature; fault-tolerant control

1. Introduction

Compared to two-level inverters, multi-level inverters [1–5] exert less voltage stress on switches, feature smaller change rates in their output voltages (*dv/dt*), and can be applied in high power scenarios. As multi-level inverters incorporate multiple power transistors that are connected in both series and parallel, thus making the output line-to-ground voltage waves embedded with many step-like waveforms, their waveforms become step-like voltage waveforms which are better approximated in the form of sine waves, thus being suitable for reducing harmonic components.

Multi-level inverters are generally divided as diode-clamped, neutral-point clamped (NPC, I-type), cascaded H-bridge (CHB), and flying capacitor inverters. Of these types, three-level diode-clamped inverters [6] are widely used due to their simple circuit and ease of control. The inverter works by dividing, using capacitors, the voltage on DC side equally into three voltage levels ($+V_{dc}/2$, 0, and $-V_{dc}/2$), allowing the output voltage to change in three modes, thus, with diodes and switches, clamping the output voltage at a neutral point on the DC side. Yet, as the switches may be damaged when operating in high-current or high-temperature conditions over extended periods of time, components aging, faulty drive circuits, and other faults, thereby resulting in an inability to work normally, multi-level inverters should be designed with a fault detection mechanism and fault-tolerant control in the hope of operation without disruption in the event of a faulty inverter component [7–9].

In the past, faults in motor drive systems have been diagnosed based on messages from measurement instruments that are shown directly by numbers or waveforms, which allows easy and fast indication for fault occurrence. Such a practice requires on-site operators to determine (directly and from their own experience) what fault a message indicates for subsequent repairs and component replacement as needed. Such a conventional method of fault diagnosis [10,11] tends to lead to misjudgment for fault points, causing unnecessary wastes of effort and time for repairs and component replacement. Hence, a new technology for diagnosing system faults deserves study. In this regard, there has been much effort made in both Taiwan and overseas [12–15]. The methods that have resulted from these efforts combine more than two algorithms to form an artificial intelligence (AI) system for motor fault diagnosis, which can increase the efficiency in fault diagnosis for motor drive systems and thus helps save labor and time costs. These AI systems are only suitable for motor fault diagnosis and cannot be directly applied to fault diagnosis for multi-level inverters. As such, in the research of the same diagnosis, some researchers have used artificial neural networks (ANNs) to create fault diagnosis systems [16–19]. ANN methods perform diagnosis faster, yet they require massive data quantities for learning. Therefore, the accuracy of the fault diagnosis is based on the amount of learning data and the correctness of the learning process. In [20], an effective control strategy including fault detection, localization, and tolerant operation was proposed for a modular multi-level converter with a transistor open-circuit fault; however, this method is only applicable to a large number of series-connected two-level converter modules, and only for AC-to-DC converters. A novel fault-tolerant control strategy was proposed in [21] for bypassing converter submodules (SM) with faults and re-regulating the SM capacitor voltage and carrier phase-shift angle to maintain the main components of circulating current, additionally reducing the total harmonic distortion (THD) of the grid-connected current to enable the stable operation of the photovoltaic inverter; however, this method is also only suitable for a large number of series-connected two-level inverters. A detailed analysis of a single open-switch fault for a NPC active rectifier was presented in [22]. The study proposed a fault-tolerant control method to reduce the effect of the open-switch fault via compensation with a distorted reference voltage. Although this method could be used with a three-level NPC converter, it can only be used in AC-to-DC converters, but not in inverters. A fast fault diagnosis scheme in a grid-connected photovoltaic (PV) system based on applying the combination of wavelet multi-resolution singular spectrum entropy and a support vector machine was proposed in [23] to identify different types of grid faults in a three-phase grid-tied PV system; however, the calculation procedure of this fault diagnosis method is quite complex and it is aimed at fault detection for a PV system, rather than the diagnosis of the inverter. A fault prognostics method which makes full use of the similarities between inverter clusters was proposed in [24]. Although the divergence of inverter clusters could be used to predict inverter faults, the process for this fault diagnosis method is quite complicated and the calculation procedure is also very complicated. A diagnostic method for an open-circuit transistor failure in a hybrid active neutral-point clamped (HANPC) inverter was proposed in [25]. By analyzing the individual characteristics of each transistor failure, it is possible to detect the exact location of the failed transistors in a short period; however, this method can only diagnose transistor faults in the same leg, and other legs will have the same characteristics, so it is impossible to determine exactly which leg contains the faulty transistor.

In contrast, extension theory-based methods do not require large data quantities for training as they feature models of matter-elements, the coupling of qualification and quantification, and non-closure [26–30], while the distances and rank values makes it possible to establish correlation functions for fault types, and the correlation degrees can be used to identify fault types, thus reducing the learning process and making determination easier, faster, and more accurate. Hence, this study proposes an intelligent fault-tolerant control system based on extension theory for locating faults in a three-level NPC-type inverter. First, the frequency spectra of output line current waveforms are extracted during a power transistor failure event and used as feature values for fault detection. Therefore, the proposed extension theory-based method can be applied to accurately locate a faulty power transistor. Subsequently, a fault-tolerant control strategy is used to maintain the three-phase balanced line voltage output if any power semiconductor switch of the inverter fails. Figure 1 illustrates the overall architecture of the extension theory-based fault-diagnosis and fault-tolerant control system proposed herein for NPC-type inverters.

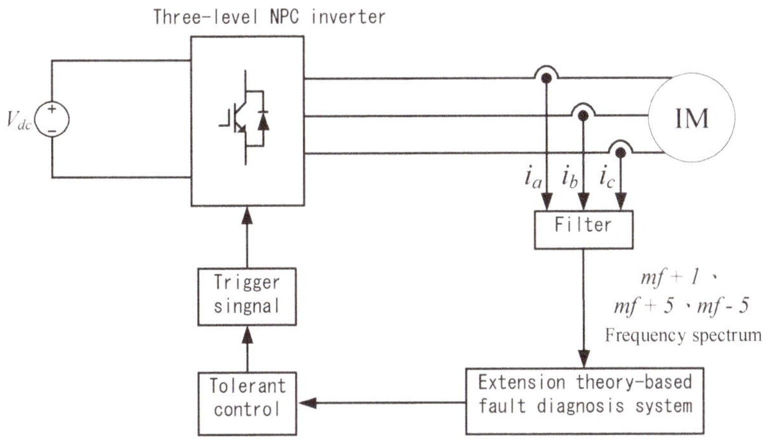

Figure 1. Framework for the extension theory-based fault diagnosis and tolerant control system for neutral-point clamped-type (NPC-type) inverters.

This paper is organized as follows: First, the characteristics of switch faults for three-level NPC-type inverters are discussed in Section 2. Then, in Section 3, the concept of extension theory is described in detail. The procedure for using the extension theory-based fault diagnosis method for a three-level NPC-type inverter is explained in Section 4. Practical tests are detailed in Section 5 to demonstrate the effectiveness of the proposed fault diagnosis method. Finally, in Section 6, fault-tolerant control in the event of power switch failure in the three-level NPC-type inverter is analyzed, and the simulation and experimental results are used to prove the feasibility of the method.

2. Characteristics of Faults for Three-Level Inverters

Considering research on inverter fault diagnosis, this paper targets a three-level NPC inverter as shown in Figure 2. In general, faults for inverters come in three types: short-circuit faults, open-circuit faults, and trigger signal faults. A short-circuit fault arises when a switch component is blown due to an overly high voltage across both ends of the switch. An open-circuit fault refers to a power transistor without a trigger signal to actuate conductivity through it. A trigger signal fault is the receipt, by a switch component, of an incorrect trigger signal for command.

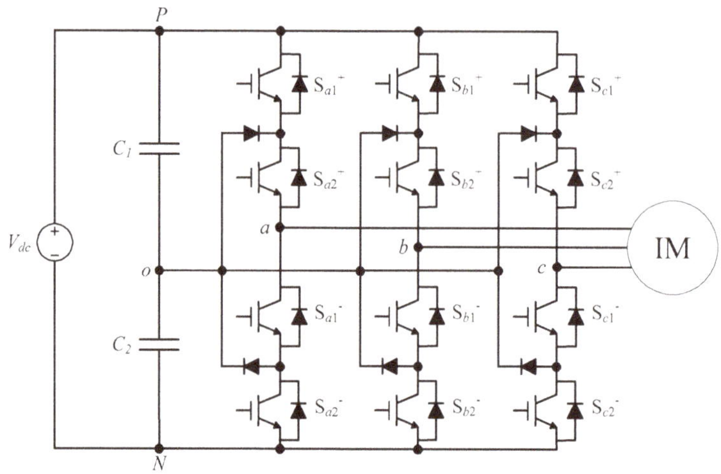

Figure 2. Framework for three-level NPC inverters.

The PSIM software package is the ultimate simulation environment for power electronics and motor control and was developed by PowerSIM. Using the PSIM software package, a simulation environment for three-level NPC inverters was created and used to investigate the diagnosis of faults occurring on any switch at any time. It was observed from the simulations and analysis that the output waveform measured on an inverter in normal working conditions was a balanced three-phase waveform. For example, when the inverter has a working frequency at 60 Hz without a faulty power transistor, the output waveforms of its line currents and their frequency spectra are given as shown in Figures 3 and 4, respectively. Figure 3 reveals that the waveforms for the line currents are of the same magnitude and all are sine waves, where each is different in phase by 120°, which is typical of the balanced three-phase characteristic. Regarding the parameter setting of the motor drive, a 300 V DC inverter with a switching frequency of 18 kHz was connected to an induction motor. The frequency spectra for line currents i_a, i_b, and i_c for an inverter working at 60 Hz without a faulty switch are shown in Figure 4. We can see from Figure 4 that the frequency spectra at $m_f - 5$, $m_f + 1$, and $m_f + 5$, multiplied by the working frequency, are very small. Hence, the values of the frequency spectra at such frequencies were used in this paper as the feature spectra for faults, where m_f is defined as the frequency modulation index:

$$m_f \triangleq \frac{f_{carrier}}{f_{reference}} = \frac{f_{tri}}{f_{\sin}} \tag{1}$$

where f_{tri} is the frequency of a triangular carrier wave, i.e., the switching frequency of inverter (18 kHz here) and $f_{\sin}$ is a sine wave frequency (60 Hz here) and also the working frequency of the inverter. Therefore, its frequency modulation index m_f is 300. Based on this, the frequency spectra for line currents i_a, i_b, and i_c for an inverter working at 60 Hz without a faulty switch at $m_f - 5$, $m_f + 1$, and $m_f + 5$, multiplied by the working frequency, are 17.7, 18.06, and 18.3 kHz, respectively.

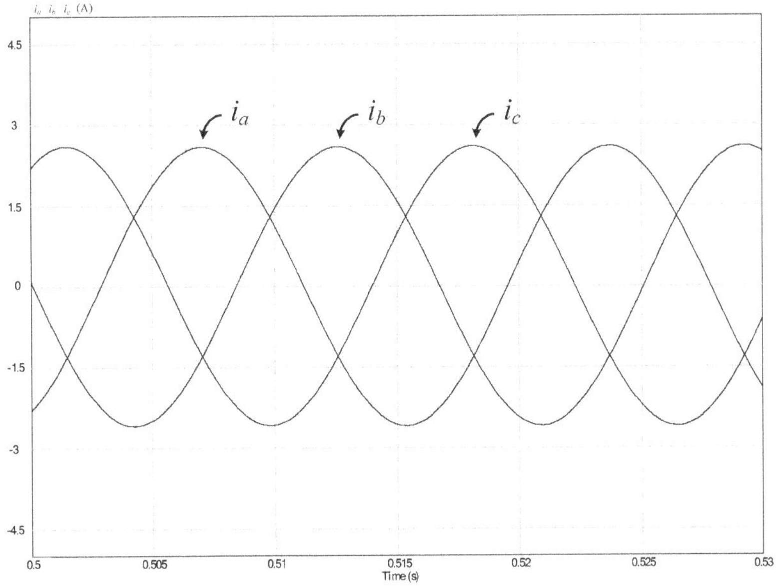

Figure 3. Output waveforms of line currents for an inverter working at 60 Hz without any faulty switches.

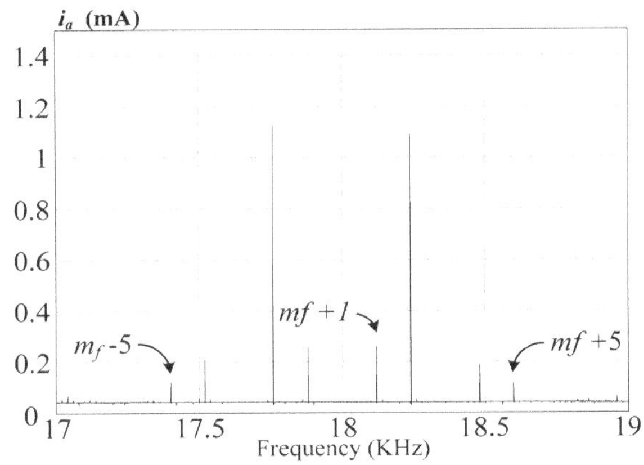

(a)

Figure 4. *Cont.*

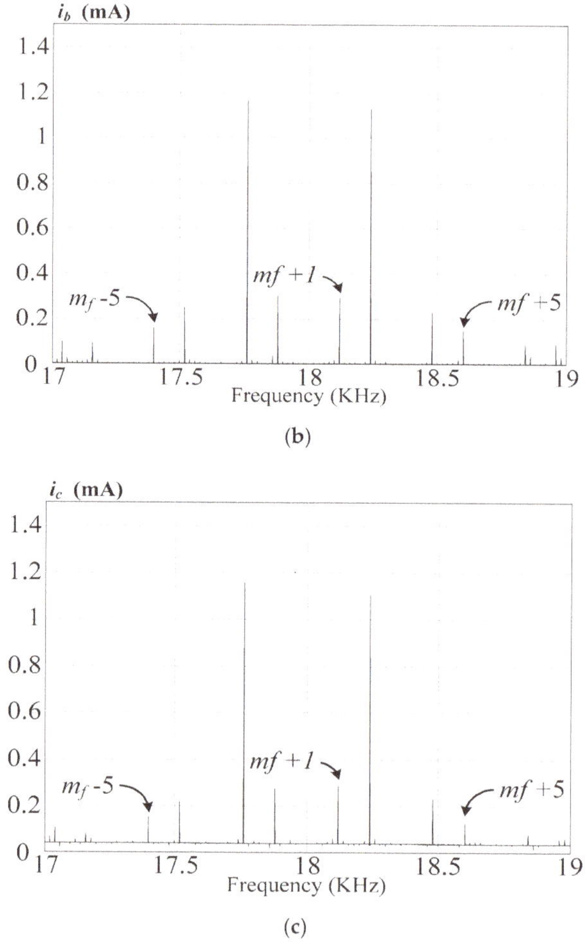

Figure 4. Frequency spectra of line currents for an inverter working at 60 Hz without any faulty switches: (**a**) line current i_a; (**b**) line current i_b; (**c**) line current i_c.

If any switch in the inverter is faulty, the characteristics of the inverter change. For instance, when a switch of the inverter $S_{a1}{}^{+}$ is faulty, a distorted waveform of its output line current i_a can be seen as shown in Figure 5. Figure 6 shows the waveform of the line current i_b with the fault inverter switch $S_{b2}{}^{-}$, where the waveform obviously differs from the one in normal working conditions. Figure 7 shows the waveform of the line current i_c with the faulty switch $S_{c2}{}^{+}$, where the distortion is apparent.

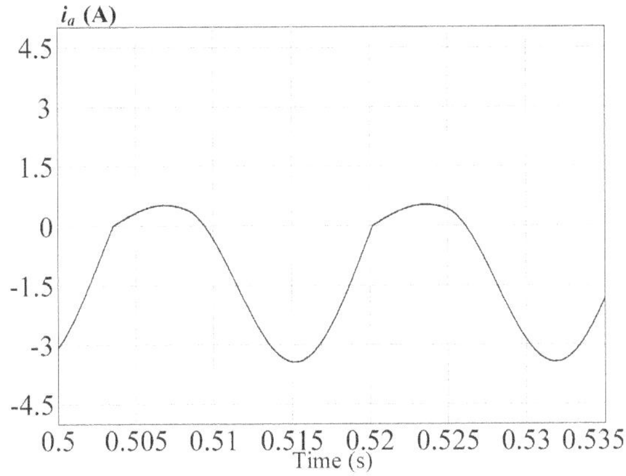

Figure 5. Waveform of line current i_a for an inverter working at 60 Hz with faulty switch $S_{a1}{}^{+}$.

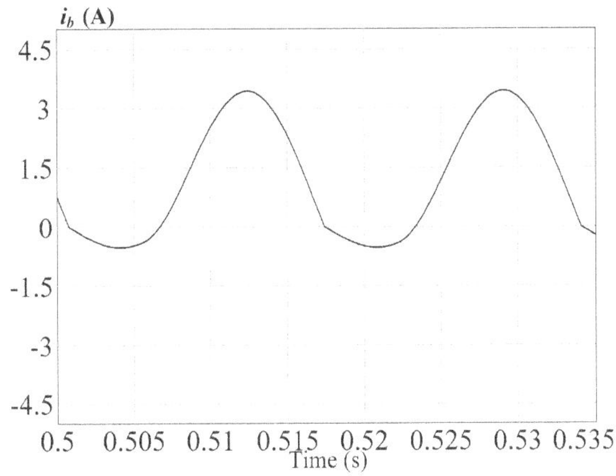

Figure 6. Waveform of line current i_b for an inverter working at 60 Hz with faulty switch $S_{b2}{}^{-}$.

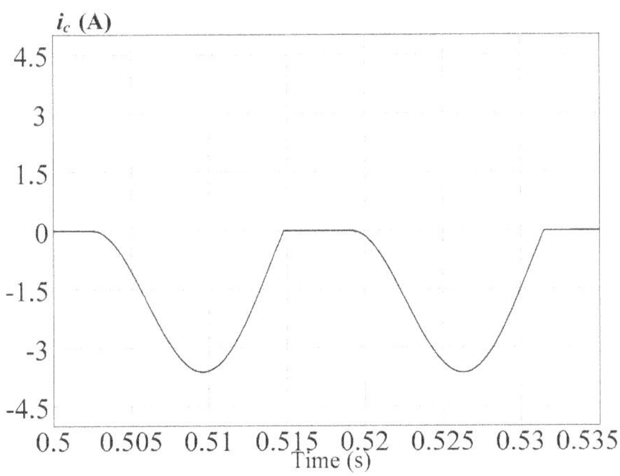

Figure 7. Waveform of line current i_c for an inverter working at 60 Hz with faulty switch $Sc_2{}^{+}$.

It is clear from the above analysis that anomalies can be observed in current frequency spectra when an inverter is faulty, as in the case of an inverter working at 60 Hz with faulty switch $S_{c2}{}^{+}$, where the resultant current frequency spectrum is given as shown in Figure 8. When compared with the frequency spectra without a fault in Figure 4, this demonstrates that the frequency spectrum of line current i_c has a feature spectrum with relatively major changes at the positions of $m_f + 1$, $m_f + 5$, and $m_f - 5$.

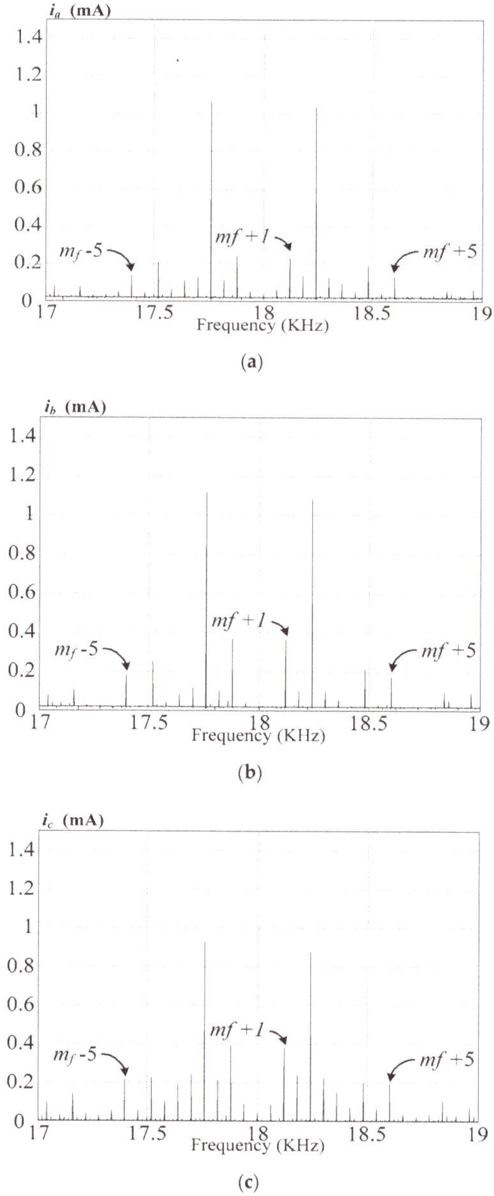

Figure 8. Frequency spectra of line currents for an inverter working at 60 Hz with faulty switch $Sc_2{}^{+}$: (**a**) line current i_a; (**b**) line current i_b; (**c**) line current i_c.

Relevant data for power transistor faults can be obtained by simulation and analysis. These data can be used with extension theory to create an inverter fault diagnosis system for detecting faults in power semiconductor switches in the main circuits of three-level NPC inverters.

3. Extension Theory

Extension theory, which was proposed by Dr. Cai Wen in 1983, is a formalized tool for studying and solving problems with qualitative and quantitative methods [29]. The theory structurally consists of matter-element theory and extension mathematics, in which the extension theory uses logic algorithms with matter elements. The matter-element theory deals with extensibility and the change of matter elements, describes them in formalized language, and details computation and reasoning. Extension mathematics builds adaptive mathematical tools from extension sets and correlation functions. As such, extension theory creates matter-element models and uses the property transformation of matter elements for qualitative and quantitative transformation, then determining qualitative and quantitative effects via correlation functions to clearly express the effects of features.

Extension theory is a science of solving contradictory problems by dealing with extensibility and rules and methods of transformation. It features (1) the idea of changing contradictory problems into compatible ones and (2) creates matter-element theory to provide new ways of solving contradictory problems with matter-element transformation based on the extensibility of matter elements. Extension theory also (3) creates extension set theory, giving quantitative descriptions of quantitative change to qualitative changes in the extension domain and critical elements, thus realizing quantitative extension mathematics in the extension theory based on extension sets [26–30].

3.1. Basic Concept of Extension Matter-Element Theory

In dealing with contradictory problems, if concepts, characteristics, and their corresponding data are brought together for consideration, it is possible to deduce problem solving methods. Hence, the concept of "matter elements" was introduced [30], which consists of "name" and "characteristic" of a matter and the "value" that corresponds to the characteristic. Matter elements are the basic elements in extension theory that describe matters, denoted by R, N (name), c (characteristic), and v (value) as an expression as follows:

$$R = (N, c, v) \tag{2}$$

Additionally, via the above definition for a matter element, the correlation between the characteristics of these three basic elements and the corresponding value can be expressed by the equation below:

$$v = c(N) \tag{3}$$

Thus, Equation (2) for a matter element is converted into the following form:

$$R = (N, c, c(N)) \tag{4}$$

To understand the relationships between basic matter elements, references can be made to the expression by a space of matter elements, as Figure 9 shows, where the name, characteristic, and value are plotted in the x-, y- and z-axes, which also displays the feature of the change of such a combination.

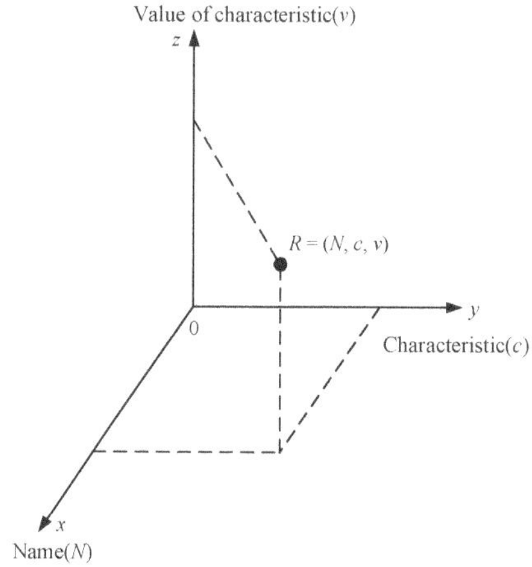

Figure 9. Space of matter elements.

Furthermore, in extension matter-element theory, if the matter elements have multiple characteristics, they can be expressed by m characteristics, i.e., $c_1, c_2, \ldots c_m$, and m corresponds to the values of characteristics, i.e., $v_1, v_2, \ldots v_m$, in an equation for their correlations as follows:

$$R = \begin{bmatrix} N, & c_1, & v_1 \\ & c_2, & v_2 \\ & \cdots & \cdots \\ & c_m, & v_m \end{bmatrix} = \begin{bmatrix} R_1 \\ R_2 \\ \cdots \\ R_m \end{bmatrix} \tag{5}$$

In Equation (5), R is called an m-dimensional matter element, whose components are expressed by $R_k = (N, c_k, v_k)$ $(k = 1, 2, \ldots, m)$. Thus, Equation (5) can be rewritten as $R = (N, C, V)$, where matrix $C = [c_1, c_2, \ldots c_m]^T$ and matrix $V = [v_1, v_2, \ldots v_m]^T$. Therefore, by the way of a multi-dimensional definition, it enables describing any single thing in the real world.

The characteristic value corresponding to a characteristic can be a single point or a single range. In the latter case, such a range is referred to as a classical domain which is contained in a neighborhood domain. Assume point f is any point in the interval $F = < a_q, b_q >$, and $F_0 \in F$, then, the matter element, R_0, corresponded by $F_0 = < a_p, b_p >$ can be expressed by Equation (6), where C is a characteristic and V_p is the value for C, i.e., its classical domain:

$$\begin{aligned} R_0 &= (F_0, C, V_p) \\ &= \begin{bmatrix} F_0, & c_1, & V_{p1} \\ & c_2, & V_{p2} \\ & \vdots & \vdots \\ & c_m, & V_{pm} \end{bmatrix} \\ &= \begin{bmatrix} F_0, & c_1, & < a_{p1}, b_{p1} > \\ & c_2, & < a_{p2}, b_{p2} > \\ & \vdots & \vdots \\ & c_m, & < a_{pm}, b_{pm} > \end{bmatrix} \end{aligned} \tag{6}$$

A matter element, R_F, corresponded by F, can be expressed by Equation (7). Similarly, C is the value of the characteristic for F and V_q is the value for C, i.e., its neighborhood domain:

$$R_F = (F, C, V_q) = \begin{bmatrix} F, & c_1, & V_{q1} \\ & c_2, & V_{q2} \\ & \vdots & \vdots \\ & c_m, & V_{qm} \end{bmatrix} = \begin{bmatrix} F, & c_1, & < a_{q1}, b_{q1} > \\ & c_2, & < a_{q2}, b_{q2} > \\ & \vdots & \vdots \\ & c_m, & < a_{qm}, b_{qm} > \end{bmatrix} \tag{7}$$

3.2. Correlation Function

If $F_0 \in F$, then the correlation function $K(f)$ can be defined as follows:

$$K(f) = \frac{\rho(f, F_0)}{D(f, F_0, F)} \tag{8}$$

If set $K(f) \leq 1$, then:

$$\rho(f, F_0) = \left| f - \frac{a_p + b_p}{2} \right| - \frac{b_p - a_p}{2} \tag{9}$$

$$D(f, F_0, F) = \begin{cases} \rho(f, F) - \rho(f, F_0) & f \notin F_0 \\ -\frac{|(a_p - b_p)|}{2} & f \in F_0 \end{cases} \tag{10}$$

where:

$$\rho(f, F) = \left| f - \frac{a_q + b_q}{2} \right| - \frac{b_q - a_q}{2} \tag{11}$$

Correlation functions can be used to determine the membership grade between f and F_0. Extension correlation functions are given in Figure 10. The figure indicates that if $K(f) > 0$, it means that point f currently lies within domain F_0. If $K(f) < -1$, it means point f will not be within either of the two domains, and if $-1 \leq K(f) \leq 0$, it means that the point lies not within domain F_0, but within domain F. In extension domains, it is possible to make use of transformation of conditions such that f can belong to the domain of F_0.

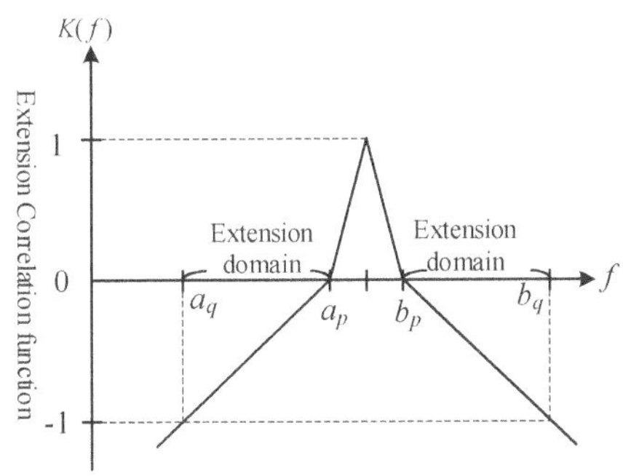

Figure 10. The curve for the correlation function.

4. Extension Theory-Based Fault Diagnosis for Inverters

The inverter fault diagnosis system operates when a switch fails to actuate normally as a result of having worked in over-current or high-temperature conditions for an extended period of time or after aging, among other causes. Hence, the present research considers testing switch faults with the framework for three-level NPC inverters as shown in Figure 2, where the positions of faulty transistors are identified by extension theory.

To begin with, the waveforms of the line currents in three phases on the inverter under simulation were subjected to fast Fourier transformation (FFT) to generate the feature frequency spectra for the line currents at $m_f + 1$, $m_f + 5$, and $m_f - 5$ as input signals in the fault diagnosis system that were built based on extension theory. The faults were divided into 12 types by switches $S_{a1}{}^{+}$, $S_{a2}{}^{+}$, $S_{a1}{}^{-}$, $S_{a2}{}^{-}$, $S_{b1}{}^{+}$, $S_{b2}{}^{+}$, $S_{b1}{}^{-}$, $S_{b2}{}^{-}$, $S_{c1}{}^{+}$, $S_{c2}{}^{+}$, $S_{c1}{}^{-}$, and $S_{c2}{}^{-}$.

Additionally, from the 84 records for the feature frequency spectra that were measured with twelve different switch faults with the inverter working at a frequency range of 30–90 Hz, and additionally based on the values for the feature frequency spectra of the twelve fault types, the upper and lower limits for the neighborhood domains in each fault type were identified. The classical domains were further set by the feature frequency spectra of each fault type.

The diagnosis and identification process for inverter faults with extension theory is delineated as follows:

Step 1: Create a matter-element model for fault characteristics C_1, C_2, and C_3 on the frequency spectra for the line current for each fault type.

$$R_g = (F, C, V_p) = \begin{bmatrix} F_0 & C_1 & < x_1, y_1 > \\ & C_2 & < x_2, y_2 > \\ & C_3 & < x_3, y_3 > \end{bmatrix}, \ g = 1, 2, \ldots, 12 \tag{12}$$

Step 2: Input uncategorized fault characteristics C_1, C_2, and C_3 with the frequency spectra of the line current to create a matter-element model.

$$R_{new} = \begin{bmatrix} F_{new} & C_1 & V_{new1} \\ & C_2 & V_{new2} \\ & C_3 & V_{new3} \end{bmatrix} \tag{13}$$

Step 3: Assign weights to the characteristics of each fault type (W_1, W_2, and W_3) which represent the significance of these characteristics. Set $W_1 = W_2 = W_3 = 1/3$ here.

Step 4: Calculate the correlation between uncategorized fault characteristics on the frequency spectra of the line current and each existing fault type.

$$\lambda_g = \sum_{j=1}^{3} W_j K_{g\ j}, g = 1, 2, \ldots, 12 \tag{14}$$

Step 5: The maximum correlation value from the calculation determines the category of the uncategorized fault characteristics C_1, C_2, and C_3 on the frequency spectra of the line current. Thus, the fault type is identified.

5. Test Results

In order to identify faulty power transistors, all the fault types were divided into 12 switch fault conditions, i.e., $S_{a1}{}^{+}$, $S_{a2}{}^{+}$, $S_{a1}{}^{-}$, $S_{a2}{}^{-}$, $S_{b1}{}^{+}$, $S_{b2}{}^{+}$, $S_{b1}{}^{-}$, $S_{b2}{}^{-}$, $S_{c1}{}^{+}$, $S_{c2}{}^{+}$, $S_{c1}{}^{-}$, and $S_{c2}{}^{-}$, as shown in Table 1.

Table 1. Types of fault diagnosis.

Fault Condition	Type
A faulty S_{a1}^{+} switch	F_1
A faulty S_{a2}^{+} switch	F_2
A faulty S_{a1}^{-} switch	F_3
A faulty S_{a2}^{-} switch	F_4
A faulty S_{b1}^{+} switch	F_5
A faulty S_{b2}^{+} switch	F_6
A faulty S_{b1}^{-} switch	F_7
A faulty S_{b2}^{-} switch	F_8
A faulty S_{c1}^{+} switch	F_9
A faulty S_{c2}^{+} switch	F_{10}
A faulty S_{c1}^{-} switch	F_{11}
A faulty S_{c2}^{-} switch	F_{12}

Tables 2 and 3 show the data for the characteristic frequency spectra of the line currents with a faulty switch working at 40 and 80 Hz. These measured data were used as inputs in the fault diagnosis system, generating the results of identification shown in Tables 4 and 5. The two tables indicate that the data for every fault were accurately identified. For example, considering F_5 in the test data in Table 2, it is clear from the identification results in Table 4 that F_5 was determined according to the highest output correlation at 0.78784, while the correlation F_{12} was the lowest at −0.33376, which suggests that F_{12} was least likely.

Table 2. Data for feature frequency spectra for faulty switches in an inverter working at 40 Hz.

Fault Type	i_a Characteristic Frequency Spectrum (mA)			i_b Characteristic Frequency Spectrum (mA)			i_c Characteristic Frequency Spectrum (mA)		
	m_f+1	m_f+5	m_f-5	m_f+1	m_f+5	m_f-5	m_f+1	m_f+5	m_f-5
F_1	0.59	0.14	0.15	0.26	0.11	0.11	0.33	0.03	0.06
F_2	0.43	0.26	0.23	0.17	0.13	0.14	0.26	0.14	0.10
F_3	0.37	0.28	0.27	0.22	0.13	0.10	0.15	0.16	0.18
F_4	0.56	0.21	0.22	0.31	0.06	0.07	0.26	0.15	0.15
F_5	0.26	0.10	0.11	0.58	0.23	0.21	0.30	0.16	0.13
F_6	0.21	0.10	0.11	0.32	0.31	0.33	0.12	0.22	0.21
F_7	0.13	0.20	0.20	0.36	0.28	0.28	0.24	0.08	0.08
F_8	0.32	0.13	0.12	0.56	0.20	0.21	0.26	0.10	0.13
F_9	0.31	0.08	0.07	0.28	0.10	0.11	0.59	0.19	0.18
F_{10}	0.19	0.14	0.14	0.18	0.13	0.15	0.37	0.27	0.27
F_{11}	0.20	0.11	0.17	0.19	0.11	0.14	0.39	0.22	0.28
F_{12}	0.22	0.14	0.14	0.25	0.11	0.09	0.48	0.26	0.29

Table 3. Data for feature frequency spectra for faulty switches in an inverter working at 80 Hz.

Fault Type	i_a Characteristic Frequency Spectrum (mA)			i_b Characteristic Frequency Spectrum (mA)			i_c Characteristic Frequency Spectrum (mA)		
	m_f+1	m_f+5	m_f-5	m_f+1	m_f+5	m_f-5	m_f+1	m_f+5	m_f-5
F_1	0.59	0.13	0.14	0.28	0.08	0.08	0.3	0.05	0.06
F_2	0.25	0.24	0.24	0.12	0.12	0.12	0.13	0.12	0.11
F_3	0.19	0.28	0.30	0.11	0.12	0.13	0.07	0.16	0.16
F_4	0.55	0.19	0.19	0.28	0.08	0.08	0.26	0.11	0.10
F_5	0.26	0.09	0.10	0.55	0.20	0.20	0.29	0.12	0.11
F_6	0.11	0.12	0.13	0.17	0.31	0.31	0.07	0.19	0.19
F_7	0.07	0.19	0.18	0.17	0.31	0.31	0.11	0.12	0.13
F_8	0.29	0.13	0.11	0.55	0.20	0.20	0.26	0.09	0.10
F_9	0.30	0.08	0.07	0.26	0.10	0.11	0.56	0.18	0.17
F_{10}	0.10	0.15	0.14	0.10	0.13	0.15	0.19	0.27	0.29
F_{11}	0.09	0.14	0.17	0.10	0.13	0.14	0.19	0.27	0.32
F_{12}	0.23	0.12	0.14	0.27	0.09	0.09	0.51	0.21	0.23

Table 4. Results for the identification of different switch faults in an inverter working at 40 Hz.

Fault Type	F_1	F_2	F_3	F_4	F_5	F_6	F_7	F_8	F_9	F_{10}	F_{11}	F_{12}	Identification Result
F_1	0.559732	−0.2341	−0.30163	−0.25791	−0.35161	−0.28997	−0.28827	−0.37557	−0.13128	−0.19571	−0.09069	0.122352	F_1
F_2	−0.31894	0.529985	0.19564	0.101215	−0.24656	−0.39278	−0.13254	−0.18231	−0.38297	−0.06586	−0.03606	−0.39277	F_2
F_3	−0.32544	0.153303	0.480904	−0.28298	−0.35341	−0.24157	−0.23005	−0.45123	−0.15916	−0.30098	−0.31193	−0.30453	F_3
F_4	−0.19607	−0.02785	−0.05336	0.669501	−0.25911	−0.38318	−0.191	−0.29812	−0.35435	−0.31991	−0.25472	−0.35481	F_4
F_5	−0.30139	−0.1844	−0.11302	−0.27134	0.78784	−0.14891	−0.31367	0.170666	−0.27814	−0.28528	−0.1707	−0.33376	F_5
F_6	−0.5183	−0.40867	−0.23671	−0.5169	−0.1806	0.410636	−0.05613	−0.31277	−0.3506	−0.35322	−0.1498	−0.15425	F_6
F_7	−0.35259	−0.12849	−0.3183	−0.11627	−0.33122	−0.09984	0.386716	−0.25002	−0.54508	−0.20199	−0.1635	−0.48438	F_7
F_8	−0.19771	−0.04102	−0.25002	−0.07334	0.370901	−0.22944	−0.12948	0.729028	−0.28932	−0.14899	−0.15479	−0.23812	F_8
F_9	−0.10774	−0.24794	−0.13496	−0.26971	−0.19914	−0.13057	−0.42272	−0.2614	0.729557	−0.40752	−0.27084	−0.06767	F_9
F_{10}	−0.21181	−0.04626	0.010629	−0.39599	−0.31255	−0.06477	−0.34739	−0.33145	−0.31563	0.35106	0.337094	0.172031	F_{10}
F_{11}	−0.20393	−0.17468	−0.14651	−0.38417	−0.24391	0.13807	−0.3556	−0.2593	−0.17523	0.138607	0.454056	0.202084	F_{11}
F_{12}	−0.11796	−0.17041	−0.18303	−0.36593	−0.38078	−0.13336	−0.31591	−0.39715	−0.15802	0.096222	0.373375	0.5391	F_{12}

Table 5. Results for the identification of different switch faults in an inverter working at 80 Hz.

Fault Type	F_1	F_2	F_3	F_4	F_5	F_6	F_7	F_8	F_9	F_{10}	F_{11}	F_{12}	Identification Result
F_1	0.622689	−0.34653	−0.43611	−0.06351	−0.32677	−0.28556	−0.30873	−0.27985	−0.34084	−0.23062	−0.18383	−0.13332	F_1
F_2	−0.38422	0.592683	0.4343	−0.09045	−0.25795	−0.38591	−0.06168	−0.31945	−0.35681	−0.21063	−0.05411	−0.33311	F_2
F_3	−0.58218	−0.28377	0.547845	−0.56388	−0.3993	−0.3059	−0.50629	−0.56016	−0.41968	−0.28528	−0.13195	−0.44254	F_3
F_4	0.06017	0.043765	−0.23491	0.561843	−0.33158	−0.35213	0.19983	−0.17871	−0.29778	−0.28035	−0.17302	−0.29794	F_4
F_5	−0.36619	−0.07315	−0.25264	−0.24308	0.585479	−0.27545	−0.20206	0.325634	−0.21915	−0.27345	−0.18597	−0.38598	F_5
F_6	−0.46206	−0.39656	−0.16292	−0.52598	−0.31768	0.599934	0.122139	−0.29312	−0.27273	−0.14926	−0.08569	−0.31517	F_6
F_7	−0.50241	−0.04344	−0.06313	−0.13996	−0.31732	0.130542	0.403082	−0.39168	−0.55918	−0.37664	−0.19821	−0.52746	F_7
F_8	−0.18896	−0.09863	−0.27318	−0.18287	0.249617	−0.28389	−0.10745	0.686704	−0.36196	−0.19957	−0.1727	−0.25936	F_8
F_9	−0.17866	−0.19605	−0.01026	−0.32171	−0.17182	−0.15735	−0.38808	−0.29794	0.773997	−0.39863	−0.25987	0.095117	F_9
F_{10}	−0.39289	−0.05756	−0.16112	−0.47144	−0.42423	−0.19086	−0.21326	−0.40214	−0.47384	0.664821	0.341783	−0.13587	F_{10}
F_{11}	−0.34599	−0.00483	−0.19398	−0.4578	−0.50207	−0.32113	−0.22635	−0.47022	−0.51012	0.347249	0.458702	−0.19029	F_{11}
F_{12}	0.1033	−0.28286	−0.18499	−0.22448	−0.27963	0.12889	−0.33649	−0.22669	0.053569	−0.15907	0.002915	0.446516	F_{12}

6. Fault-Tolerant Control for Three-Level NPC Inverters

6.1. Analysis of Tolerant Control

The switch faults of a three-level NPC inverter are divided into two types: external switch faults and internal switch faults. In the framework of a three-level NPC inverter in Figure 2, if any of the external switches $Sa1^+$, $Sa2^-$, $Sb1^+$, $Sb2^-$, $Sc1^+$, and $Sc2^-$ are faulty because of an open circuit, then it is necessary to alter both the switching condition of the inverter and the phase angle of sine reference voltage in the pulse width modulation (PWM) control signals to keep the output voltage of the inverter in a balanced three-phase form. The parameters of the induction motor are listed in Table 6.

Table 6. Parameter of the three-phase inductor motor.

Horsepower (Hp)	Rotor Resistance (Ω)	Rotor Leakage Inductance (H)	Stator Resistance (Ω)	Stator Leakage Inductance (H)	Magnetization Inductance (H)	Moment of Inertia (kg-m^2)
1	10.4	0.04	11.6	0.04	0.557	0.004

6.1.1. Tolerant Control Strategy for External Switch Faults

When an external switch, for example $S_{a1}{}^+$ or $S_{a2}{}^-$, encounters an open-circuit fault, the *a*-phase half-bridge switches ($S_{a1}{}^+$ and $S_{a2}{}^-$) must be deactivated to activate the internal (neutral-point) switches ($S_{a1}{}^-$ and $S_{a2}{}^+$). Specifically, point *a* is connected to the neutral point and *b*- and *c*-phase are still switched normally. The fault-tolerant control for the occurrence of an open-circuit fault in $S_{a1}{}^+$ is shown in Figure 11. The voltage phasor diagram corresponding to this situation is illustrated in Figure 12a. As illustrated in the figure, the phasor positions of the line voltages **Vab** and **Vca** without any switch fault in Figure 11 become those of **Vab1** and **Vca1**, with the voltage magnitude decreasing by 0.577 times relative to the original line voltage. The line voltage **Vbc1** remains unchanged; however, because the voltage v_{ao} is 0, the phase angle of the *b*-phase voltage should be simultaneously adjusted to be 150° behind that of the *a*-phase voltage, and the phase angle of the *c*-phase voltage should be 150° ahead of that of the *a*-phase voltage. After the occurrence of a fault, the three-phase voltage can still maintain the operation of the balanced three-phase system and the corresponding voltage phasor diagram of the system is shown in Figure 12b. The phasor positions of the line voltages **Vab1** and **Vca1** presented in Figure 12a shift to those of **Vab2** and **Vca2**, with the magnitude of the line voltage **Vbc2** decreasing by 0.577 times relative to that of the original line voltage **Vbc1**. Considering switch $S_{a1}{}^+$ with an open-circuit fault as an example, the waveforms of the sine reference voltages in three phases are shown in Figure 13.

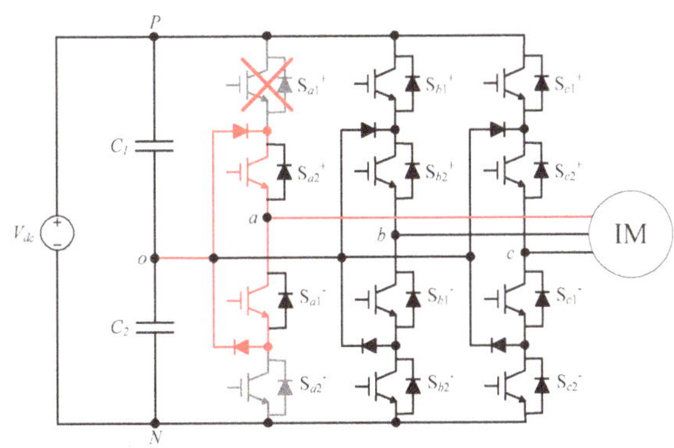

Figure 11. Fault-tolerant control for the occurrence of an open-circuit fault in $S_{a1}{}^+$.

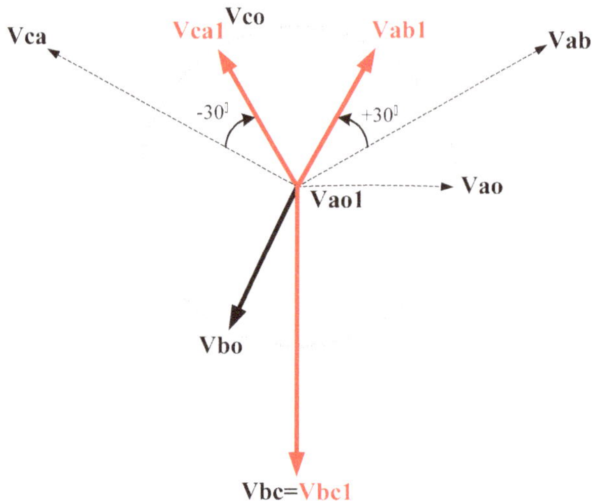

(a)

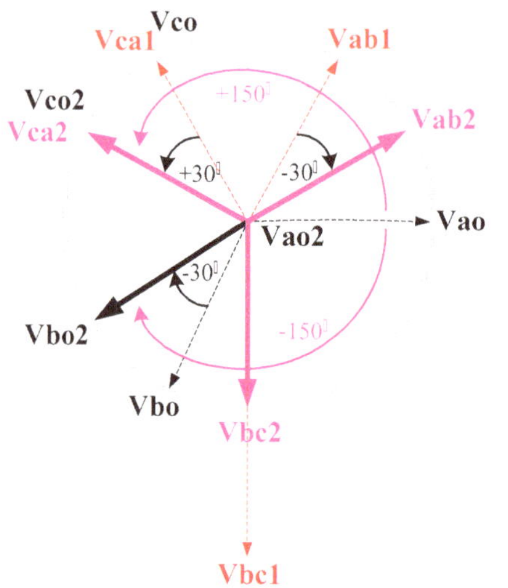

(b)

Figure 12. Voltage phasor diagram of fault-tolerant control when a fault occurs in the switch $S_{a1}{}^{+}$: (**a**) unadjusted voltage phase angle; (**b**) adjusted phase angle of the b-phase voltage such that it is 150° behind that of the *a*-phase voltage; adjusted phase angle of the c-phase voltage such that it is 150° ahead of that of the *a*-phase voltage.

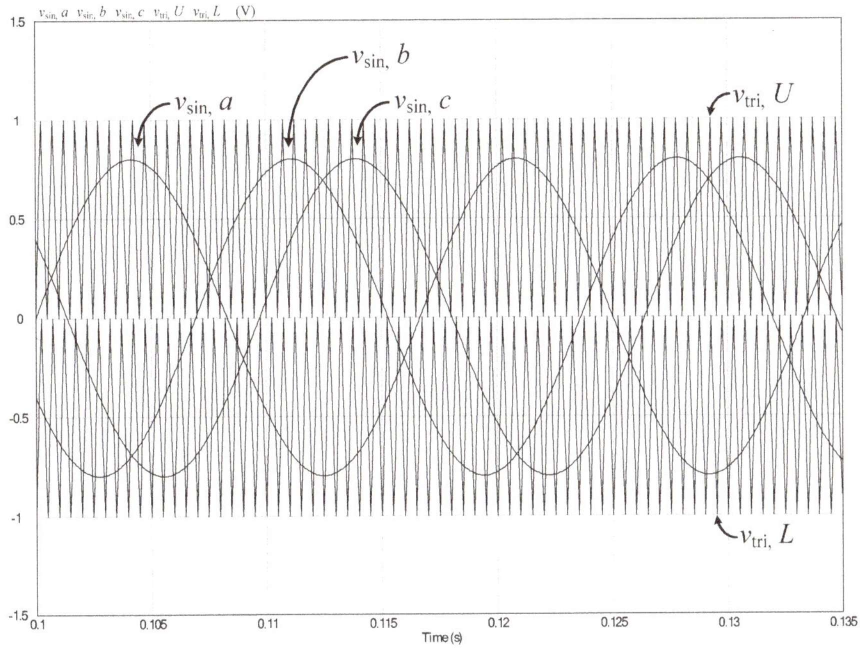

Figure 13. The waveform of pulse width modulation (PWM) control signals for the case of inverter switch $Sa1^+$ with an open-circuit fault.

Analysis of the tolerant control strategy for other external switch failures shows the same results and is not be repeated here for the sake of brevity.

6.1.2. Tolerant Control Strategy for Internal Switch Faults

In the framework of a three-level NPC inverter, in order to exercise the tolerant control of internal (neutral point) switches, it is necessary to connect a serial-connected H-bridge switch in parallel to the neutral-point switches in each arm ($Sa2^+$, $Sa1^-$; $Sb2^+$, $Sb1^-$; and $Sc2^+$, $Sc1^-$) as shown in Figure 14. If any of the internal (neutral point) switches $Sa2^+$, $Sa1^-$, $Sb2^+$, $Sb1^-$, $Sc2^+$, and $Sc1^-$ in the inverter encounter an open-circuit fault, in order for the inverter to remain in operation, it is necessary to make every parallel-connected tolerant control switch operate at once while altering the switching condition of the inverter. For example, if switch $Sc2^+$ has an open-circuit fault, where, in this case, the *c* phase output of the faulty inverter cannot connect to neutral point *o*, the *c* phase voltage, v_{co}, becomes erroneous and causes current distortion; hence, it is necessary to activate the tolerant control, i.e., disconnecting H-bridge switches in the *c* phase ($Sc1^-$, $Sc2^+$), maintaining the tolerant bypass switches (Sc^+, Sc^-), and allowing the H-bridge switches ($Sa1^+$, $Sa2^-$, $Sb1^+$, $Sb2^-$, $Sc1^+$, and $Sc2^-$) to actuate two-level switching of voltage P and voltage N, that is, changing the three-level inverter output voltage into two-level output where three-phase balanced output is still maintained. Figures 15 and 16 show the fault-tolerant control and the PWM control signal waveforms for the case of inverter internal switch $Sc2^+$ with an open-circuit fault, respectively.

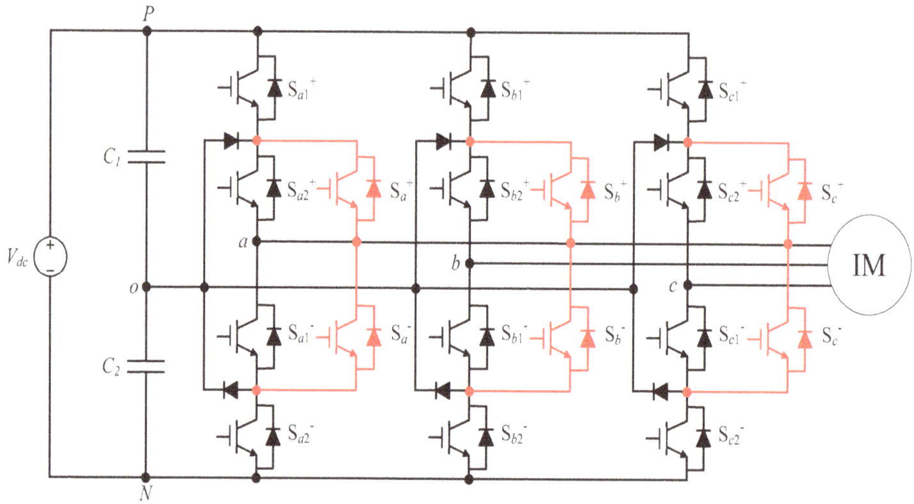

Figure 14. Framework of three-level NPC inverter with additional tolerant control switches.

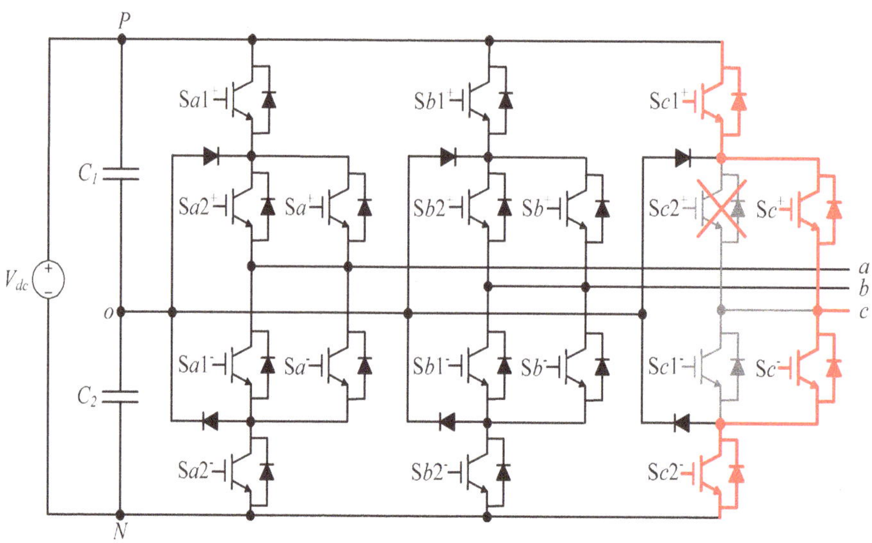

Figure 15. The fault-tolerant control for the occurrence of an open-circuit fault in $Sc2^+$.

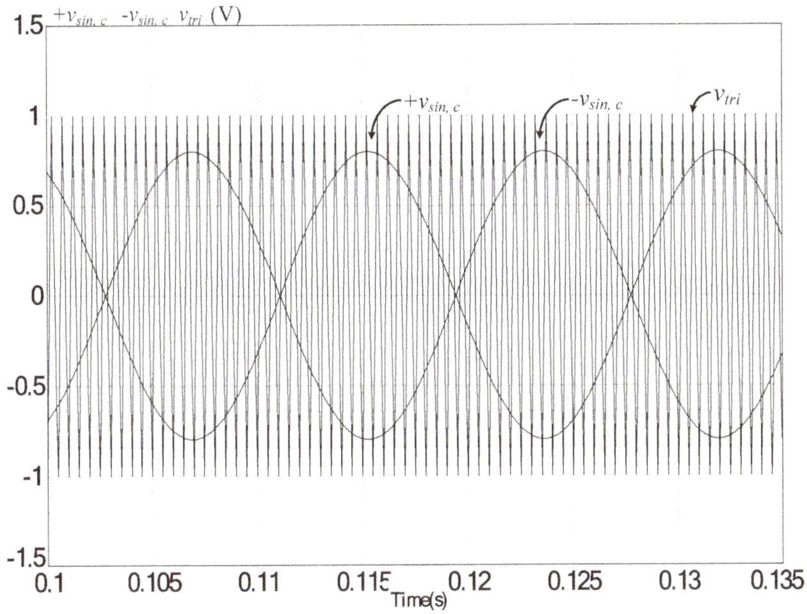

Figure 16. The PWM control signal waveforms for the case of inverter switch $Sc2^+$ with an open-circuit fault.

6.2. Tolerant Control Simulations

When applying tolerant control to an open-circuit fault with external switch $Sa1^+$, for example, its output line voltage and line current is given as shown in Figure 17. In the case of internal (neutral point) switch $Sc2^+$ with an open-circuit fault, its output line voltage and line current are also shown in Figure 18. From Figures 17 and 18, we can observe that if an open-circuit fault occurs at 0.12 s, which would cause the line current of the corresponding phase and three-phase output line voltage to be distorted, particularly so in the affected phase, and the tolerant control strategy is enacted at 0.18 s, then it may observed in the figures that once the tolerant control strategy is enacted that the three-phase output line voltage is downgraded from three-level to two-level output, but it remains a balanced three-phase system despite the fault. Moreover, although the three-phase line current displays a small amount of lag in the beginning of tolerant control due to inductive loads, it maintains a balanced three-phase system output.

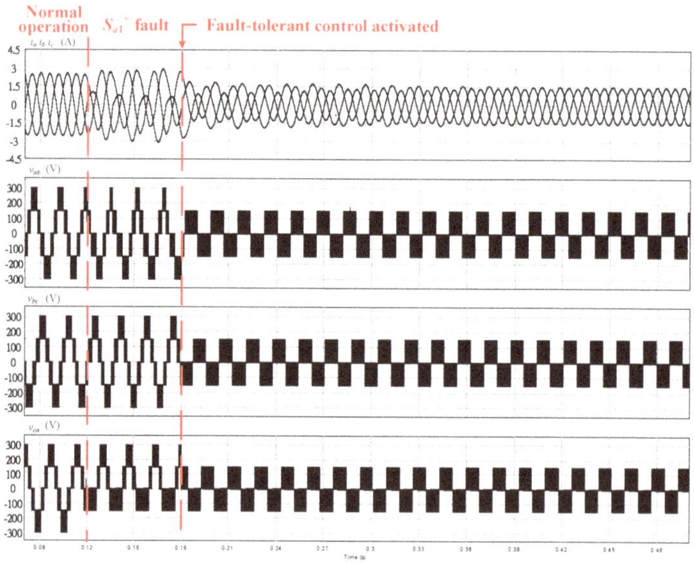

Figure 17. Waveforms of the output line voltage and line current of external switch $S_{a1}{}^{+}$ with an open-circuit fault and tolerant control.

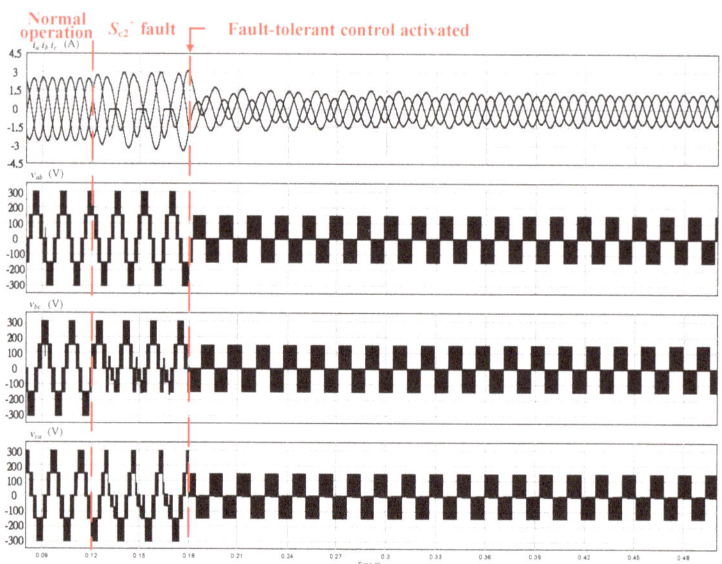

Figure 18. Waveforms of the output line voltage and line current of internal switch $S_{c2}{}^{+}$ with an open-circuit fault and tolerant control.

According to the analysis of the tolerant control in Section 6.1, other switch failures in different phases will also produce the same results. As such, additional results are not presented here for the sake of brevity.

6.3. Tolerant Control Experiments

Figure 19 shows the three-level neutral-point clamped (NPC) inverter used in this study. To verify the experimental results, this study used the digital signal processor TMS320F28335 as the control core and considered the occurrence of open-circuit faults in the switches to test the fault-tolerant control strategy. Figure 20 shows that the occurrence of an open-circuit fault in the switch $S_{a1}{}^{+}$ would distort the three-phase output line current and voltage. The fault is particularly severe in the i_a, v_{ab}, and v_{ca} phases, and the fault-tolerant control strategy can be implemented 0.015 s after the occurrence of the fault. In this situation, the *a* phase half-bridge switches ($S_{a1}{}^{+}$ and $S_{a2}{}^{-}$) are deactivated and the neutral-point switches ($S_{a1}{}^{-}$ and $S_{a2}{}^{+}$) are activated. The *b*- and *c*-phase switches still operate normally. The phase angle of the *b* phase voltage is simultaneously adjusted such that it is 150° behind the phase angle of the *a* phase reference voltage, and the phase angle of the *c* phase reference voltage is adjusted such that it is 150° ahead of that of the *a* phase voltage. After the implementation of the fault-tolerant control strategy, the three-phase output line voltage is reduced from five to three levels (Figure 20b); however, after the occurrence of the fault, the output line current and voltage are still maintained via the operation of the balanced three-phase system. Therefore, the motor can still operate normally with a reduced load.

Based on the tolerance control analysis detailed in Section 6.1, the same results can be obtained when other external switches fail.

If any of the internal switches $Sa2^{+}$, $Sa1^{-}$, $Sb2^{+}$, $Sb1^{-}$, $Sc2^{+}$, and $Sc1^{-}$ in the inverter experience an open-circuit fault, in order for the inverter to remain in operation, it is necessary to make every parallel-connected tolerant control switch (as shows in Figure 14) operate at once. If internal (neutral point) switch $Sc2^{+}$ experiences an open-circuit fault, its output line current and voltage are same as shown in Figure 21. The output line current and voltage still maintain the operation of the balanced three-phase system. Hence, the motor can also operate normally with a reduced load.

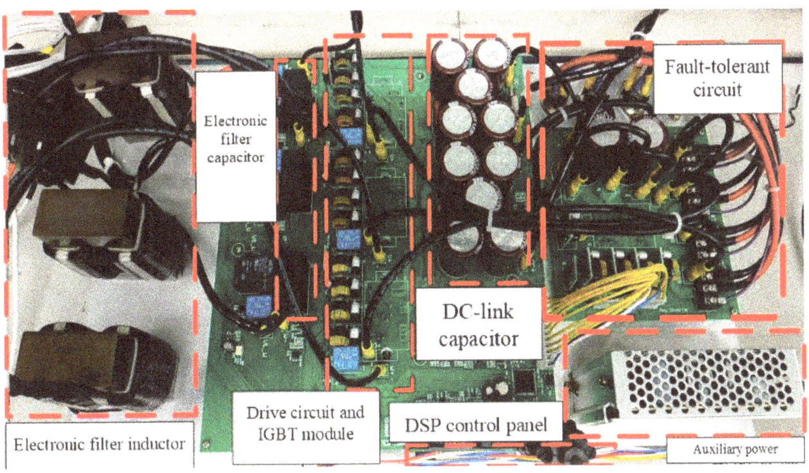

Figure 19. Experimental hardware circuits for the three-level NPC inverter.

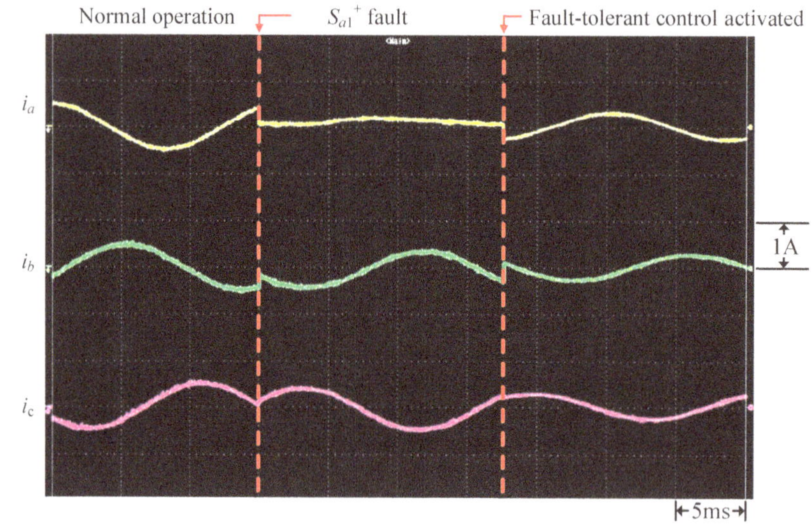

(a)

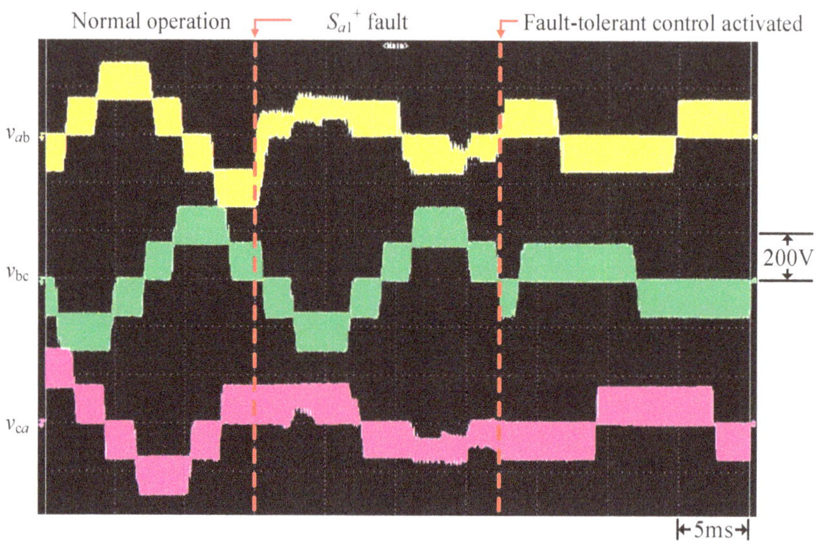

(b)

Figure 20. Measured waveforms of switch $S_{a1}{}^+$ with an open-circuit fault and tolerant control: (**a**) output line current; (**b**) output line voltage.

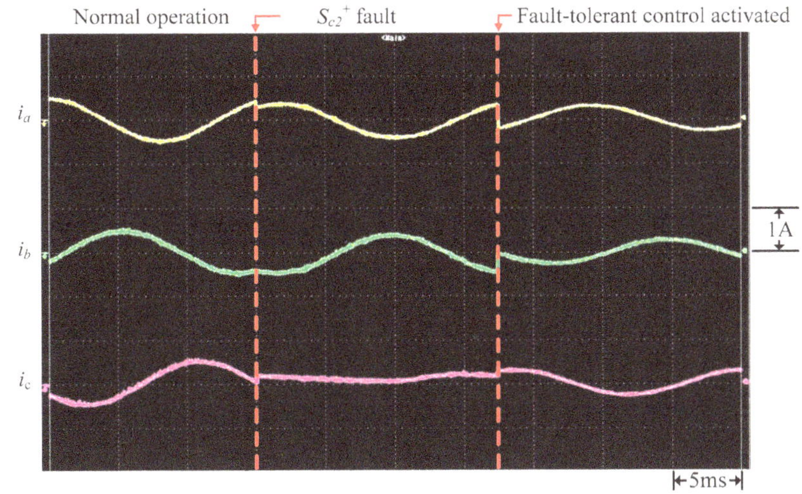

(a)

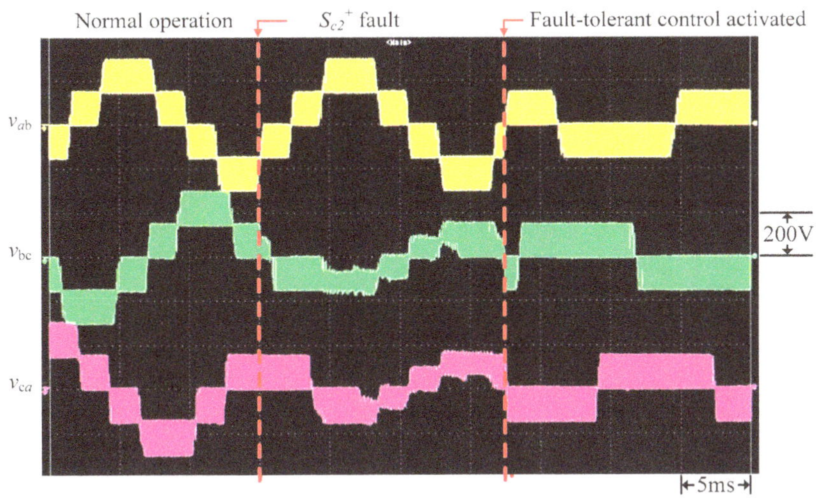

(b)

Figure 21. Measured waveforms of switch $S_{c2}{}^+$ with an open-circuit fault and tolerant control: (**a**) output line current; (**b**) output line voltage.

According to the previous analysis in Section 6.1, other internal switch failures will also have the same results.

7. Conclusions

This paper has presented a fault diagnosis system for inverters based on extension theory. The system can identify the positions of faulty power transistors in a three-level NPC inverter. The extension theory-based method, as applied here, can be implemented without requiring massive data quantities for training, thus being able to reduce the data volume and enable faster identification with higher accuracy. In addition, the tolerant control strategy of the system is able to be implemented as soon as any switch in the inverter becomes faulty, where the inverter is enabled to continue supplying power, enhancing the reliability of power supply by the three-level NPC inverter. Finally, the simulation and experimental results suggest that the system is able to correctly pinpoint the positions of faulty power transistors, and, when any switch becomes faulty, it is able to maintain the output line voltage and line current with three-phase balance by means of the tolerant control strategy, which verifies that the method proposed herein is viable.

Author Contributions: The conceptualization was proposed by K.-H.C., who was also responsible for writing (review and editing) this paper. K.-H.C. completed the formal analysis of the fault diagnosis system and tolerant control algorithm. C.-H.K. carried out the data curation, software programming, simulation, and experimental validation. K.-H.C. was in charge of project administration. All authors have read and agreed to the published version of the manuscript.

Funding: This research was funded by Ministry of Science and Technology, Taiwan, under grant number MOST 109-2221-E-167-016-MY2.

Conflicts of Interest: The authors of the manuscript declare that there are no conflicts of interest with any of the commercial identities mentioned in the manuscript.

References

1. Rodriguez, J.I.; Leeb, S.B. A multilevel inverter topology for inductively coupled power transfer. *IEEE Trans. Power Electron.* **2006**, *21*, 1607–1617. [CrossRef]
2. Escalante, M.F.; Vannier, J.C.; Arzande, A. Flying capacitor multilevel inverters and DTC motor drive applications. *IEEE Trans. Ind. Electron.* **2002**, *49*, 809–815. [CrossRef]
3. Tourkhani, F.; Viarouge, P.; Meynard, T.A. Optimal design and experimental results of a multilevel inverter for an UPS application. In Proceedings of the International Conference on Power Electronics and Drive Systems, Singapore, 26–29 May 1997; pp. 340–343.
4. Daher, S.; Schmid, J.; Antunes, F.L.M. Multilevel inverter topologies for stand-alone PV systems. *IEEE Trans. Ind. Electron.* **2008**, *55*, 2703–2712. [CrossRef]
5. Naik, R.L.; Udaya, K.R.Y. A novel technique for control of cascaded multilevel inverter for photovoltaic power supplies. In Proceedings of the European Conference on Power Electronics and Applications, Dresden, Germany, 11–14 September 2005; pp. 1–9.
6. Nabae, A.; Takahashi, I.; Akagi, H. A new neutral-point clamped PWM inverter. *IEEE Trans. Ind. Appl.* **1981**, *17*, 518–523. [CrossRef]
7. Chen, A.L.; Hu, L.; Chen, L.F.; Deng, Y.; He, X.N. A multilevel converter topology with fault-tolerant ability. *IEEE Trans. Power Electron.* **2005**, *20*, 405–415. [CrossRef]
8. Khomfoi, S.; Tolbert, L.M. Fault diagnostic system for a multilevel inverter using a neural network. *IEEE Trans. Power Electron.* **2007**, *22*, 1062–1069. [CrossRef]
9. Choi, U.; Lee, K.; Blaabjerg, F. Diagnosis and tolerant strategy of an open-switch fault for T-type three-level inverter systems. *IEEE Trans. Ind. Appl.* **2014**, *50*, 495–508. [CrossRef]
10. Altug, S.; Chen, M.Y.; Trussell, H.J. Fuzzy inference systems implemented on neural architectures for motor fault detection and diagnosis. *IEEE Trans. Ind. Electron.* **2002**, *46*, 1069–1079. [CrossRef]
11. Peng, Z. An integrated approach to fault diagnosis of machinery using wear debris and vibration analysis. *Wear* **2003**, *255*, 1221–1232. [CrossRef]
12. Han, T.; Yang, B.S.; Lee, J.M. A new condition monitoring and fault diagnosis system of induction motors using artificial intelligence algorithms. In Proceedings of the IEEE International Conference on Electric Machines and Drives, San Antonio, TX, USA, 15–18 May 2005; pp. 1967–1974.
13. Pedrayes, F.; Rojas, C.H.; Cabanas, M.F.; Melero, M.G.; Orcajo, G.A.; Cano, J.M. Application of a dynamic model based on a network of magnetically coupled reluctances to rotor fault diagnosis in induction motors. In Proceedings of the IEEE International Symposium on Diagnostics for Electric Machines, Power Electronics and Drives, Cracow, Poland, 6–8 September 2007; pp. 241–246.

14. Lehtoranta, J.; Koivo, H.N. Fault diagnosis of induction motors with dynamical neural networks. In Proceedings of the IEEE International Conference on Systems, Man and Cybernetics, Waikoloa, HI, USA, 10–12 October 2005; pp. 2979–2984.
15. He, Q.; Du, D.M. Fault diagnosis of induction motor using neural networks. In Proceedings of the International Conference on Machine Learning and Cybernetics, Hong Kong, China, 19–22 August 2007; pp. 1090–1095.
16. Martin-Diaz, I.; Morinigo-Sotelo, D.; Duque-Perez, O.; Romero-Troncoso, R.J. An experimental comparative evaluation of machine learning techniques for motor fault diagnosis under various operating conditions. *IEEE Trans. Ind. Appl.* **2018**, *54*, 2215–2224. [CrossRef]
17. Torabi, N.; Sundaram, V.M.; Toliyat, H.A. On-line fault diagnosis of multi-phase drives using self-recurrent wavelet neural networks with adaptive learning rates. In Proceedings of the IEEE Applied Power Electronics Conference and Exposition (APEC), Tampa, FL, USA, 26–30 March 2017; pp. 570–577.
18. Chowdhury, D.; Bhattacharya, M.; Khan, D.; Saha, S.; Dasgupta, A. Wavelet decomposition based fault detection in cascaded H-bridge multilevel inverter using artificial neural network. In Proceedings of the 2nd IEEE International Conference Recent Trends in Electronics, Information & Communication Technology (RTEICT), Bangalore, India, 19–20 May 2017; pp. 1931–1935.
19. Xu, J.; Song, B.; Zhang, J.; Xu, L. A new approach to fault diagnosis of multilevel inverter. In Proceedings of the Chinese Control and Decision Conference (CCDC), Shenyang, China, 9–11 June 2018; pp. 1054–1058.
20. Li, W.; Li, G.Y.; Zeng, R.; Ni, K.; Hu, Y.H.; Wen, H.Q. The fault detection, localization, and tolerant operation of modular multilevel converters with an insulated gate bipolar transistor (IGBT) open circuit fault. *Energies* **2018**, *11*, 837. [CrossRef]
21. Liu, Y.Q.; Li, D.H.; Jin, Y.; Wang, Q.B.; Song, W.L. Research on unbalance fault-tolerant control strategy of modular multilevel photovoltaic grid-connected inverter. *Energies* **2018**, *11*, 1368. [CrossRef]
22. Jung, J.H.; Ku, H.K.; Son, Y.D.; Kim, J.M. Open-switch fault diagnosis algorithm and tolerant control method of the three-phase three-level NPC active rectifier. *Energies* **2019**, *12*, 2495. [CrossRef]
23. Ahmadipour, M.; Hizam, H.; Othman, M.L.; Radzi, M.A.M.; Chireh, N. A fast fault identification in a grid-connected photovoltaic system using wavelet multi-resolution singular spectrum entropy and support vector machine. *Energies* **2019**, *12*, 2508. [CrossRef]
24. He, Z.; Zhang, X.C.; Liu, C.; Han, T. Fault prognostics for photovoltaic inverter based on fast clustering algorithm and Gaussian mixture model. *Energies* **2020**, *13*, 4901. [CrossRef]
25. Kwon, B.H.; Kim, S.H.; Kim, S.M.; Lee, K.B. Fault diagnosis of open-switch failure in a grid-connected three-level Si/SiC hybrid ANPC inverter. *Electronics* **2020**, *9*, 399. [CrossRef]
26. Li, M.; Li, G. Based on theory of extenics research assessment and classification of soil erosion. In Proceedings of the World Automation Congress, Puerto Vallarta, Mexico, Mexico, 24–28 June 2012; pp. 1–6.
27. Chen, P.Y.; Chao, K.H.; Wu, Z.Y. An optimal collocation strategy for the key components of compact photovoltaic power generation systems. *Energies* **2018**, *11*, 2523. [CrossRef]
28. Wang, M.H.; Huang, M.L.; Jiang, W.J. Maximum power point tracking and harmonic reducing control method for generator-based exercise equipment. *Energies* **2016**, *9*, 103. [CrossRef]
29. Ju, Y.; Yu, Y.; Ju, G.; Cai, W. Extension set and restricting qualifications of matter-elements' extension. In Proceedings of the 3rd International Conference on Information Technology and Applications, Sydney, NSW, Australia, 4–7 July 2005; pp. 395–398.
30. Xiao, L.; Li, X. Supplier selection based on matter element analysis. In Proceedings of the fourth International Conference on Transportation Engineering, Chengdu, China, 19–20 October 2013; pp. 551–557.

energies

Article

Experimental Comparison of Preferential vs. Common Delta Connections for the Star-Delta Starting of Induction Motors

José Augusto Itajiba [1,*], Cézar Armando Cunha Varnier [2], Sergio Henrique Lopes Cabral [1], Stéfano Frizzo Stefenon [3,4], Valderi Reis Quietinho Leithardt [5], Raúl García Ovejero [6], Ademir Nied [3] and Kin-Choong Yow [4]

1 Electrical Engineering Graduate Program, Electrical Engineering Department, Regional University of Blumenau, R. São Paulo 3250 (Itoupava Seca), Blumenau 89030-000, Brazil; scabral@furb.br
2 WEG Industries, R. Venâncio da Silva Pôrto 399 (Nova Brasília), Jaraguá do Sul 89252-230, Brazil; cezarv@weg.net
3 Electrical Engineering Graduate Program, Electrical Engineering Department, Santa Catarina State University (UDESC), R. Paulo Malschitzki 200 (North Industrial Zone), Joinville 89219-710, Brazil; stefano.stefenon@udesc.br (S.F.S.); ademir.nied@udesc.br (A.N.)
4 Software Systems Engineering Department, Faculty of Engineering and Applied Science, University of Regina, 3737 Wascana Parkway, Regina, SK S4S 0A2, Canada; kin-choong.yow@uregina.ca
5 VALORIZA, Research Center for Endogenous Resources Valorization, Instituto Politécnico de Portalegre, 7300-555 Portalegre, Portugal; valderi@ipportalegre.pt
6 Expert Systems and Applications Lab., E.T.S.I.I of Béjar, University of Salamanca, 37008 Salamanca, Spain; raulovej@usal.es
* Correspondence: joseitagiba@gmail.com

Citation: Itajiba, J.A.; Varnier, C.A.C.; Cabral, S.H.L.; Stefenon, S.F.; Leithardt, V.R.Q.; Ovejero, R.G.; Nied, A.; Yow, K.-C. Experimental Comparison of Preferential vs. Common Delta Connections for the Star-Delta Starting of Induction Motors. *Energies* **2021**, *14*, 1318. https://doi.org/10.3390/en14051318

Academic Editor: Anouar Belahcen and Tian-Hua Liu

Received: 25 January 2021
Accepted: 23 February 2021
Published: 1 March 2021

Abstract: Although this is a fact that is not very explored in the literature, there are two possible forms to connect the stator winding of an induction motor in the delta. The choice for one of these forms defines the amplitude of the stator transient current during the switching from star to delta connection when the motor is driven by a star-delta starting system, which is the most widely used and diffused method for starting an induction motor. One of the possible forms of the delta connection gives rise to a switching current with a relatively small amplitude, which gives it the denomination of preferential. The other form has a relatively higher amplitude of switching current, but it is the most recommended and indicated in diagrams of catalogues and motor plates. Therefore, it is here called "common". With the aim of evidencing how the differences between these two forms of delta connection are manifested, this paper approaches the issue experimentally, through a methodology with statistical support, for a better characterization of the performance of each of these forms of delta connection, in the case of the widely popular star-delta starting method.

Keywords: induction motor; delta connections; star-delta starting

1. Introduction

Despite the growing availability and the consequent increase in the use of electronic devices for the soft starting of the induction motor, such as the soft starter and the VFD —Variable frequency drive—the traditional star-delta starting method, mainly composed of contactors and timing relays, remains one of the most usual and preferred options for most drive systems in the industry [1]. This is especially true in the case of relatively small automation systems, be it new or an already running one [2], due to its relatively low cost as well as the wide availability of services for its installation and maintenance, besides its robustness as a whole [3].

Currently, many electrical drives perform with equipment that has switched converters; these have a flexible application and become economically advantageous as the power of the motors has increased. However, this type of starter has disadvantages such as noise from its switching and high cost for low power motors. The use of static converters is gaining more and more space as they are robust equipment and have become economically viable with developments in semiconductor materials; this expansion means that fewer and fewer starts are used that need to change the way the motor is connected, such as the star-delta starting, compensating start and parallel start [4]. The definition of a preferred way to start engines is rarely addressed in the literature, considering that there may be variations when engines of different powers are evaluated.

The preferred start is rarely addressed in the literature, although some works address the variations of possible delta connections and their consequences in the electrical system [4]. Some surveys are focused on starting with drives [5] and in a large part of the research the improvement of the engine efficiency is evaluated through the use of more modern materials [6] and optimization of machine parameters [7–9].

In fact, the star-delta starting is one of the most classic methods of starting an induction motor [10]. As with many other starting methods, it is based on the reduction in the voltage amplitude applied to the stator winding at the start, to decrease the stator current [11]. In this case, the value of the amplitude of the applied voltage is $1/\sqrt{3}$ of the rated voltage at the start, because of the star connection of the stator winding, which causes an equal reduction in the value of the amplitude of the stator current [12]. Thus, a while after the start of the motor, which is adjusted by the operator, the connection of the stator winding is automatically switched to the delta, and so it remains in the steady-state [13].

For this purpose, there is a wide variety of models and manufacturers on the market for the star-delta switch, which can bring together several devices to perform this type of start in a single device. In addition to contactors, these switches should have overcurrent relays, fuses, and timing relays, which allow the operator to adjust the time interval during which the motor will run as star-connected. After this interval has elapsed, the star connection of the stator winding is undone, and the motor runs idle for a short period, usually Δt = 100 ms, adopted by most manufacturers of these switches, before the delta connection of the stator winding is performed. Then, at the instant that this connection is made, there is a sudden increase in the amplitude of the stator current, which must be accepted by the motor protection system, so as not to unduly take it as a fault current [3].

Figure 1 shows a graphic with the expected performance of the stator current of the motor along this process, with an indication of the sudden increase in the current. Figure 1 is a representation of the variation that occurs in the transient. Right after receiving the command to break the circuit, the contactor opens its contacts by generating an arc per each phase that is properly quenched within a chamber until it passes to zero. Therefore, the three-phase current continues to flow for some milliseconds, approximately.

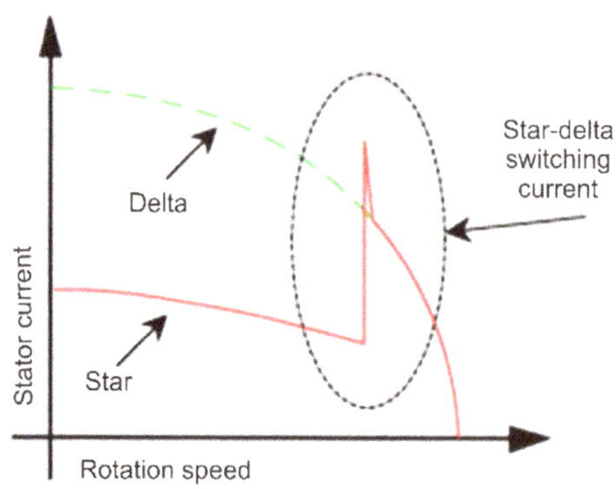

Figure 1. Stator current behavior in star-delta start-switching overcurrent.

Although the graphic of Figure 1 shows that the sudden increase in the stator current may be a natural consequence of switching from the star to the delta connection, the real reasons for this sudden increase are not so obvious. In fact, the amplitude of such an overcurrent is strongly influenced by the form of connection adopted for the delta connection of the stator winding [14].

According to [15] and [16] in the choice between delta connection forms, one of them will present a relatively smaller amplitude of switching current and is therefore considered as a delta connection in the preferential form. Consequently, the other form presents an amplitude of switching current with higher values but, curiously, it is the most commonly found form in motor plates and/or recommended by installation diagrams.

Therefore, it is called the common form. The difference between both forms of delta connection is only in the inversion of the connection of the terminals of one of the stator windings, without which there may be a reversal of the rotation direction of the motor shaft since the phase sequence of the power supply is not inverted [17]. Thus, with a few exceptions in the literature, as in [18] and [19], the importance of the correct choice between these two forms of connecting the stator winding in the delta is not properly diffused or explored, and so this is the essence of this work. The diagrams in Figure 2 show the two possible forms of delta connection.

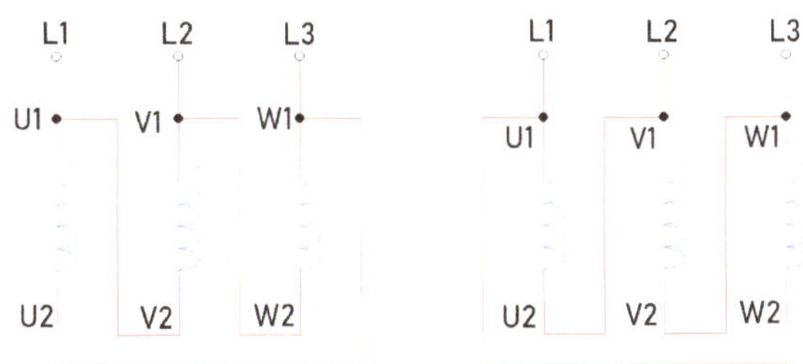

Figure 2. The two possibilities of connection in delta. Preferential on the left and common on the right.

According to [18], this difference in the amplitude of the switching current occurs because, invariably, there is a residual magnetic field, both in the rotor magnetic circuit and in the stator magnetic circuit. Therefore, the rotor movement causes both residual fields to mutually induce a voltage in the respective circuits during the short time interval in which the motor runs idle, Δt. Thus, depending on whether the delta connection is made in the preferential form or in the common one, the combination of the mutually induced voltage will give different results, causing the amplitude of the switching current to have significantly different values at the right instant of the delta connection.

As can be seen in Figure 2, there are two ways to connect a star-delta starter in an induction motor; however, this difference is little discussed considering that the industry determines a preferred way to start the induction motor. In this way, this paper addresses an important subject for the study of induction motors, evaluating the differences in the alteration of the preferred form of connection for induction motors.

In [18] there is a proposing explanation for this fact, with the help of phasor diagrams, which does not consider any residual magnetic field and thus requires some refinement. On the other hand, it is important to consider that the influence of the form of the delta connection superimposes on the effect of the inertia of the rotor, which naturally causes the reduction in its speed. This reduction in the speed is due to the time elapsed (Δt) since the star connection of the stator winding is undone until the instant they are connected in delta [20].

Therefore, since the motor runs idle along with this time interval, an additional amount of stator current will be necessary to accelerate the rotor and compensate for the loss of its kinetic energy [21], which represents a part of the total switching current. Thus, since this part of the switching current does not depend on the form of delta connection, for the same load condition of a given motor, the reason for the common connection to give a total switching current with a higher amplitude relies on the residual magnetic field of the magnetic circuit of the motor. Specific projects are needed to improve the efficiency of electrical machines [22–24], and thus meet the development of the industrial sector in a sustainable manner [25].

To improve energy quality and increase machine efficiency, specific projects have been developed [26], as highlighted by Orosz et al. [27], who present a study highlighting open problems related to electric machines and highlighting how optimization can help in the development of emerging technologies. Optimization is undergoing this great evolution due to the higher analysis capabilities and can be used in various applications, such as electrical transformers [28] and autotransformers [29], among other applications [30]. Research that applies artificial intelligence and modern methods of evaluation of the consequences of parameter changes is promising for addressing the optimization problems [31,32].

However, since there is an influence of the residual value of the magnetic field that varies sinusoidally over time, this value itself is not constant but depends on several factors that are difficult to evaluate or to assess, especially the instantaneous values of the current and voltage on the stator before and after the switching, as well as the motor temperature, among others [33]. Thus, the occurrence of a greater amplitude of the switching current for the common form of star connection is of a statistical nature. Consequently, several starts of the same motor are required, with each of the forms of delta connections to evaluate the behavior of the switching current.

It is important to mention that the peak current does not depend on the residual magnetic field and the residual voltage induced by this field in the stator winding only, but also significantly on the phase difference between the residual voltage and the supply voltage in the instant of the reconnection of the motor to the supply in the delta connection [34]. This phase difference, as well as the residual voltage, is not only influenced by the switching time Δt and the alternative chosen for delta connection, but also by parameters like inertia, load torque, and stator leakage [35].

For this purpose, an experimental arrangement was elaborated to, firstly, check the occurrence of differences in the amplitude of the switching current according to the form of the delta connection, and then to characterize this difference in amplitude.

2. Materials and Methods

In view of the analysis of the differences in amplitude of the switching current for the common and the preferential forms of delta connection, a prototype of a drive system was assembled in the laboratory to enable the sequential use of each of the two forms of delta connection, with the start of the motor in star connection of the stator winding. As a sample, a standard three-phase induction motor was used. It was a 380-Y/220-ΔV, 4 poles, 0.75 kW, 1730 rpm, squirrel cage, SF = 1.15 and continuous duty S1 type of motor. The power circuit of the drive system, with an indication of some contactors providentially taken for the proposed type of drive, is shown in the diagram in Figure 3.

The channels shown in Figure 3 represent the connection of the experiment for measuring voltage and current that was used in the laboratory analysis. As shown in the diagram, four contactors are used. Initially, the contactor K1 energizes the terminals 1, 2, and 3 of the stator winding. Then, the contactor K4 short-circuits the other terminals, 4, 5, and 6, making the connection in star. After the time interval (Δt) is previously set to the timing relay, the star connection is undone through the opening of the contactor K4, and the stator windings are then connected in delta. By doing so, the contactor K2 allows the delta connection in the common form, whereas the contactor K3 allows the preferential form.

Regarding the control circuit, although there are many commercial models of star-delta starting solutions, fitted with timers or not, Misir et al. [36] developed a specific solution, since most of them do not allow the choice between the common or the preferential form. Moreover, commercial switches do not allow adjustment of the time interval between the star and the delta connections (Δt) which hinders the analysis of the influence of the inertia of the motor rotor on the amplitude of the switching current to the delta connection. For example, this time interval could then be set to 50 ms, besides the usual 100 ms. To do so, a

PLC-Programmable Logic Controller was chosen for use in the control circuit, allowing significant simplicity in implementation of the desired actions and easy communication through a serial interface.

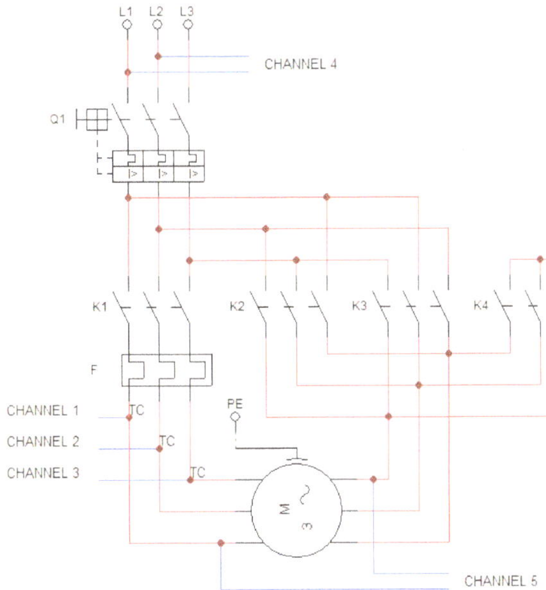

Figure 3. Diagram of the power circuit (main) of the proposed drive system.

The PLC model adopted was Siemens® S7-200, Blumenau, Brazil, and Figure 4 shows a photo of the complete arrangement, containing the command and power circuits, assembled on a laboratory bench, proper for induction motor tests. In the foreground are the contactors and the fuses in one of the boards, whereas on the right is the adopted PLC. Between the motor, taken at the idle mode, and the contactors/fuses board there is an autotransformer to adjust the available mains voltage from 380 V to 220 V as required by the windings of the motor stator.

Figure 4. Arrangement of star-delta starting tests for analysis.

A star-delta starter is a low voltage starter used to reduce the starting current of the motor. The great advantage of the star-delta starter is its need to use 1/3 of the peak current compared to a direct starting. For this consideration, having V_L = supply line voltage, I_{LS} = supply line current and, I_{PS} = winding current per phase and Z = impedance per phase winding at stand still condition.

When the winding is star connected, the I_{PS} is equal to the I_{LS}. Then, the voltage in each phase of the winding is $V_L/\sqrt{3}$. Since here, the I_{PS} equals to the I_{LS}, it can be written

$$I_{PS} = \frac{V_L}{\sqrt{3}\,Z} \leftrightarrow I_{LS} = \frac{V_L}{\sqrt{3}\,Z}. \tag{1}$$

When the winding is delta connected, the (I_{LD}) supply line current is root three times of the winding (I_{PD}) current per phase, the voltage in each phase of the winding is V_L. Thus, the winding current per phase is V_L/Z. Rewriting the equation

$$I_{LD} = \sqrt{3}\,I_{PD} = \frac{\sqrt{3}\,V_L}{Z}. \tag{2}$$

In this way comparing the equations

$$\frac{I_{LD}}{I_{LS}} = \frac{\frac{\sqrt{3}\,V_L}{Z}}{\frac{V_L}{\sqrt{3}\,Z}} = 3. \tag{3}$$

In this way, it is possible to say that the starting current of the network in the case of delta star is one third of the direct switching in the delta.

The voltage equation for a motor consists of the evaluation of the following components: the voltage drop of the winding resistance R, the induced voltage proportional to the rate of change over time of the winding flow connection λ, as follows:

$$v(t) = Ri(t) + \frac{d\lambda(t)}{dt} = Ri(t) + \frac{d[L(\theta_r)i(t)]}{dt}, \tag{4}$$

wherein $\theta_r = \omega_r t$.

Considering a three-phase and bipolar induction motor, whereas the stator windings have a number of effective turns N_s, resistance R_s, leakage inductance L_{1s} and self-inductance L_s. Likewise, the equivalent rotor windings have a number of effective turns Nr, resistance Rr, leakage inductance L_{1r} and self-inductance. θ_r is the angular displacement between the stator and rotor axes, disregarding nonideal features such as groove effect, saturation of the iron core and toothed torque [37].

The winding voltage is given by the sum of the voltage drop of the winding resistance and the voltage induced by the variation of the winding flow connection. Thus, by deriving the voltage range of an induction motor, the six voltage equations for the stator and rotor windings can be calculated by

$$v_{as} = R_s i_{as} + \frac{d\lambda_{as}}{dt} \tag{5}$$

$$v_{bs} = R_s i_{bs} + \frac{d\lambda_{bs}}{dt} \tag{6}$$

$$v_{cs} = R_s i_{cs} + \frac{d\lambda_{cs}}{dt} \tag{7}$$

$$v'_{ar} = R' i'_{ar} + \frac{d\lambda'_{ar}}{dt} \tag{8}$$

$$v'_{br} = R' i'_{br} + \frac{d\lambda'_{br}}{dt} \tag{9}$$

$$v'_{cr} = R'^{i'_{cr}} + \frac{d\lambda'_{cr}}{dt} \tag{10}$$

wherein v_{as}, v_{bs}, v_{cs} are the stator voltages, i_{as}, i_{bs}, i_{vs} are the stator currents, v'_{ar}, v'_{br}, v'_{cr} are the rotor voltages, i'_{ar}, i'_{cr} are the rotor currents, λ_{as}, λ_{bs}, λ_{cs} are the stator flux linkages, λ_{ar}, λ_{br}, λ_{cr} are the rotor flux linkages.

To solve these voltage equations, it is necessary to know the flow connection of each winding. Considering that there are six windings, each of them will be influenced by the flow produced by the current flowing in the other five windings. Therefore, the full flow connection λ from the phase as the winding has the following six components:

$$\lambda_{as} = \lambda_{asas} + \lambda_{asbs} + \lambda_{ascs} + \lambda_{asar} + \lambda_{asbr} + \lambda_{ascr} \tag{11}$$

Therefore,

$$\lambda_{as} = L_{asas}i_{as} + L_{asbs}i_{bs} + L_{ascs}i_{cs} + L_{asar}i_{ar} + L_{asbr}i_{br} + L_{ascr}i_{cr} \tag{12}$$

where the flow connection λ_{xsys} represents the flow, which is produced by the current i_{ys} flowing in winding ys and connecting winding xs. The inductance L_{xsys} is determined as the ratio of the flux that connects the winding xs to the current i_{ys} that generates the flux. Thus, the six flux linkages in the stator and rotor windings can be calculated by:

$$\begin{bmatrix} \lambda_{as} \\ \lambda_{bs} \\ \lambda_{cs} \\ \lambda_{ar} \\ \lambda_{br} \\ \lambda_{cr} \end{bmatrix} = \begin{bmatrix} \lambda_{asas} & \lambda_{asbs} & \lambda_{ascs} & \lambda_{asar} & \lambda_{asbr} & \lambda_{ascr} \\ \lambda_{bsas} & \lambda_{bsbs} & \lambda_{bscs} & \lambda_{bsar} & \lambda_{bsbr} & \lambda_{bscr} \\ \lambda_{csas} & \lambda_{csbs} & \lambda_{cscs} & \lambda_{csar} & \lambda_{csbr} & \lambda_{cscr} \\ \lambda_{aras} & \lambda_{arbs} & \lambda_{arcs} & \lambda_{arar} & \lambda_{arbr} & \lambda_{arcr} \\ \lambda_{bras} & \lambda_{brbs} & \lambda_{brcs} & \lambda_{brar} & \lambda_{brbr} & \lambda_{brcr} \\ \lambda_{cras} & \lambda_{crbs} & \lambda_{crcs} & \lambda_{crar} & \lambda_{crbr} & \lambda_{crcr} \end{bmatrix} \tag{13}$$

wherein,

$$\begin{bmatrix} \lambda_{as} \\ \lambda_{bs} \\ \lambda_{cs} \\ \lambda_{ar} \\ \lambda_{br} \\ \lambda_{cr} \end{bmatrix} = \begin{bmatrix} L_{asas} & L_{asbs} & L_{ascs} & L_{asar} & L_{asbr} & L_{ascr} \\ L_{bsas} & L_{bsbs} & L_{bscs} & L_{bsar} & L_{bsbr} & L_{bscr} \\ L_{csas} & L_{csbs} & L_{cscs} & L_{csar} & L_{csbr} & L_{cscr} \\ L_{aras} & L_{arbs} & L_{arcs} & L_{arar} & L_{arbr} & L_{arcr} \\ L_{bras} & L_{brbs} & L_{brcs} & L_{brar} & L_{brbr} & L_{brcr} \\ L_{cras} & L_{crbs} & L_{crcs} & L_{crar} & L_{crbr} & L_{crcr} \end{bmatrix} \begin{bmatrix} i_{as} \\ i_{bs} \\ i_{cs} \\ i_{ar} \\ i_{br} \\ i_{cr} \end{bmatrix}. \tag{14}$$

Writing in compact form the inductances can be divided into four groups, according to:

$$\begin{bmatrix} \boldsymbol{\lambda}_{abcs} \\ \boldsymbol{\lambda}_{abcr} \end{bmatrix} = \begin{bmatrix} \boldsymbol{L}_s & \boldsymbol{L}_{sr} \\ [\boldsymbol{L}_{sr}]^T & \boldsymbol{L}_r \end{bmatrix} \begin{bmatrix} \boldsymbol{i}_{abcs} \\ \boldsymbol{i}_{abcr} \end{bmatrix} \tag{15}$$

thereby, $\boldsymbol{L}_s$ is the inductance matrix of the stator windings, $\boldsymbol{L}_r$ is the inductance matrix of the rotor windings, and $\boldsymbol{L}_{sr}$ is the mutual-inductance matrix between the stator and rotor windings.

The $\boldsymbol{L}_s$ stator inductance matrix consists of the mutual inductance between the stator windings and the self-inductances of each stator winding. The stator self-inductances consist of the sum of the leakage L_{ls} inductance and L_{ms} magnetizing inductance, according to

$$L_{asas} = L_{bsbs} = L_{cscs} = L_{ls} + L_{ms} \tag{16}$$

where,

$$L_{ms} = \mu_0 N_s^2 \left(\frac{rl}{g}\right)\left(\frac{\pi}{4}\right) \tag{17}$$

wherein, μ_0 is the permeability of air, r is the radius of the air gap, and l is the axial length of the air gap.

The mutual inductances between the two stator windings are all equal and related to the inductance of magnetization as:

$$L_{asbs} = L_{ascs} = L_{bsas} = L_{bscs} = L_{csas} = L_{csbs} = L_{ms}\cos\left(\frac{2\pi}{3}\right) = -\frac{1}{2}L_{ms}. \quad (18)$$

The inductances of the stator windings are given by

$$\boldsymbol{L}_s = \begin{bmatrix} L_{asas} & L_{asbs} & L_{ascs} \\ L_{bsas} & L_{bsbs} & L_{bscs} \\ L_{csas} & L_{csbs} & L_{cscs} \end{bmatrix} = \begin{bmatrix} L_{ls} + L_{ms} & -\frac{L_{ms}}{2} & -\frac{L_{ms}}{2} \\ -\frac{L_{ms}}{2} & L_{ls} + L_{ms} & -\frac{L_{ms}}{2} \\ -\frac{L_{ms}}{2} & -\frac{L_{ms}}{2} & L_{ls} + L_{ms} \end{bmatrix}. \quad (19)$$

For the inductance of the rotor windings $\boldsymbol{L}_r$, the rotor self-inductance L_{arar}, L_{brbr}, L_{crcr} consist of the leakage inductance L_{lr} and the magnetizing inductance L_{mr}, so it has:

$$L_{arar} = L_{brbr} = L_{crcr} = L_{lr} + L_{mr} \quad (20)$$

where,

$$L_{mr} = \mu_0 N_r^2\left(\frac{rl}{g}\right)\left(\frac{\pi}{4}\right) = \left(\frac{N_r}{N_s}\right)^2 L_{ms}. \quad (21)$$

The mutual inductances between the two rotor windings are all equal and related to the inductance of magnetization as:

$$L_{arbr} = L_{arcr} = L_{brar} = L_{brcr} = L_{crar} = L_{crbr} = L_{mr}\cos\left(\frac{2\pi}{3}\right) = -\frac{1}{2}L_{mr} = -\frac{1}{2}\left(\frac{N_r}{N_s}\right)^2 L_{ms} \quad (22)$$

where Nr/Ns is the turns ratio of the stator and rotor windings n, inductances of the rotor windings are given by

$$\boldsymbol{L}_r = \begin{bmatrix} L_{arar} & L_{arbr} & L_{arcr} \\ L_{brar} & L_{brbr} & L_{brcr} \\ L_{crar} & L_{crbr} & L_{crcr} \end{bmatrix} = \begin{bmatrix} L_{ls} + n^2 L_{ms} & -n^2\frac{L_{ms}}{2} & -n^2\frac{L_{ms}}{2} \\ -\frac{L_{ms}}{2} & L_{ls} + n^2 L_{ms} & -n^2\frac{L_{ms}}{2} \\ -n^2\frac{L_{ms}}{2} & -n^2\frac{L_{ms}}{2} & L_{ls} + n^2 L_{ms} \end{bmatrix} \quad (23)$$

The same analysis is performed to inductance between the stator and rotor windings $\boldsymbol{L}_{sr}$, considering the rotating at a speed w_r, the relative position θ_r, which is the displacement angle [37]. Considering the turns ratio between the two windings, we have:

$$L_{asar} = L_{mr}\left(\frac{N_s}{N_r}\right)\cos(\theta_r) = \left(\frac{N_r}{N_s}\right)L_{ms}\cos(\theta_r), \quad \left(\theta_r = \int \omega_r dt\right). \quad (24)$$

So we have the other mutual-inductances

$$L_{asar} = L_{bsbr} = L_{cscr} = \left(\frac{N_r}{N_s}\right)L_{ms}\cos(\theta_r) \quad (25)$$

$$L_{asbr} = L_{bscr} = L_{csar} = \left(\frac{N_r}{N_s}\right)L_{ms}\cos\left(\theta_r + \frac{2\pi}{3}\right) \quad (26)$$

$$L_{ascr} = L_{bsar} = L_{csbr} = \left(\frac{N_r}{N_s}\right)L_{ms}\cos\left(\theta_r - \frac{2\pi}{3}\right). \quad (27)$$

So the matrix $\boldsymbol{L}_{sr}$ is given by:

$$\boldsymbol{L}_{sr} = \begin{bmatrix} L_{asar} & L_{asbr} & L_{ascr} \\ L_{bsar} & L_{bsbr} & L_{bscr} \\ L_{csar} & L_{csbr} & L_{cscr} \end{bmatrix} = nL_{ms}\begin{bmatrix} \cos(\theta_r) & \cos\left(\theta_r + \frac{2\pi}{3}\right) & \cos\left(\theta_r - \frac{2\pi}{3}\right) \\ \cos\left(\theta_r - \frac{2\pi}{3}\right) & \cos(\theta_r) & \cos\left(\theta_r + \frac{2\pi}{3}\right) \\ \cos\left(\theta_r + \frac{2\pi}{3}\right) & \cos\left(\theta_r - \frac{2\pi}{3}\right) & \cos(\theta_r) \end{bmatrix} \quad (28)$$

Finally, from Equations (19), (23), and (28) for the calculation inductance of the stator and rotor windings, and mutual-inductance the full flow connections of the induction motor are obtained.

As an analytical approach for trying to elucidate this problem, the authors consider the hysteresis model suggested by [38] that was applied to transformers. In this case, minor but providential adaptations were added to that model by including the leakage inductance, for its inherent and significant influence, as well as by considering the value of the circuit parameters of the stator and the rotor as being the equivalent and referred to the primary, that is the stator.

The basic principle of hysteresis can be described through two principles; the first is related to the magnetization curve, with the presence of saturation, which is mathematically fitted with any function that adds signal coherence. In the second, the magnetization current will always present the static and dynamic components, described by:

$$i(\lambda,t) = i(\lambda(t)) + k\frac{d\lambda(t)}{dt}, \tag{29}$$

where $\lambda(t)$ is the magnetic flux linkage along the time and k is a constant of proportionality.

Considering that the Equation (29) must be derived for Ohm's and applying Lenz's law to the winding circuit we obtain:

$$V(t) = r\,i(t) + k\frac{d\lambda(t)}{dt}. \tag{30}$$

Based on this, it seems to be clear how effective the influence of the residual magnetism is on the instantaneous value of the current by including the fact that it contributes for not allowing us to affirm that the initial value of this current should be null. Moreover, it shows how the instantaneous voltage has similar influence on the instantaneous values of the current. Nonetheless, the influence of the choice for the preferential or common connection does not seem clear, since any choice does not represent a change in the phasor voltage sequence that would cause the rotor to invert its rotating direction.

At last, due to its key influence, the observed difference in the behavior of the residual magnetism for the common and preferential connections seems to be the most important item to be researched. For its turn, its initial value and behavior also have not electrical or magnetic influences but mechanical, as the percent of the rated load, instantaneous speed and even thermal conditions of the magnetic circuit of the inductions motor. This challenging scenario has made the authors investigate this problem experimentally in view of contributing to addressing this very unclear issue.

In the analysis, the current of 10 starts are evaluated, for which the average and standard deviation are calculated. The standard deviation is given by:

$$std\_Dev = \frac{1}{n-1}\sum_{p=1}^{n}(y_{i,p} - \overline{y_i})^2, \tag{31}$$

wherein, $y_{i,p}$ is the value of the predicted output i in object p and $\overline{y_i}$ is the average of the variable i.

3. Results

To check under what conditions the difference between the values of the switching current to the delta commutation occurs in the common and preferential forms, ten starts were performed for each of these two forms. In each of the starts, the motor was in the resting state without any load on its shaft and was run for only a few minutes. In order to record the transient current, the current probe Tektronix® model A-622, Blumenau, Brazil, was connected to an oscilloscope and applied to the same phase.

First, the model was applied to the common form of delta connection, to serve as a reference. Ten starts were performed and the waveform of the switching current of the delta connection was recorded (see Table 1). The presented results show that, except for two of the occurrences, with 9.4 and 11 A, the amplitude of the switching current is characterized by an average and homogeneous value of about 18 A. However, the most important result is that the value of the amplitude of the switching current is not repetitive, which confirms the statistical nature of the event and the influence of the residual magnetism, as previously mentioned.

Table 1. Amplitude of switching current—Δt = 100 ms.

Start	Amplitude (A) for the Common Form	Amplitude (A) for the Preferential Form
1	18.2	13.0
2	11.4	10.2
3	17.8	12.2
4	18.0	13.0
5	18.0	7.8
6	18.0	9.2
7	17.8	13.0
8	18.2	9.8
9	9.4	14.4
10	11.0	13.6
Average	15.8	11.6
Std Deviation	3.4	2.1

For the sake of illustration, Figure 5 shows a typical waveform of the switching current of the delta connection for the ten different starts performed.

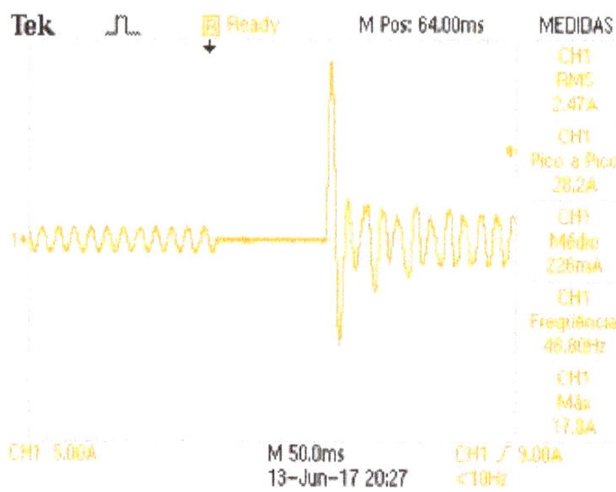

Figure 5. Typical current waveform for the common form of delta connection—Δt = 100 ms. (Vertical scale—5 A/division; Horizontal scale—50 ms per division).

In the sequence, ten starts were performed for the same motor, under the same conditions, except for the delta connection in the preferential form. Table 1 shows the experimentally obtained values for the amplitude of the switching current for each of the ten starts, with the time interval between the star and delta connections (Δt) set to 100 ms.

As expected, for the preferential form of the delta connection, the average value for the amplitude of the switching current is significantly lower than that obtained for the common form, which justifies it being given the designation. In addition, it can be noticed that the variation of values around the average value, or the mean deviation, is also smaller than in the common form.

Thus, for the sake of illustration, Figure 6 shows a typical waveform of the switching current for the preferential form of delta connection, with the same value of Δt, 100 ms.

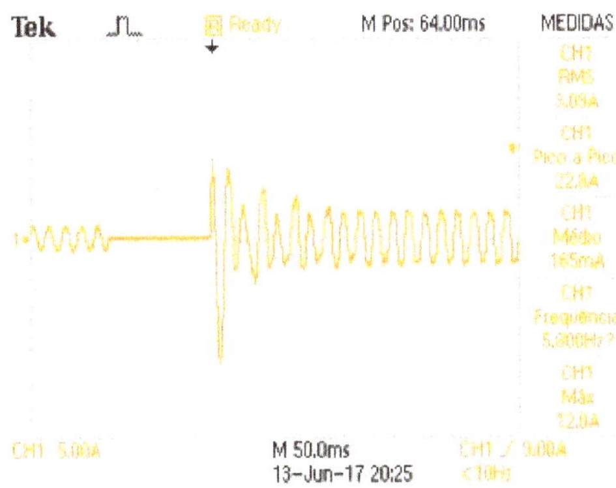

Figure 6. Typical current waveform for the preferential form of delta connection—Δt = 100 ms (Vertical scale—5 A/division; Horizontal scale—50 ms per division).

In this case, the value of the amplitude of the switching current was about 13 A, with negative polarity. In the sequence, ten other starts of the same motor were performed, in the same conditions as before, with the connection in the common and preferential forms, but with a value of Δt set to 50 ms.

The expectation for the results of these tests was that the average value of the amplitude would be significantly smaller than that for the similar conditions of before, since a smaller time interval (Δt) does not allow a significant loss of the rotor speed due to its inertia, and thus, a smaller amount of power is required from the power grid in the form of part of the switching current to compensate for the unavoidable loss of speed (see Table 2).

Table 2. Amplitude of the switching current—Δt = 50 ms.

Start	Amplitude (A) For the Common Form	Amplitude (A) for the Preferential Form
1	10.8	9.6
2	11.4	9.4
3	15.2	6.2
4	11.4	11.2
5	15.2	7.2
6	15.2	7.0
7	15.2	12.0
8	9.2	6.8
9	15.6	11.0
10	11.4	11.6
Average	13.1	9.2
Std Deviation	2.3	2.1

From a comparison with the results obtained with those of this same form of delta connection, but with Δt set in 100 ms, shown in Table 1, it is observed that, on average, the amplitude of the switching current becomes significantly smaller, although the value of the mean deviation is the same, which confirms the expectations. On the other hand, it is interesting to note that, on average, the amplitude of the switching current for the common form of delta connection with Δt = 50 ms is still significantly higher than that for the preferential form with a higher value of Δt, 100 ms.

A typical waveform of the switching current, in the function of time, is shown in Figure 7 for the common form and Δt set to 50 ms.

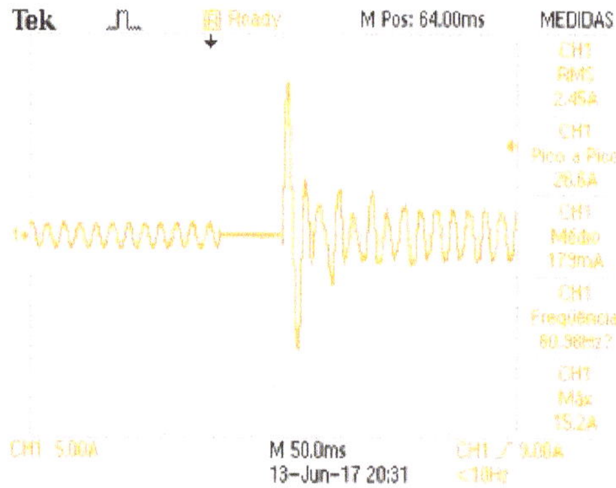

Figure 7. Typical current for the common form of delta connection—Δt = 50 ms. (Vertical scale—5 A/division; Horizontal scale—50 ms per division).

For the tests with the delta connection in the preferential form and common form, Δt with a value of 50 ms, the results of the values of the amplitude of the switching current are shown in Table 2.

From the results of Table 2 and a comparison with all the previous results, it is noted that the reduction of the value of Δt causes a significant reduction in the amplitude of the switching current, regardless of the form of connection, as expected. On the other hand, the results confirm that, as also expected, for the same form of delta connection the reduction of Δt tends to decrease the amplitude of the switching current.

In Figure 8 the waveform of the switching current, in the function of time, is shown in one of the ten starts.

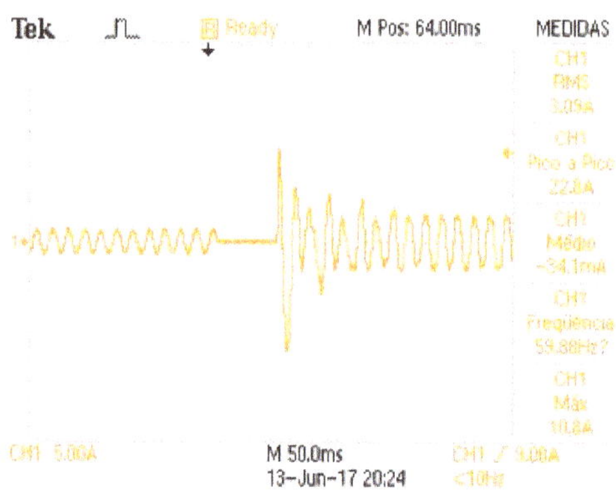

Figure 8. Typical current waveform for the preferential form of delta connection—Δt = 50 ms (Vertical scale—5 A/division; Horizontal scale—50 ms per division).

In this case, the current peak also has negative polarity.

4. Conclusions

The results presented in this work show that for the widely spread option of the star-delta start of induction motors, for which there are two possible forms of the delta connection of the stator winding, one has a switching current with a lower average value of amplitude and is, therefore, considered the preferential form. The other form, on average, has a greater value of the amplitude of the switching current and is, curiously, the most commonly indicated and recommended form.

Thus, the choice between one of these forms has important consequences, since the amplitude of the switching current is one of the most recognized disadvantages of this starting option due to the need for careful analysis and setting of the relaying system for the motor's protection.

This situation becomes even more critical since the star-delta start shall remain for a long time as the very first choice for drive motor systems despite the undeniable growing competition offered by the other start options, based on power electronics components. Thus, the methodology of comparison of the options available for the star-delta start proposed and implemented here, based on an adequately developed arrangement of tests for an induction motor of common use, running idle, experimentally indicates that the influence of the connection effectively exists, which is significant and overlaps with the effect of the loss of speed during the switching time interval, along with the inertia of the motor shaft rules.

The results also confirm the influence of the residual magnetism of the magnetic circuit of the motor, which explains why the peak of the switching current may have a statistical nature, by including the negative polarity. The evaluation in different phases did not result in a variation in the results, so the results of the three-phase system were not presented, since the results are equivalent for each phase.

Lastly, it is important to reinforce that, although this analyzed issue is not so widely present in the literature, the obtained results favor the deepening in studies looking for economically feasible ways to mitigate this switching current, even for the case of the preferential form. This paper shows that there must be an evaluation of the preferred connection depending on the conditions in which the motor is used; using a standard form in all conditions can cause greater losses. For this reason, the evaluation of the motor connection type is a matter that must be checked in order to improve the efficiency of electric motors. Other time intervals can be evaluated to check their influence on the shape of the star-delta starting.

Author Contributions: Writing—original draft preparation, J.A.I.; conceptualization, formal analysis, S.H.L.C. and C.A.C.V.; methodology, S.F.S.; writing—review and editing, V.R.Q.L. and R.G.O.; writing—review and editing, A.N.; supervision, K.-C.Y. All authors have read and agreed to the published version of the manuscript.

Funding: This work was supported by Supported by project PLATAFORMA DE VEHÍCULOS DE TRANSPORTE DE MATERIALES Y SEGUIMIENTO AUTÓNOMO—TARGET. 463AC03, SA063G19. Project co-financed with Junta Castilla y León, Consejería de Educación and FEDER funds.

Institutional Review Board Statement: Not applicable.

Informed Consent Statement: Not applicable.

Acknowledgments: Al Proyeto: Uso de algoritmos y protocolos de comunicación en dispositivos con énfasis en la privacidad de los datos, and Laboratório de Telecomunicações de Portugal IT—Branch Universidade da Beira Interior, Covilhã.

Conflicts of Interest: The authors declare no conflict of interest.

References

1. Udovichenko, A.V. New Energy Saving Multizone Alternating-Voltage Soft Starters of Induction Machines. In Proceedings of the International Conference and Seminar on Micro/Nanotechnologies and Electron Devices Proceedings, Erlagol, Russia, 30 June–4 July 2011; pp. 415–419. [CrossRef]
2. Mallick, T.C.; Dhar, S.; Khan, J. Artificial Neural Network Based Soft-Starter for Induction Motor. In Proceedings of the 2nd International Conference on Electrical Information and Communication Technologies (EICT), Khulna, Bangladesh, 10–12 December 2015; pp. 228–233. [CrossRef]
3. Cistelecan, M.V.; Ferreira, F.J.T.E.; Popescu, M. Adjustable Flux Three-Phase AC Machines with Combined Multiple-Step Star-Delta Winding Connections. *IEEE Trans. Energy Convers.* **2010**, *25*, 348–355. [CrossRef]
4. Mircevski, S.A.; Andonov, Z. An Approach to Laboratory Training in Electric Drives. *IFAC Proc. Vol.* **1997**, *30*, 211–215. [CrossRef]
5. Rathore, T.S. A Systematic Method for Finding the Input Impedance of Two-terminal Networks. *IETE J. Educ.* **2017**, *58*, 83–89. [CrossRef]

6. Yan, B.; Wang, X.; Yang, Y. Starting Performance Improvement of Line-Start Permanent-Magnet Synchronous Motor Using Composite Solid Rotor. *IEEE Trans. Magn.* **2018**, *54*, 1–4. [CrossRef]
7. Rabbi, S.F.; Zhou, P.; Rahman, M.A. Design and Performance Analysis of a Self-Start Radial Flux-Hysteresis Interior Permanent Magnet Motor. *IEEE Trans. Magn.* **2017**, *53*, 1–4. [CrossRef]
8. Hu, Y.; Chen, B.; Xiao, Y.; Shi, J.; Li, L. Study on the Influence of Design and Optimization of Rotor Bars on Parameters of a Line-Start Synchronous Reluctance Motor. *IEEE Trans. Ind. Appl.* **2020**, *56*, 1368–1376. [CrossRef]
9. Yan, B.; Wang, X.; Yang, Y. Comparative Parameters Investigation of Composite Solid Rotor Applied to Line-Start Permanent-Magnet Synchronous Motors. *IEEE Trans. Magn.* **2018**, *54*, 1–5. [CrossRef]
10. Albert, E.; Clayton, D. A Mathematical Development of the Theory of the Magnetomotive Force of Windings. *J. Inst. Electr. Eng.* **1923**, *61*, 749–787.
11. Ferreira, F.J.T.E. On the Star, Delta and Star-Delta Stator Winding Connections Tolerance to Voltage Unbalance. In Proceedings of the IEEE International Electric Machines & Drives Conference (IEMDC), Coeur d'Alene, ID, USA, 10–13 May 2015; pp. 1888–1894. [CrossRef]
12. Goh, H.H.; Looi, M.S.; Kok, B.C. Comparison between Direct-On-Line, Star-Delta and Auto-Transformer Induction Motor Starting Method in Terms of Power Quality. In Proceedings of the International Multiconference of Engineers and Computer Scientists (IMECS), Hong Kong, China, 18–20 March 2009; pp. 1–6.
13. Misir, O.; Raziee, S.M.; Hammouche, N.; Klaus, C.; Kluge, R.; Ponick, B. Prediction of Losses and Efficiency for Three-Phase Induction Machines Equipped with Combined Star–Delta Windings. *IEEE Trans. Ind. Appl.* **2017**, *53*, 3553–3587. [CrossRef]
14. Abdel-Hamid, M.N. Improved Starting Process Alternative to Conventional Star-Delta Switching of A-C Meters. *IEEE Trans. Appl. Ind.* **1963**, *82*, 52–60. [CrossRef]
15. Vansompel, H.; Sergeant, P.; Dupre, L.; Bossche, A. A Combined Wye-Delta Connection to Increase the Performance of Axial-Flux PM Machines with Concentrated Windings. *IEEE Trans. Energy Convers.* **2012**, *27*, 403–410. [CrossRef]
16. Vercelli, L. Rechts-und Linkslauf der Motoren bei Y-D-Anlauf. *Elektron. Ch* **1978**, *53*, 1–10.
17. Koņuhova, M.; Ketners, K.; Ketnere, E.; Klujevska, S. Research of the Effect of the Rotor Constant upon the Attenuation Characteristic if the Induction Motor Residual Voltage under the Switching Regime. In Proceedings of the Problems of Present-Day Electrotechnics-2010 (PPE-2010): Conference Proceedings, Kiev, Ukraine, 1–3 June 2010; National Academy of Sciences of Ukraine: Kiev, Ukraine, 2010; pp. 152–155.
18. Ferreira, F.J.T.E.; Ge, B.; Quispe, E.C.; de Almeida, A.T. Star-and Delta-Connected Windings Tolerance to Voltage Unbalance in Induction Motors. In Proceedings of the International Conference on Electrical Machines (ICEM), Berlin, Germany, 2–5 September 2014; pp. 2045–2054. [CrossRef]
19. Siemens, A.G. Grundlagen der Niederspannungs-Schalttechnik, Seite 21 B-1 Zum 21 B-4. 2008. Available online: https://cache.industry.siemens.com/dl/files/099/34973099/att70195/v1/Grundlagen_der_Niederspannungs_Schalttechnik.pdf (accessed on 6 December 2020).
20. Bradley, A.; Low—Voltage Switchgear and Controlgear. Technical Document. 3–10, Rockwell Automation. 2009. Available online: http://literature.rockwellautomation.com/idc/groups/literature/documents/rm/lvsam-rm001-en-p.pdf (accessed on 6 December 2020).
21. Moros, O.; Gerling, D. Geometrical and Electrical Optimization of Stator Slots in Electrical Machines with Combined Wye-Delta Winding. In Proceedings of the 2014 International Conference on Electrical Machines (ICEM), Berlin, Germany, 2–5 September 2014; pp. 2026–2030. [CrossRef]
22. Stefenon, S.F.; Seman, L.O.; Schutel Furtado Neto, C.; Nied, A.; Seganfredo, D.M.; da Luz, F.G.; Sabino, P.H.; Torreblanca González, J.; Quietinho Leithardt, V.R. Electric Field Evaluation Using the Finite Element Method and Proxy Models for the Design of Stator Slots in a Permanent Magnet Synchronous Motor. *Electronics* **2020**, *9*, 1975. [CrossRef]
23. Kasburg, C.; Stefenon, S.F. Deep Learning for Photovoltaic Generation Forecast in Active Solar Trackers. *IEEE Lat. Am. Trans.* **2019**, *17*, 2013–2019. [CrossRef]
24. Stefenon, S.F.; Kasburg, C.; Nied, A.; Klaar, A.C.R.; Ferreira, F.C.S.; Branco, N.W. Hybrid Deep Learning for Power Generation Forecasting in Active Solar Trackers. *IET Gener. Transm. Distrib.* **2020**, *14*, 5667–5674. [CrossRef]
25. Ninno Muniz, R.; Stefenon, S.F.; Gouvêa Buratto, W.; Nied, A.; Meyer, L.H.; Finardi, E.C.; Marino Kühl, R.; de Sá, J.A.S.; da Rocha, B.R.P. Tools for Measuring Energy Sustainability: A Comparative Review. *Energies* **2020**, *13*, 2366. [CrossRef]
26. Stefenon, F.S.; Ademir, N. FEM Applied to Evaluation of the Influence of Electric Field on Design of the Stator Slots in PMSM. *IEEE Lat. Am. Trans.* **2019**, *17*, 590–596. [CrossRef]
27. Orosz, T.; Rassõlkin, A.; Kallaste, A.; Arsénio, P.; Pánek, D.; Kaska, J.; Karban, P. Robust Design Optimization and Emerging Technologies for Electrical Machines: Challenges and Open Problems. *Appl. Sci.* **2020**, *10*, 6653. [CrossRef]
28. Orosz, T. Evolution and Modern Approaches of the Power Transformer Cost Optimization Methods. *Period. Polytech. Electr. Eng. Comput. Sci.* **2019**, *63*, 37–50. [CrossRef]
29. Orosz, T.; Sleisz, Á.; Tamus, Z.Á. Metaheuristic Optimization Preliminary Design Process of Core-Form Autotransformers. *IEEE Trans. Magn.* **2016**, *52*, 1–10. [CrossRef]
30. Pánek, D.; Orosz, T.; Karban, P. Artap: Robust Design Optimization Framework for Engineering Applications. *arXiv* **2019**, arXiv:1912.11550.

31. Stefenon, S.F.; Kasburg, C.; Freire, R.Z.; Silva Ferreira, F.C.; Bertol, D.W.; Nied, A. Photovoltaic Power Forecasting Using Wavelet Neuro-Fuzzy for Active Solar Trackers. *J. Intell. Fuzzy Syst.* **2021**, *40*, 1083–1096. [CrossRef]
32. Rolim, C.O.; Schubert, F.; Rossetto, A.G.; Leithardt, V.R.; Geyer, C.F.; Westphall, C. Comparison of a Multi Output Adaptative Neuro-Fuzzy Inference System (MANFIS) and Multi Layer Perceptron (MLP) in Cloud Computing Provisioning. In Proceedings of the 29th Brazilian Symposium on Computer Networks and Distributed Systems, Campo Grande, Brazil, 30 May–3 June 2011.
33. Ferreira, F.J.T.E.; de Almeida, A.T. Novel Multiflux Level, Three-Phase, Squirrel-Cage Induction Motor for Efficiency and Power Factor Maximization. *IEEE Trans. Energy Convers.* **2008**, *23*, 23–109. [CrossRef]
34. Furqani, J.; Kawa, M.; Kiyota, K.; Chiba, A. Current Waveform for Noise Reduction of a Switched Reluctance Motor Under Magnetically Saturated Condition. *IEEE Trans. Ind. Appl.* **2018**, *54*, 213–222. [CrossRef]
35. Misir, O.; Ponick, B. Analysis of Three-Phase Induction Machines with Combined Star-Delta Windings. In Proceedings of the 2014 IEEE 23rd International Symposium on Industrial Electronics (ISIE), Istanbul, Turkey, 1–4 June 2014; pp. 756–761. [CrossRef]
36. Misir, O.; Raziee, S.M.; Hammouche, N.; Klaus, C.; Kluge, R.; Ponick, B. Calculation Method of Three-Phase Induction Machines Equipped with Combined Star-Delta Windings. In Proceedings of the 2016 XXII International Conference on Electrical Machines (ICEM), Lausanne, Switzerland, 4–7 September 2016; pp. 166–172. [CrossRef]
37. Sang-Hoon, K. Modeling of Alternating Current Motors and Reference Frame Theory. In *Electric Motor Control*; Elsevier: Amsterdam, The Netherlands, 2017. [CrossRef]
38. Cabral, S.H.L.; Matos, J. Simplified Modelling of Hysteresis for Power System Transformers Studies. In *International Conference on Power Systems Transients: Conference Proceedings*; IPST 2001 (Federal University of Rio de Janeiro): Rio de Janeiro, Brazil, 2001; pp. 61–64.

energies

Article

Implementation of an FPGA-Based Current Control and SVPWM ASIC with Asymmetric Five-Segment Switching Scheme for AC Motor Drives

Ming-Fa Tsai *, Chung-Shi Tseng and Po-Jen Cheng

Department of Electrical Engineering, Minghsin University of Science and Technology, No. 1, Xinxing Rd., Xinfeng, Hsinchu 30401, Taiwan; cstseng@must.edu.tw (C.-S.T.); pjcheng3255@gmail.com (P.-J.C.)
* Correspondence: mftsai@must.edu.tw; Tel.: +886-3-5593142 (ext. 3070); Fax: +886-3-5573895

Abstract: This paper presents the design and implementation of an application-specific integrated circuit (ASIC) for a discrete-time current control and space-vector pulse-width modulation (SVPWM) with asymmetric five-segment switching scheme for AC motor drives. As compared to a conventional three-phase symmetric seven-segment switching SVPWM scheme, the proposed method involves five-segment two-phase switching in each switching period, so the inverter switching times and power loss can be reduced by 33%. In addition, the produced PWM signal is asymmetric with respect to the center-symmetric triangular carrier wave, and the voltage command signal from the discrete-time current control output can be given in each half period of the PWM switching time interval, hence increasing the system bandwidth and allowing the motor drive system with better dynamic response. For the verification of the proposed SVPWM modulation scheme, the current control function in the stationary reference frame is also included in the design of the ASIC. The design is firstly verified by using PSIM simulation tool. Then, a DE0-nano field programmable gate array (FPGA) control board is employed to drive a 300W permanent-magnet synchronous motor (PMSM) for the experimental verification of the ASIC.

Keywords: SVPWM ASIC; asymmetric five-segment switching; AC motor drives; current control; FPGA control

Citation: Tsai, M.-F.; Tseng, C.-S.; Cheng, P.-J. Implementation of an FPGA-Based Current Control and SVPWM ASIC with Asymmetric Five-Segment Switching Scheme for AC Motor Drives. *Energies* **2021**, *14*, 1462. https://doi.org/10.3390/en14051462

Academic Editor: Mario Marchesoni

Received: 1 February 2021
Accepted: 27 February 2021
Published: 7 March 2021

1. Introduction

The AC motor drives, including induction motors (IM), permanent-magnet synchronous motor (PMSM) drives, synchronous reluctance motors (SynRM), and others, have been very popular, being applied to electric vehicles, railway traction engines, and industrial applications such as CNC tools and robots, because they have higher power density and efficiency as compared to DC motor drives [1–4]. Since the AC motor is a time-varying, multi-variable, and nonlinear control system with very complicated dynamic characteristics, it can be reduced to a simpler linear control system by using the field-oriented vector control method [5–11]. In vector-controlled AC motor drives, various pulse-width modulation (PWM) techniques exist to determine the inverter switch-on and switch-off instants from the control output modulating signals. Popular examples include sinusoidal PWM (SPWM), hysteresis PWM, and space-vector PWM (SVPWM). Among them, the SPWM method generates the inverter switching signals by comparing the modulating signals with a common triangular carrier wave, in which the intersection points determine the switching points of the inverter power devices [12,13]. The hysteresis PWM method controls the inverter output currents to track the reference inputs within the hysteresis band, but with various switching frequencies [14,15]. The SVPWM method is based on the concept of voltage vector space, which can be divided into six sectors, and calculates the switch dwelling or firing time in each sector to generate the PWM signals [16–40]. The relationship between SVPWM and SPWM was presented in [18], which indicated that the SVPWM can be viewed as a particular form of asymmetric regular-sampled PWM. In [24], the relationship between SVPWM and three-phase SPWM was also analyzed and the implementation of them in a closed-loop feedback converter was discussed as well. The SVPWM technique is the preferred approach in most applications due to the higher voltage utilization and lower harmonic distortion (THD) than the other two methods, and hence it can extend the linear range of the vector-controlled AC motor drives. Furthermore, sensorless and direct torque control (DTC) cooperating with an SVPWM scheme for a flux-modulated permanent-magnet wheel motor and induction motor has been described owing to several advantages, such as low torque/flux ripples in motor drive and reduced direct axis current, over the conventional hysteresis direct torque control method when the motor is operated at a light or a sudden increased load [34,37–40]. The SVPWM scheme has also been extensively applied to the control of a five-phase permanent-magnet motor and multilevel inverters [35,36].

Different SVPWM techniques can be used for three-phase inverters according to the choice of the null vector on the voltage space and the number of samplings during a PWM switching period. They can be divided into seven-segment, five-segment, and three-segment techniques for the switching sequences and the choice of null vector, and classified into symmetric and asymmetric techniques according to the number of samplings during a PWM switching period as well. For the symmetric technique, the sampling period is equal to the switching period, while the sampling period is one half of the switching period for the asymmetric technique. So, the asymmetric methods have only 50% switching action in each switching period as compared with symmetric methods, and hence the sampling frequency of the current control can be doubled at a certain switching frequency. In [28], three SVPWM strategies, including symmetric seven-segment technique, symmetric five-segment technique, and asymmetric three-segment technique, were compared through simulation analysis. The switching loss of the three-segment scheme is the lowest and the seven-segment scheme performs better in terms of the THD of the output line voltage.

Figure 1 shows the rotor flux-oriented control (RFOC) structure of a PMSM motor drive with a SVPWM modulation scheme, in which an asymmetric five-segment switching is proposed rather than the conventional symmetric seven-segment switching so as to double the sampling frequency and reduce the switching loss. This control structure, from the inner to the outer loops, includes the SVPWM modulation, current control loop, torque and flux control loops, and speed control loop. Conventionally, the execution of all the control tasks is performed by using a high-performance microprocessor or dual digital signal processor (DSP) [21,41,42]. It may take a lot of computation time for the microprocessor or DSP because all the computation of the tasks is very complicated. If the computation of the current control and the SVPWM tasks can be executed by a field programmable gate array (FPGA) device, the computation load can be greatly reduced [16,17,22,25,30]. In such cases, the microprocessor or DSP can have enough time to process higher level tasks, such as position control, motion control, adaptive, fuzzy, or neural-network learning and intelligent control [43]. In [17], an eight-bit SVPWM control IC was realized with a symmetrical five-segment switching scheme, but without current control function included in the chip. In [22], a microcoded machine with a 16-bit computational ALU unit and control sequencer was designed in an FPGA for the execution of the current control and SVPWM modulation of an induction motor with symmetrical seven-segment switching scheme. However, the microcoded machine structure is very complicated.

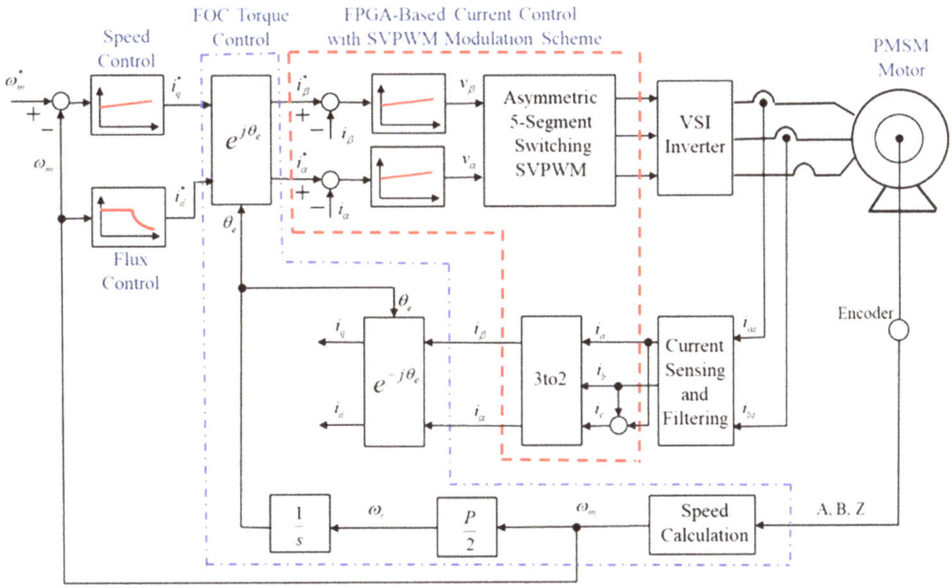

Figure 1. The rotor flux-oriented control (RFOC) structure of a permanent-magnet synchronous motor (PMSM) AC motor drive.

This paper describes the design and implementation of an application-specific integrated circuit (ASIC) for a discrete-time current control and space-vector pulse-width modulation (SVPWM) with asymmetric five-segment switching scheme AC motor drives. It can not only reduce the computation load of a microprocessor or DSP, but also reduce the power transistor switching loss of the inverter. Traditionally, the current control loop is performed in the synchronously *d-q* rotating reference frame [3–7]. However, as indicated in Figure 1, the presented current control loop is performed in the stationary reference frame rather than the synchronously rotating reference frame in order to simplify the computation complexity in the FPGA device. Furthermore, because the current reference commands can be updated two times during one switching period, the current control system can also increase the sampling frequency two times so as to increase the bandwidth.

The algorithm of the proposed current control and asymmetric five-segment switching SVPWM scheme was firstly verified by using PSIM simulation tool as applied to the current control of a PMSM AC motor. The proposed SVPWM scheme and the current controller function were implemented on a DE0-nano control board with Altera Cyclone IV E FPGA device for the experimental verification.

This paper is organized as follows. The principle of the asymmetric five-segment switching SVPWM modulation scheme is described in Section 2. The simulation verification in PSIM simulation tool is analyzed in Section 3 and the FPGA implementation and the experimental results are presented in Section 4. A discussion is then given in Section 5. Finally, the conclusion is given in Section 6.

2. The Principle of the Asymmetrical Five-Segment Switching SVPWM Modulation

Firstly, as shown in Figure 2, for a three-phase PWM inverter of Y-connected AC motor with neutral point n, the three-phase stator voltages with respect to the inverter ground can be written as

$$v_{a0} = v_{an} + v_{n0} \tag{1}$$

$$v_{b0} = v_{bn} + v_{n0} \tag{2}$$

$$v_{c0} = v_{cn} + v_{n0} \tag{3}$$

where v_{an}, v_{bn}, and v_{cn} are the three-phase voltages with respect to the neutral point of the motor, and v_{n0} is the neutral-point voltage with respect to the inverter ground.

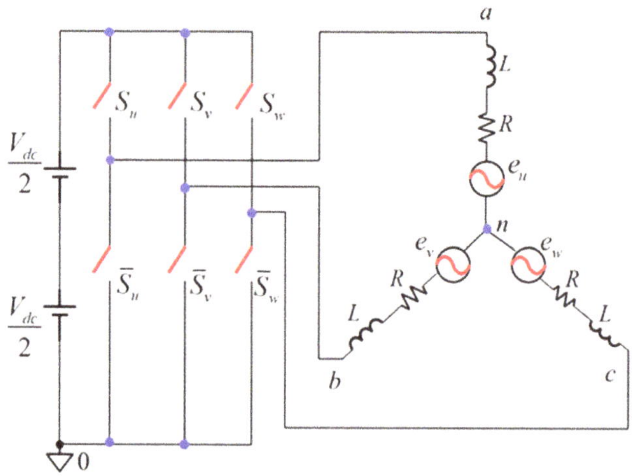

Figure 2. The pulse-width modulation (PWM) inverter circuit for a Y-connected AC motor.

For balanced three-phase supply, it can be written as

$$v_{an} + v_{bn} + v_{cn} = 0 \tag{4}$$

Thus, by adding (1)–(3) with the condition of (4), the neutral-point voltage with respect to the inverter ground, v_{n0}, can be written as

$$v_{n0} = \frac{1}{3}(v_{a0} + v_{b0} + v_{c0}) \tag{5}$$

Substituting (5) into (1)–(3) yields

$$\begin{bmatrix} v_{an} \\ v_{bn} \\ v_{cn} \end{bmatrix} = \begin{bmatrix} \frac{2}{3} & -\frac{1}{3} & -\frac{1}{3} \\ -\frac{1}{3} & \frac{2}{3} & -\frac{1}{3} \\ -\frac{1}{3} & -\frac{1}{3} & \frac{2}{3} \end{bmatrix} \begin{bmatrix} v_{a0} \\ v_{b0} \\ v_{c0} \end{bmatrix} \tag{6}$$

Secondly, because the three-phase stator voltages are dependent upon (4), the inputs of the SVPWM circuit can be reduced from the three-phase variables into α-β two-axes components with the Clarke transformation given by (7).

$$\begin{bmatrix} v_{\alpha} \\ v_{\beta} \end{bmatrix} = \begin{bmatrix} 1 & 0 & 0 \\ 0 & \frac{1}{\sqrt{3}} & \frac{-1}{\sqrt{3}} \end{bmatrix} \begin{bmatrix} v_{an} \\ v_{bn} \\ v_{cn} \end{bmatrix} \tag{7}$$

where v_{α} and v_{β} are the α-β axes component voltages in the stationary reference frame. Substituting (6) into (7) yields

$$\begin{bmatrix} v_{\alpha} \\ v_{\beta} \end{bmatrix} = \begin{bmatrix} \frac{2}{3} & -\frac{1}{3} & -\frac{1}{3} \\ 0 & \frac{1}{\sqrt{3}} & -\frac{1}{\sqrt{3}} \end{bmatrix} \begin{bmatrix} v_{a0} \\ v_{b0} \\ v_{c0} \end{bmatrix} \tag{8}$$

Thus, using (8), one can get the eight output voltage vectors, $V_0 - V_7$, corresponding to the eight switching states of the inverter, as shown in Table 1. The resulted space vector diagram is shown in Figure 3, which is divided into six sectors from Sector I to Sector VI.

As can be seen, among the eight voltage vectors, V_0 and V_7 are the null vectors and $V_1 - V_6$ are located on the six corners of the hexagonal diagram.

Table 1. The eight space voltage vectors.

Inverter Switch Status (S_u, S_v, S_w)	**Three-Phase Stator Voltages with Respect to Ground** $[v_{a0}, v_{b0}, v_{c0}]$	**Space Vectors** $[v_\alpha, v_\beta]$
(1, 0, 0)	$[V_{dc}, 0, 0]$	$V_1 = \left[\frac{2V_{dc}}{3}, 0\right]$
(0, 1, 0)	$[0, V_{dc}, 0]$	$V_3 = \left[\frac{-V_{dc}}{3}, \frac{V_{dc}}{\sqrt{3}}\right]$
(1, 1, 0)	$[V_{dc}, V_{dc}, 0]$	$V_2 = \left[\frac{V_{dc}}{3}, \frac{V_{dc}}{\sqrt{3}}\right]$
(0, 0, 1)	$[0, 0, V_{dc}]$	$V_5 = \left[\frac{-V_{dc}}{3}, \frac{-V_{dc}}{\sqrt{3}}\right]$
(1, 0, 1)	$[V_{dc}, 0, V_{dc}]$	$V_6 = \left[\frac{V_{dc}}{3}, \frac{-V_{dc}}{\sqrt{3}}\right]$
(0, 1, 1)	$[0, V_{dc}, V_{dc}]$	$V_4 = \left[\frac{-2V_{dc}}{3}, 0\right]$
(1, 1, 1)	$[V_{dc}, V_{dc}, V_{dc}]$	$V_7 = [0, 0]$
(0, 0, 0)	$[0, 0, 0]$	$V_0 = [0, 0]$

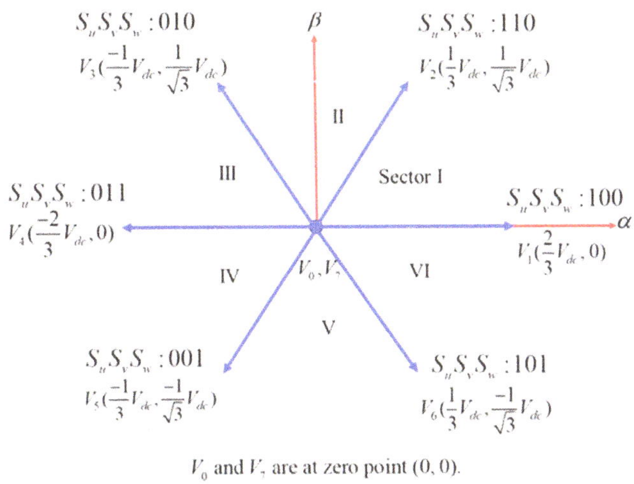

Figure 3. Inverter's eight output voltage space vectors.

The switching waveform generations of the inverter for driving an AC motor can be accomplished by rotating a reference voltage vector around the vector space. Any instance of the reference voltage vector can be produced by the two nearest adjacent vectors and a null vector in an arbitrary sector. For example, as shown in Figure 4, for providing the reference voltage vector V_s in Sector I to the motor, the voltage-second balance equation in this sector determines the time length of the two adjacent active inverter states in the following:

$$V_s T = V_A T_1 + V_B T_2 \tag{9}$$

where T_1 and T_2 are the dwelling time length for V_1 and V_2, respectively, and T is the sampling period, which is one half of the inverter switching period. In order to get the solution of T_1 and T_2 in (9), it follows that

$$\begin{bmatrix} v_\alpha \\ v_\beta \end{bmatrix} T = \frac{2}{3} V_{dc} \left(T_1 \begin{bmatrix} 1 \\ 0 \end{bmatrix} + T_2 \begin{bmatrix} \frac{1}{2} \\ \frac{\sqrt{3}}{2} \end{bmatrix} \right) \tag{10}$$

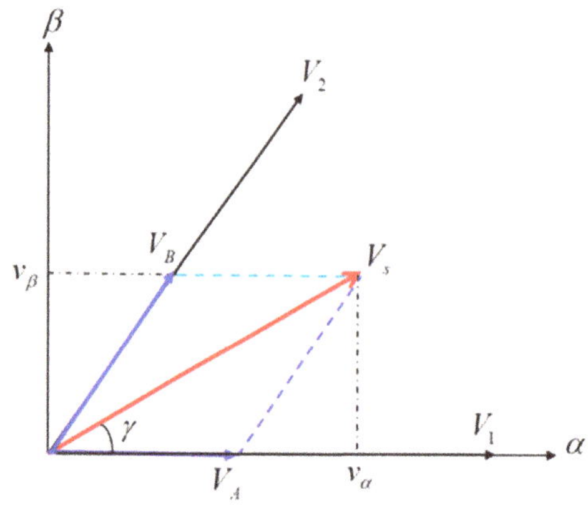

Figure 4. Determination of the switching time in Sector I.

Thus, T_1 and T_2 can be, respectively, solved as

$$T_1 = \frac{\sqrt{3}T}{V_{dc}}(\frac{\sqrt{3}}{2}v_\alpha - \frac{1}{2}v_\beta) \tag{11}$$

and

$$T_2 = \frac{\sqrt{3}T}{V_{dc}}v_\beta \tag{12}$$

Let

$$\begin{bmatrix} v_\alpha \\ v_\beta \end{bmatrix} = V_s \begin{bmatrix} \cos\gamma \\ \sin\gamma \end{bmatrix} \tag{13}$$

then (11) and (12) can be rewritten as the following equations [18]:

$$T_1 = Ta\frac{2}{\sqrt{3}}\sin(\frac{\pi}{3} - \gamma) \tag{14}$$

$$T_2 = Ta\frac{2}{\sqrt{3}}\sin(\gamma) \tag{15}$$

where

$$a = \frac{V_s}{\frac{2}{3}V_{dc}} \tag{16}$$

Similarly, the switching time interval in the other sectors can be derived. Table 2 summaries the results, in which the condition for sector number selection is also illustrated.

Table 2. Dwelling-time interval calculation in each sector.

Sector	Sector Selection	T_1	T_2
I	$v_\alpha \geq 0, 0 \leq v_\beta < \sqrt{3}v_\alpha$	$\frac{\sqrt{3}T}{V_{dc}}(\frac{\sqrt{3}}{2}v_\alpha - \frac{1}{2}v_\beta)$	$\frac{\sqrt{3}T}{V_{dc}}v_\beta$
II	$v_\beta \geq 0, v_\beta \geq \sqrt{3}\lvert v_\alpha \rvert$	$\frac{\sqrt{3}T}{V_{dc}}(\frac{-\sqrt{3}}{2}v_\alpha + \frac{1}{2}v_\beta)$	$\frac{\sqrt{3}T}{V_{dc}}(\frac{\sqrt{3}}{2}v_\alpha + \frac{1}{2}v_\beta)$
III	$v_\alpha \leq 0, 0 \leq v_\beta < -\sqrt{3}v_\alpha$	$\frac{\sqrt{3}T}{V_{dc}}v_\beta$	$\frac{\sqrt{3}T}{V_{dc}}(\frac{-\sqrt{3}}{2}v_\alpha - \frac{1}{2}v_\beta)$
IV	$v_\alpha \leq 0, \sqrt{3}v_\alpha \leq v_\beta < 0$	$\frac{-\sqrt{3}T}{V_{dc}}v_\beta$	$\frac{\sqrt{3}T}{V_{dc}}(\frac{-\sqrt{3}}{2}v_\alpha + \frac{1}{2}v_\beta)$
V	$v_\beta \leq 0, v_\beta \leq -\sqrt{3}\lvert v_\alpha \rvert$	$\frac{\sqrt{3}T}{V_{dc}}(\frac{-\sqrt{3}}{2}v_\alpha - \frac{1}{2}v_\beta)$	$\frac{\sqrt{3}T}{V_{dc}}(\frac{\sqrt{3}}{2}v_\alpha - \frac{1}{2}v_\beta)$
VI	$v_\alpha \geq 0, -\sqrt{3}v_\alpha \leq v_\beta < 0$	$\frac{\sqrt{3}T}{V_{dc}}(\frac{\sqrt{3}}{2}v_\alpha + \frac{1}{2}v_\beta)$	$\frac{-\sqrt{3}T}{V_{dc}}v_\beta$

For minimizing the number of switchings in a switching period, the asymmetrically alternating-reversing pulse sequence with five-segment ($V_0 - V_1 - V_2 - V_1 - V_0$) switching technique is employed without using the null vector V_7, as shown in Figure 5a. The pulse patterns for two consecutive sampling intervals can be configured by beginning with the null vector V_0 on the kT sampling time and also by ending with the same null vector V_0 on the $(k + 1)T$ sampling time, where $k = 1, 3, 5, \cdots$. The time length T_0 on the kT and $(k + 1)T$ sampling time intervals can be different, either for T_1 or T_2. Therefore, the PWM pulses are asymmetric with respect to the center point of the switching period. The technique benefits from one of the inverter legs not switching during a full switching period, only one inverter leg switching at a time, and only two commutations per sampling period. Thus, the number of switchings in a switching period is four and is less than the conventional symmetric seven-segment SVPWM method, in which the number of switchings in a switching period is six.

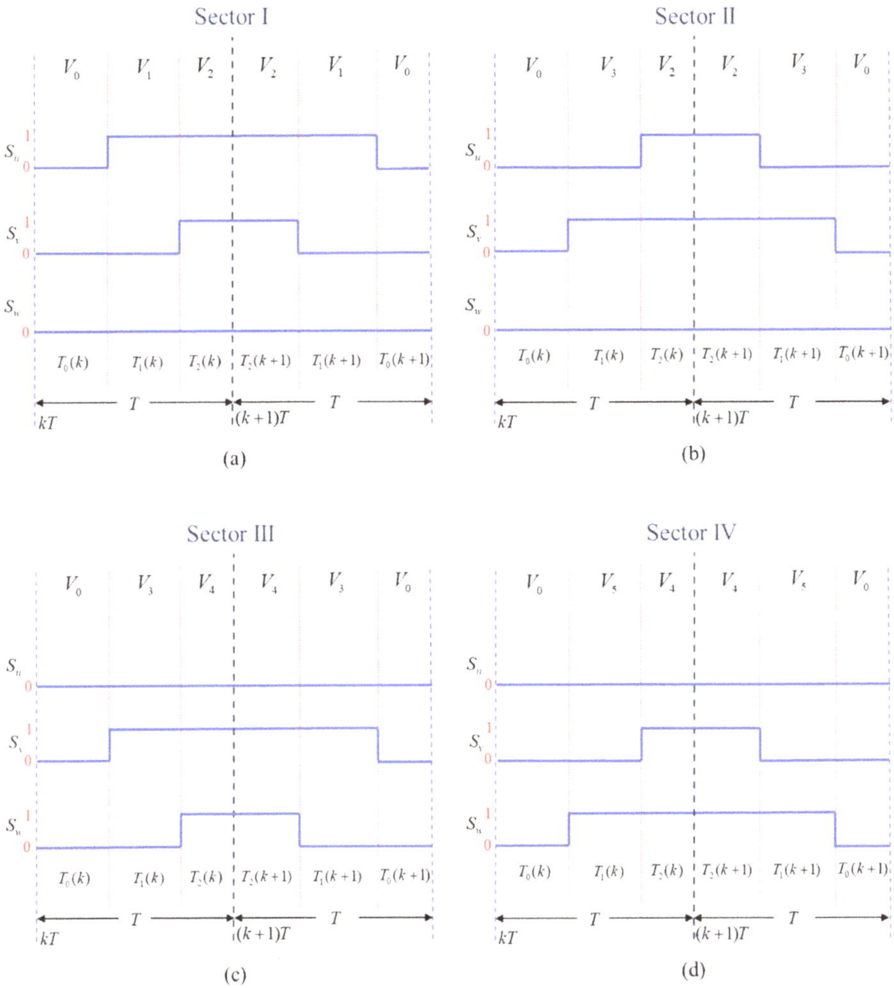

Figure 5. *Cont.*

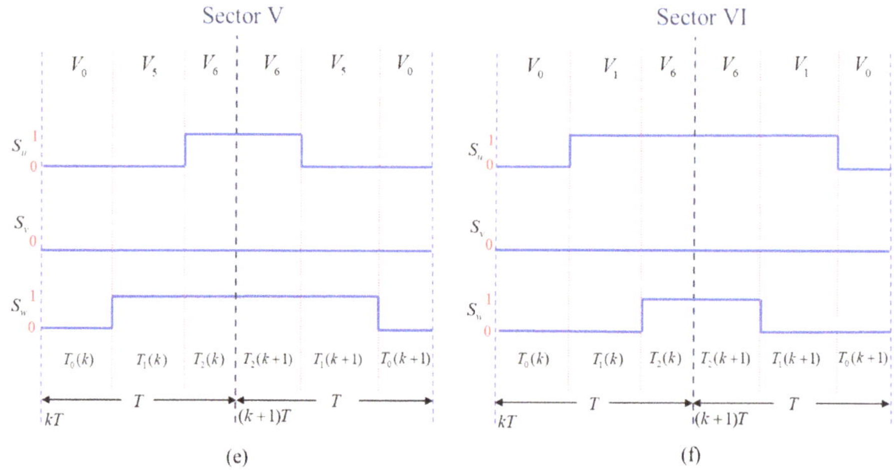

Figure 5. The asymmetric five-segment switching space-vector pulse-width modulation (SVPWM) pulse waveform in each sector: (**a**) Sector I; (**b**) Sector II; (**c**) Sector III; (**d**) Sector IV; (**e**) Sector V; (**f**) Sector VI.

Furthermore, as can be seen from Table 2, the switch dwelling time length T_1 and T_2 are a function of the component voltages, v_α and v_β, the sampling period, T, and the DC bus voltage, V_{dc}. The calculations of the functions are very simple, and hence can be easily carried out by the FPGA-based digital hardware. The pulse waveforms of the proposed asymmetric five-segment switching SVPWM scheme in other sectors are also shown in Figure 5. However, there may be the cases of the reference voltage vector travelling across the sector on the trailing sampling interval. In this case, the sector pulse pattern in the leading sampling interval must change to the next sector pulse waveforms on the trailing sampling interval. Figure 6 illustrates a pulse waveform crossing from Sector I to the other five sectors and from Sector II to I on the trailing sampling interval.

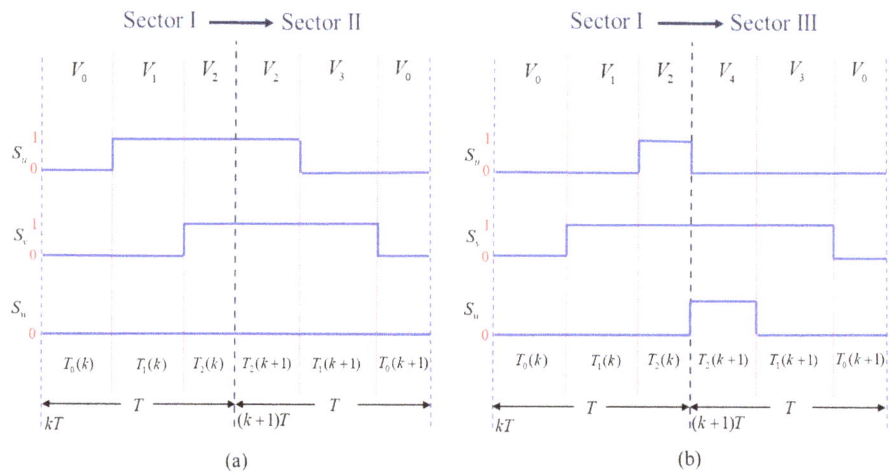

Figure 6. *Cont.*

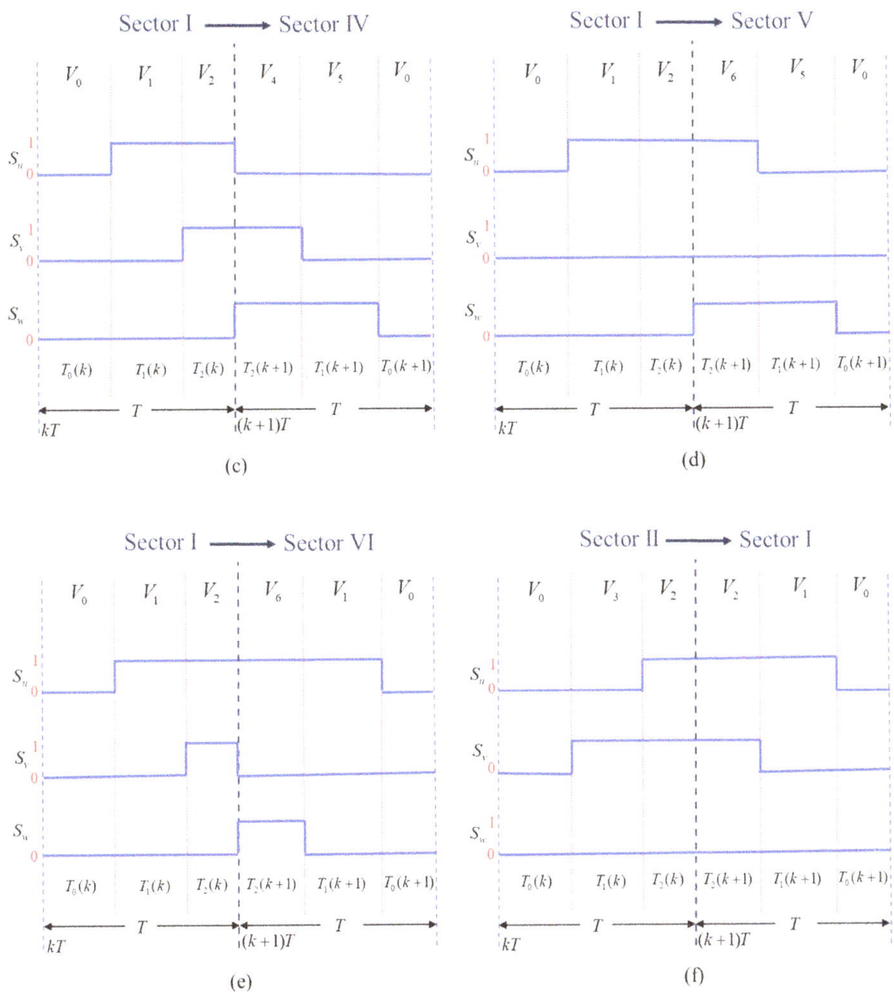

Figure 6. The pulse waveforms of sector crossing on the trailing sampling time: (**a**) Sector I to II, (**b**) Sector I to III, (**c**) Sector I to IV, (**d**) Sector I to V, (**e**) Sector I to VI, (**f**) Sector II to I.

The scaled reference commands can be routed to the circuit block for the calculation of the dwelling time duration T_1 and T_2 in each sector for generating the PWM signals [17]. However, for reducing the computation complexity, an alternative method is used by calculating the firing time, which is defined as the time interval from the start to the leading edge of the PWM pulse. For example, in Sector I, the PWM firing times for the switches S_u, S_v and S_w in the kT sampling period can be obtained as follows:

$$\begin{aligned} f_u &= T_0 \\ &= T - \frac{\sqrt{3}T}{V_{dc}}\left(\frac{\sqrt{3}}{2}v_\alpha + \frac{1}{2}v_\beta\right) \end{aligned} \quad (17)$$

$$\begin{aligned} f_v &= T_0 + T_1 = T_s - T_2 \\ &= T - \frac{\sqrt{3}T}{V_{dc}}v_\beta \end{aligned} \quad (18)$$

$$f_w = T \quad (19)$$

where $T_0 = T - T_1 - T_2$. The resulting equations for the PWM firing time in another sector can be obtained in the same way. Table 3 summarizes the results. As can be seen, the firing-time equations in Sector I are the same as in Sector II. They are the same in Sector III and IV and in Sector V and VI as well. Thus the circuit realization for the sectors with the same firing-time equation can share the same circuit as each other. This can simplify the circuit complexity for the implementation of the ASIC. As shown in Figure 7, the PWM pulse pattern on each phase can be generated by comparing the firing-time signal with a common center-symmetrical triangular wave with amplitude of T and period of $2T$, which can be implemented by an up-down counter. The pulse is low when the firing-time signal is larger than the magnitude of the triangular wave and is high otherwise.

Table 3. PWM firing time calculation in each sector.

Sector	f_u	f_v	f_w
I, II	$T - \frac{\sqrt{3}T}{V_{dc}}(\frac{\sqrt{3}}{2}v_\alpha + \frac{1}{2}v_\beta)$	$T - \frac{\sqrt{3}T}{V_{dc}}v_\beta$	T
III, IV	T	$T - \frac{\sqrt{3}T}{V_{dc}}(\frac{-\sqrt{3}}{2}v_\alpha + \frac{1}{2}v_\beta)$	$T - \frac{\sqrt{3}T}{V_{dc}}(\frac{-\sqrt{3}}{2}v_\alpha - \frac{1}{2}v_\beta)$
V, VI	$T - \frac{\sqrt{3}T}{V_{dc}}(\frac{\sqrt{3}}{2}v_\alpha - \frac{1}{2}v_\beta)$	T	$T + \frac{\sqrt{3}T}{V_{dc}}v_\beta$

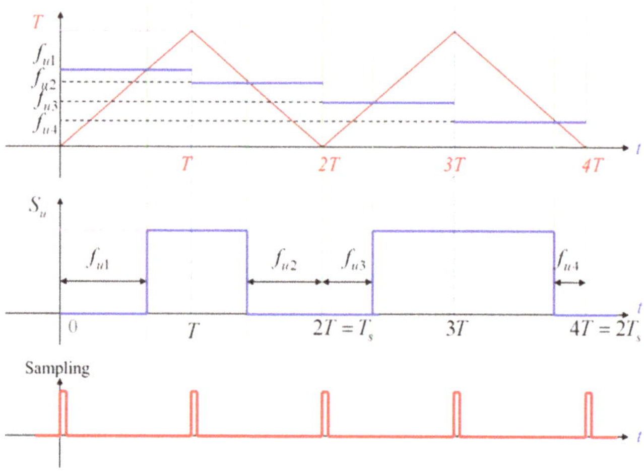

Figure 7. The SVPWM signal generation on one phase.

3. Simulation Verification Using PSIM

Figure 8 shows the simulation verification of the proposed asymmetric five-segment switching SVPWM modulation scheme to drive a PMSM motor by using the PSIM simulation tool, in which the PMSM motor model (Sinano 7CB30-2DE6FKS) was constructed, as shown in Figure 8b. In this simulation, the sampling period is $T = 40\mu s$, the DC bus voltage is $V_{dc} = 40\text{V}, v_\alpha = 23\cos(40\pi t)$, and $v_\beta = 23\sin(40\pi t)$. The SVPWM modulation scheme algorithm was written in C language inside the C block function of PSIM with the flowchart shown in Figure 9. Figure 10 shows the simulation results of current and speed responses, the three-phase firing-time signals, and the firing-time signal vector trajectory with the 20 Hz voltage reference inputs given above and 0.3 Nm load torque at 0.1 s. As can be seen, the speed is 31.4 rad/s in the steady-state, which can verify the correction of the algorithm of the proposed SVPWM modulation scheme.

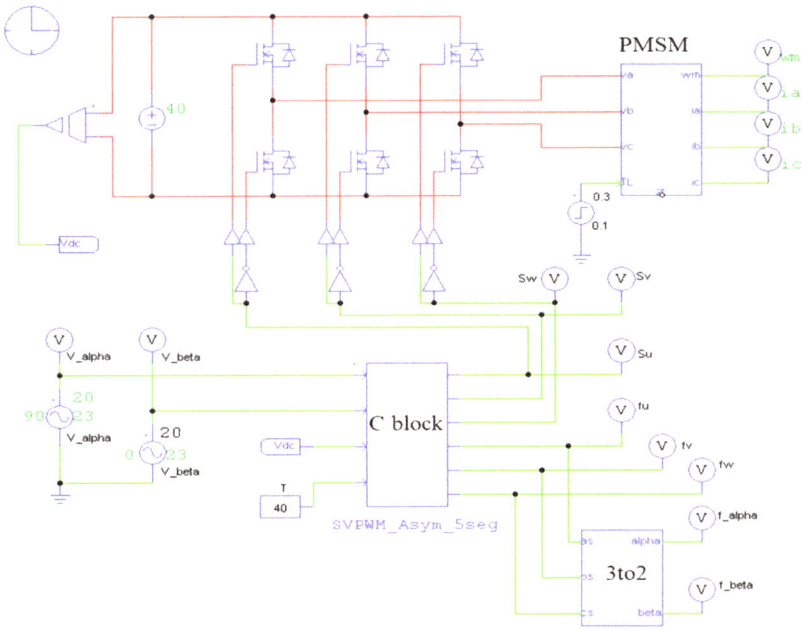

(**a**)

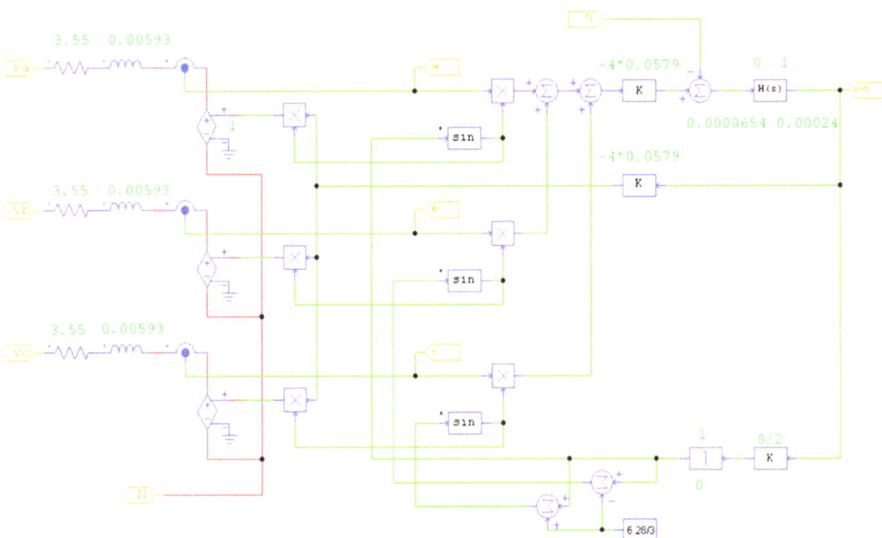

(**b**)

Figure 8. Simulation verification of the SVPWM scheme: (**a**) simulation model using C block, (**b**) PMSM motor model with pole number equal to 8.

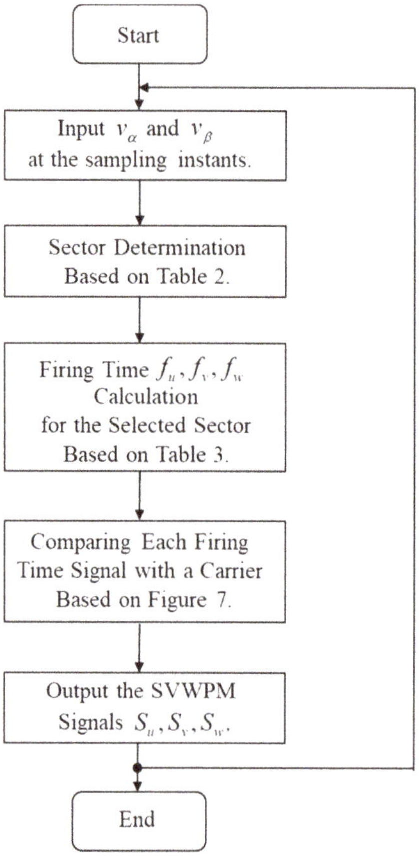

Figure 9. Flowchart of the proposed asymmetric five-segment switching SVPWM scheme algorithm.

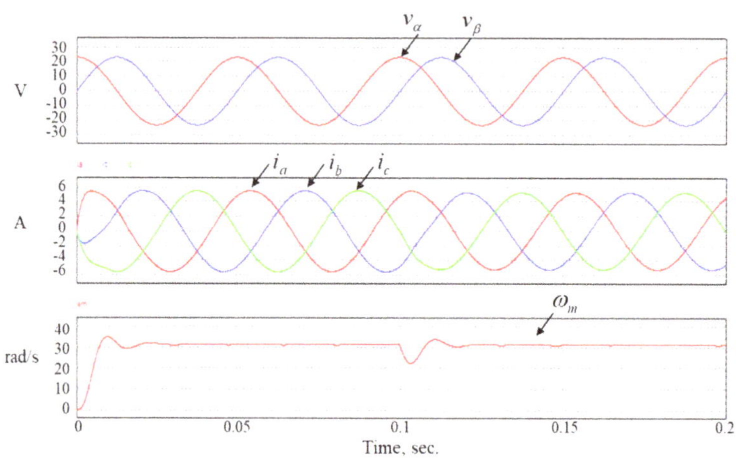

(a)

Figure 10. *Cont.*

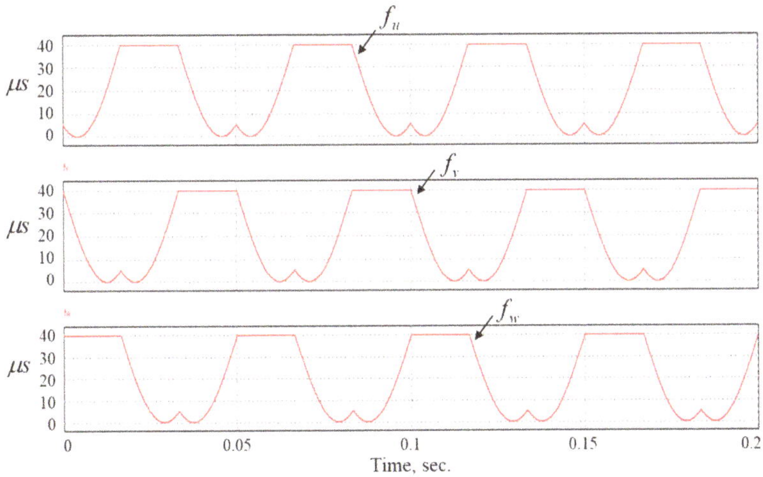

(**b**)

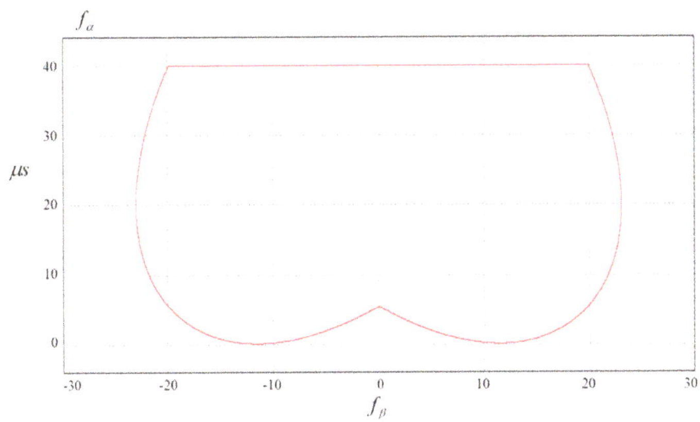

(**c**)

Figure 10. Simulation results with 0.3 Nm load torque at 0.1 s: (**a**) the reference voltages, current and speed responses, (**b**) the three-phase firing-time signals, (**c**) the firing-time signal vector trajectory.

Figure 11 shows the simulation verification of the discrete-time current control of the PMSM motor using the proposed SVPWM scheme. A discrete-time proportional-integral (PI) controller is designed with the pulse transfer function in (20) and the zero-order holder with sampling frequency of 25 kHz, which is two times of the inverter switching frequency. As can be seen, the current responses can track the current reference with the amplitude of 3A and frequency of 20 Hz.

$$H(z) = \frac{k_p + k_i T - k_p z^{-1}}{1 - z^{-1}} \tag{20}$$

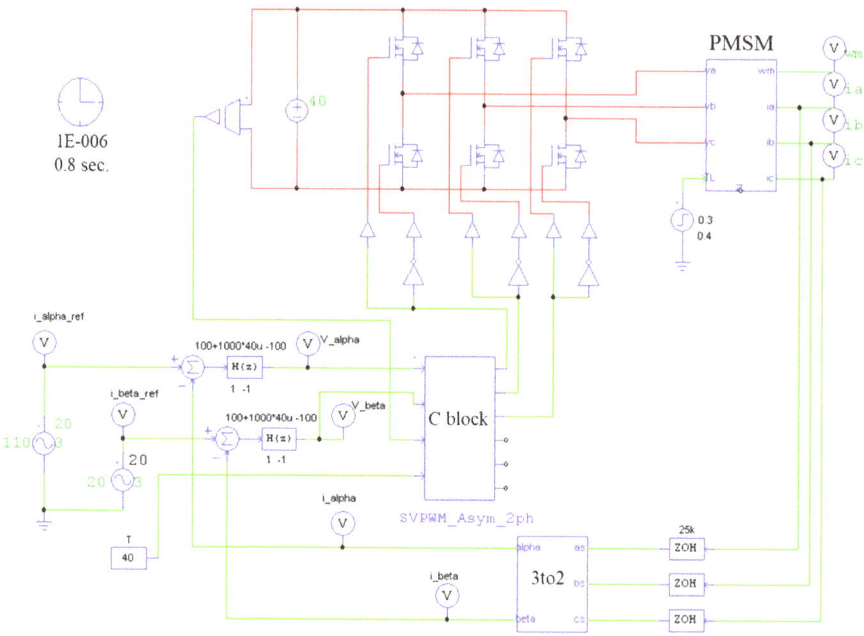

(a)

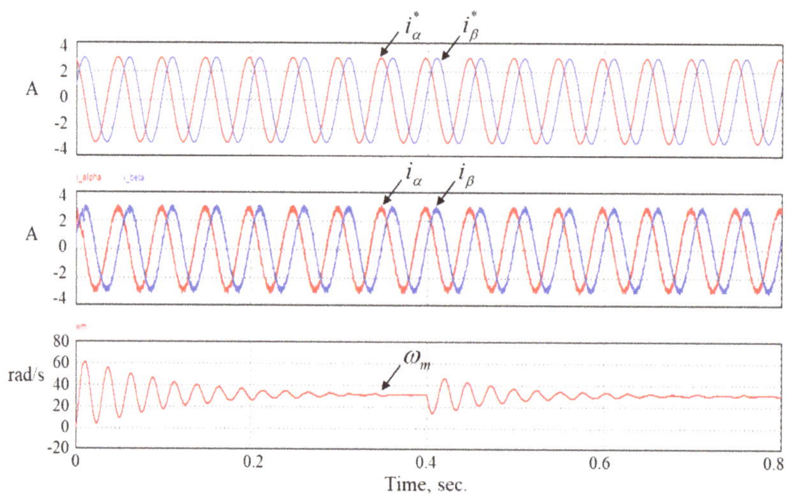

(b)

Figure 11. Simulation verification of the current control with 0.3 Nm load torque at 0.4 s: (**a**) simulation model, (**b**) the reference currents, current, and speed responses.

4. FPGA Implementation

A DE0-Nano FPGA control board by Terasic Inc. with Altera Cyclone IV-E device (EP4CE22F17C6) was employed for the implementation and experiment verification of the proposed current control and SVPWM ASIC for driving a PMSM servo motor (Sinano) with rated power of 300 W. Figure 12 shows the designed SVPWM ASIC structure with optional current control function included. The DE0-Nano board includes a 50 MHz oscillator, which can be used as a source clock to the designed multi-clock generation circuit to give various frequency clock signals, such as 2.5 MHz, 200 kHz, 100 kHz, and 10 kHz, for the design of the ASIC. A cosine and a sine look-up table, each with 500 words and 10 bits/word, were created by using LPM_ROM function to generate the voltage references, v_α and v_β, in which the frequency can be determined through the modulus-500 counter as a frequency divider. The voltage references then can be used as the inputs of the sector selection circuit according to Table 2 and the inputs of the firing time calculation circuit in each sector according to Table 3. Because some firing time equations are the same, they can be divided into three groups. The 3:1 multiplexers are used in order to select the three firing-time signals f_u, f_v, and f_w in each sector. The 50 MHz clock is used as the input of an up-and-down counter, which counts up from zero to 2000 and then counts down to zero, for the generation of a triangular carrier signal with an 80 μs period. The three firing time signals are then compared with the common triangular carrier counter value, respectively, to generate the three-phase SVPWM signals. A dead-time interval of 2 μs in each phase is also inserted to generate the six gate signals to drive the inverter. A SPI-DAC interface to a DAC7513 converter was also designed to illustrate the three-phase firing time signals on the oscilloscope.

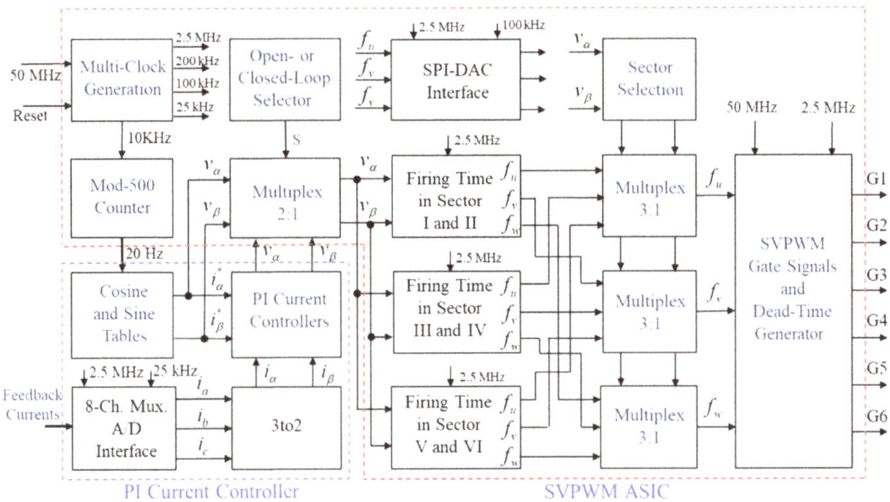

Figure 12. The SVPWM application-specific integrated circuit (ASIC) structure with optional current control function.

The SVPWM ASIC in the FPGA chip was designed, as shown in Figure 13, by using Altera Quartus II software development system of version 13.1. The experimental results of the SVPWM three-phase switching signals in each sector are shown in Figure 14. They are consistent with the expected pulse waveforms in Figure 5. The open-loop experiment result of the firing-time signal vector trajectory in a steady state is shown in Figure 15, which is also consistent with the simulation result in Figure 10c. Figure 16 shows the experimental result of the open-loop current vector trajectory in steady state with the SVPWM scheme. Two of the three-phase currents are sensed by using two LEM-55P current sensors and filtered through low-pass filters. The filtered currents are shifted up 1.65 V, being the inputs to an ADC128S022 A/D converter, in which the input voltage range is between zero and 3.3 V, for the feedback of the PI current control function with 25 kHz sampling frequency.

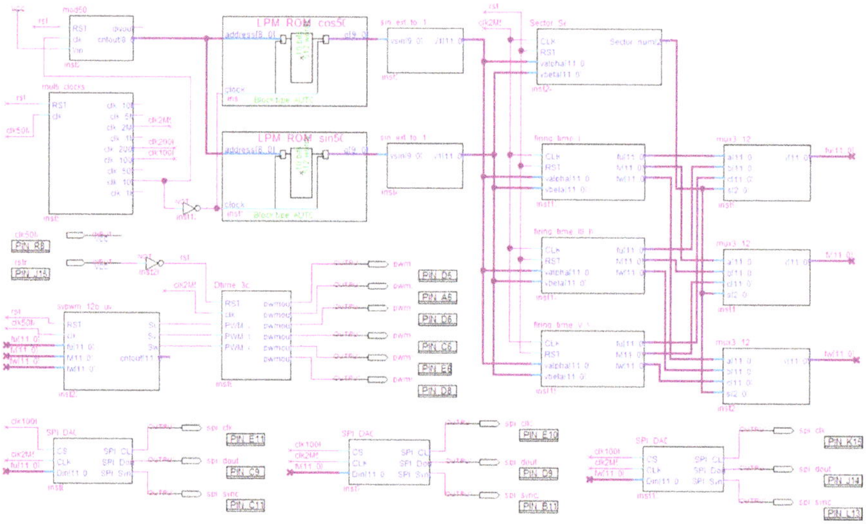

Figure 13. FPGA implementation of the proposed SVPWM scheme by using Quartus II version 13.1 tool.

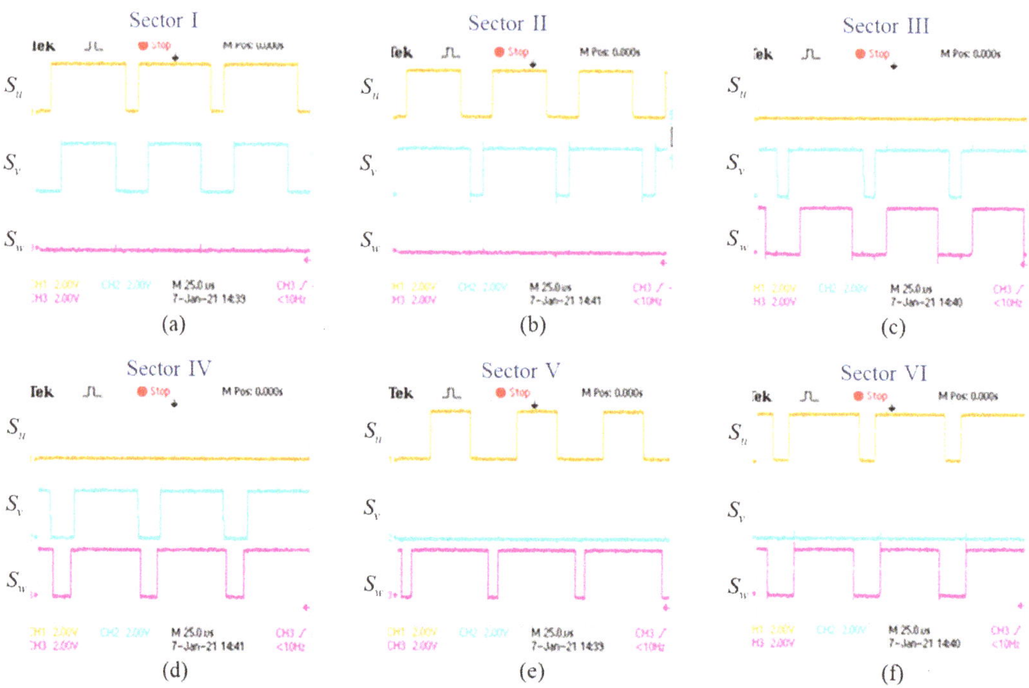

Figure 14. Experiment verification of the three-phase switching signals in each sector of the proposed SVPWM scheme: (**a**) Sector I; (**b**) Sector II; (**c**) Sector III; (**d**) Sector IV; (**e**) Sector V; (**f**) Sector VI.

Figure 15. The open-loop experimental result of the firing-time signal vector trajectory in steady state.

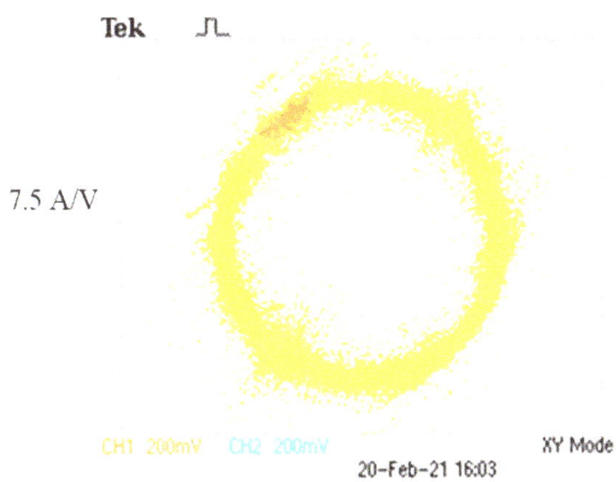

Figure 16. The open-loop current vector trajectory in steady state with the SVPWM scheme (7.5 A/V).

For saving the hardware resources, the computation architecture of the PI controller in the ASIC is shown in Figure 17, which has a control unit and a data path, which contains a 12-bit adder/subtracter, a 12-bit multiplier, and a limiter. The control unit is a finite state machine (FSM) which generates the control signals to the data path to control the computation procedures of the PI controller. The computation procedures of the PI controller according to the pulse transfer function in (20) are shown in Figure 18. There are five steps (s1–s5) for the computation of the PI control function. For the 25 kHz sampling frequency, the input clock frequency for the PI control function is 200 kHz. Therefore, there are eight clocks, in which three clocks are used for waiting state, needed to accomplish the computation of the PI controller. Figure 19 shows the experimental result for the closed-loop current vector trajectory in steady state with different DC bus voltage values. As can be seen, it can be a circular in the range from 55 to 63 V.

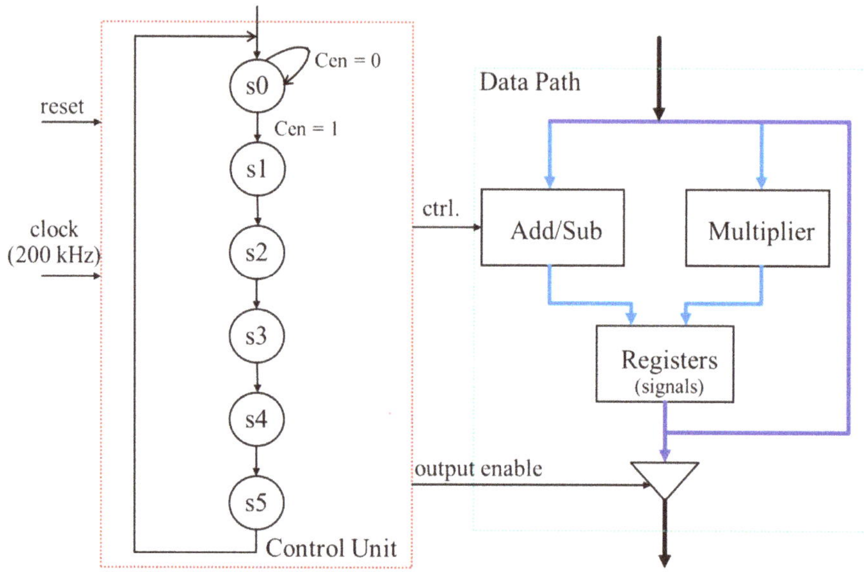

Figure 17. The computation architecture of the ASIC.

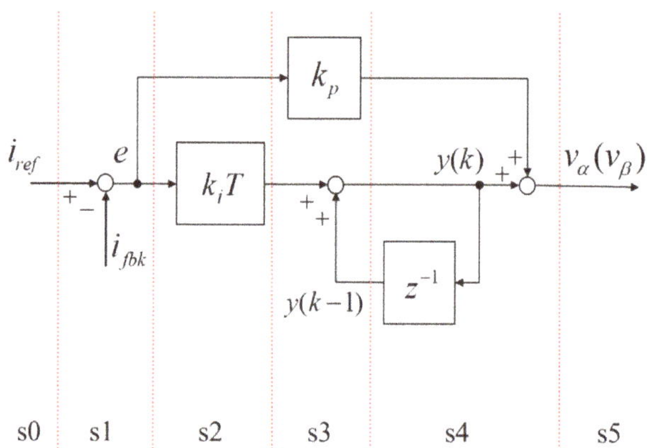

Figure 18. The computation procedures of the proportional-integral (PI) controller.

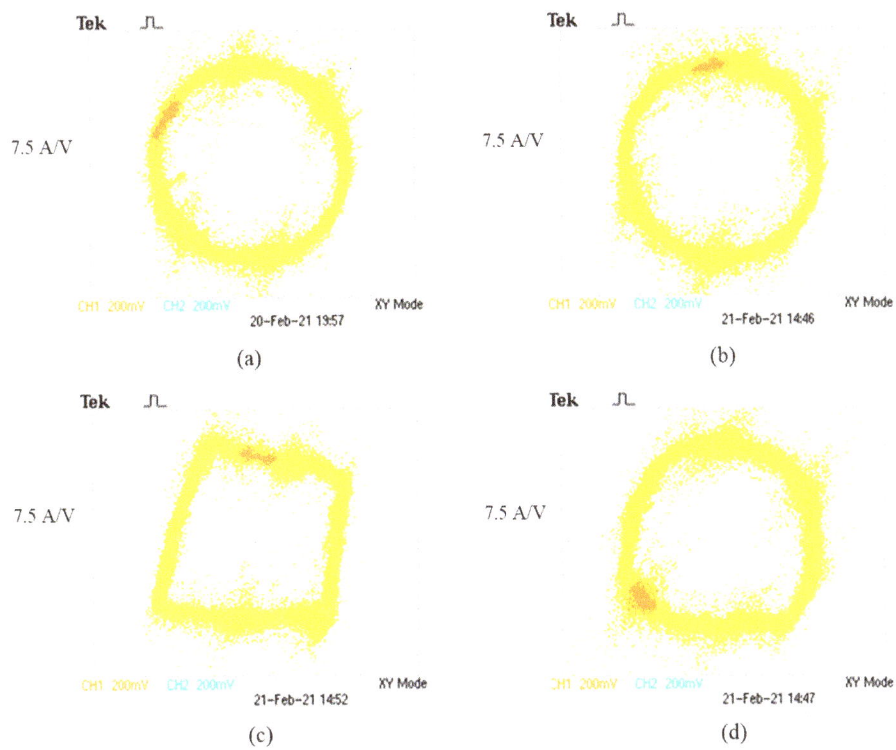

Figure 19. The closed-loop current vector trajectory in steady state with different DC bus voltage (7.5 A/V): (**a**) $V_{dc} = 55\,\text{V}$, (**b**) $V_{dc} = 63\,\text{V}$, (**c**) $V_{dc} = 45\,\text{V}$, (**d**) $V_{dc} = 50\,\text{V}$.

5. Discussion

Several practical aspects for the implementation of the proposed current control and SVPWM ASIC are discussed as follows:

Firstly, as shown in Figure 7, the firing-time signals are the modulating signals and used to compare with a common center-symmetrical triangular carrier wave for generating the PWM gating signals of the inverter in the proposed SVPWM scheme. This is different from the conventional SPWM implementation method, in which the sinusoidal control outputs are the modulating signals. For convenience, the firing time equation of f_v in Table 3 is rewritten as follows:

$$f_v = T - \frac{\sqrt{3}T}{V_{dc}} v_\beta = T\left(1 - \frac{\sqrt{3}}{V_{dc}} v_\beta\right) \tag{21}$$

The second term on the right-hand side of the equals sign has a relationship with the duty ratio of the gating signal expressed as

$$d_v = \frac{\sqrt{3}}{V_{dc}} v_\beta \tag{22}$$

Because the duty ratio is $0 \le d \le 1$, it follows that

$$v_\beta \le \frac{V_{dc}}{\sqrt{3}} \tag{23}$$

So, the maximum value of the control signals v_β or v_α is $V_{dc}/\sqrt{3}$ and is larger than $V_{dc}/2$, which is the maximum value of the modulating signals in SPWM method. Thus, as is well known, the SVPWM method has higher voltage utilization by about 115% than the SPWM modulation method.

Secondly, the term $\sqrt{3}/V_{dc}$ in (22) can be seen as the gain from the control signal v_β to the duty ratio to generate the gating signal in the SVPWM modulation scheme. Thus, this gain can be expressed as

$$K_{svpwm} = \frac{d_v}{v_\beta} = \frac{\sqrt{3}}{V_{dc}} \tag{24}$$

It was found that this SVPWM gain can be included in the PI current controller parameters. So, although the firing time equations are relative to the dc bus voltage, the multiplication of the SVPWM gain to the control output signal v_β is not necessary for the implementation of the firing-time signal. That means the small perturbation of the DC bus voltage will not affect the performance of the current control and SVPWM ASIC. As can be seen from the experimental results in Figure 19, the closed-loop current vector trajectory in steady state can be a circular for the dc bus voltage of 55 and 63 V, respectively. This finding makes the implementation circuit simpler.

Thirdly, although it is the digital hardware circuit for the computation of the current control and SVPWM scheme, the timing sequence during the sampling interval from the sampling of the feedback currents to the current control and firing-time signal output must be considered for the synchronization, as shown in Figure 20. The computation of the timing signals must be completed before the next sampling time. The resulted firing-time signal values are then loaded into the PWM signal generation circuit for the comparison with an up-down counter at the next sampling time in which the counter starts to count up from zero or count down from 2000.

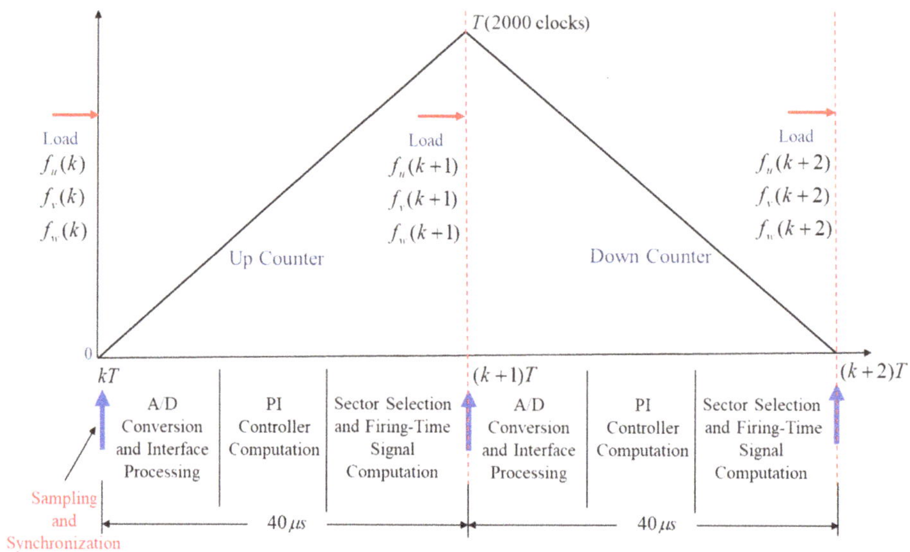

Figure 20. The timing sequence of sampling and loading for the proposed current control and SVPWM scheme.

Fourthly, the conventional SVPWM scheme including the symmetric seven-segment technique, symmetric five-segment technique, and asymmetric three-segment technique are shown in Figure 21. As can be seen, the symmetric five-segment technique has four switching times and is the lowest during a switching period, so the switching loss of the scheme is significantly reduced in comparison with the others. However, in the asymmetric three-segment technique, the sampling frequency can be doubled at a certain switching frequency. The proposed asymmetric five-segment technique can not only have the advantages of minimum switching times, but also can double the sampling frequency in the current control loop so as to improve the control performance.

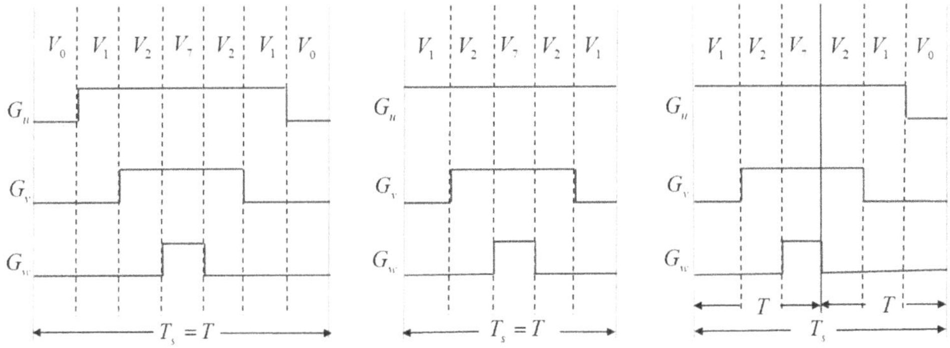

Symmetric 7-Segment Symmetric 5-Segment Asymmetric 3-Segment

(a) (b) (c)

Figure 21. SVPWM switching sequences: (**a**) symmetric 7-segment, (**b**) symmetric 5-segment, (**c**) asymmetric 3-segment.

6. Conclusions

In this work, the design and implementation of an FPGA-based SVPWM ASIC with an asymmetric five-segment switching scheme for AC motor drives have been performed. The inverter switch dwelling and firing times on each sector have been derived. It was found that the firing-time equations in Sector I and II are the same. They are the same in Sector III and IV and in Sector V and VI as well. These finding allow us to simplify the circuit complexity for the implementation of the ASIC. Compared with the conventionally symmetric seven-segment three-phase switching scheme, the inverter switching times and power loss of this proposed scheme can not only be reduced by 33%, but also the asymmetric characteristics mean that the reference voltage command signal can be given in the half period of the PWM switching time interval. Therefore, one can design the closed-loop current control while doubling the sampling frequency, hence increasing the bandwidth, and allowing the motor drive system with better dynamic response. For the verification of the proposed SVPWM modulation scheme, the closed-loop current control function in the stationary reference frame has been also included in the design of the ASIC. The ASIC function is firstly verified by using the PSIM simulation tool. Then, a DE0-nano FPGA control board has been employed to drive a 300 W PMSM AC motor for the experimental verification. The simulation and experimental results show the performance of the proposed SVPWM ASIC both in the open-loop pulse-width modulation and in the current control loop. The proposed current control and SVPWM ASIC can not only be used in PMSM motor drives, but can also be applied in other AC motor drives.

Author Contributions: Conceptualization, M.-F.T. and C.-S.T.; methodology, M.-F.T.; software, P.-J.C.; validation, M.-F.T., C.-S.T. and P.-J.C.; formal analysis, M.-F.T.; investigation, M.-F.T.; resources, M.-F.T. and P.-J.C.; data curation, M.-F.T.; writing—original draft preparation, M.-F.T.; writing—review and editing, M.-F.T. and C.-S.T.; visualization, M.-F.T.; supervision, C.-S.T.; project administration, C.-S.T.; funding acquisition, M.-F.T. All authors have read and agreed to the published version of the manuscript.

Funding: This research was funded by Project number: MUST-110-mission-3, from Ministry of Education (MOE), Taiwan.

Conflicts of Interest: The authors declare no conflict of interest.

References

1. Abouzeid, A.F.; Guerrero, J.M.; Endemaño, A.; Muniategui, I.; Ortega, D.; Larrazabal, I.; Briz, F. Control strategies for induction motors in railway traction applications. *Energies* **2020**, *13*, 700. [CrossRef]
2. Sieklucki, G. An Investigation into the Induction Motor of Tesla Model S Vehicle. In Proceedings of the IEEE 2018 International Symposium on Electrical Machines (SME 2018), Miyazaki, Japan, 10–13 June 2018.
3. Liu, T.-H.; Ahmad, S.; Mubarok, M.S.; Chen, J.-Y. Simulation and implementation of predictive speed controller and position observer for sensorless synchronous reluctance motors. *Energies* **2020**, *13*, 2712. [CrossRef]
4. Lin, C.-H. Permanent-magnet synchronous motor drive system using backstepping control with three adaptive rules and revised recurring sieved pollaczek polynomials neural network with reformed grey wolf optimization and recouped controller. *Energies* **2020**, *13*, 5870. [CrossRef]
5. Ferdiansyah, I.; Rusli, M.R.; Praharsena, B.; Toar, H.; Ridwan; Purwanto, E. Speed Control of Three Phase Induction Motor Using Indirect Field Oriented Control Based on Real-Time Control System. In Proceedings of the IEEE 2018 10th International Conference on Information Technology and Electrical Engineering (ICITEE 2018), Bali, Indonesia, 24–26 July 2018.
6. Rosa, F.C.; Bim, E. A constrained non-linear model predictive controller for the rotor flux-oriented control of an induction motor drive. *Energies* **2020**, *13*, 3899. [CrossRef]
7. Alonge, F.; Cirrincione, M.; D'Ippolito, F.; Pucci, M.; Sferlazza, A. Robust active disturbance rejection control of induction motor systems based on additional sliding mode component. *IEEE Trans. Ind. Electron.* **2017**, *64*, 5608–5621. [CrossRef]
8. De Santana, E.S.; Bim, E.; do Amaral, W.C. A predictive algorithm for controlling speed and rotor flux of induction motor. *IEEE Trans. Ind. Electron.* **2008**, *55*, 4398–4407. [CrossRef]
9. Rosa, F.C.; Lima, F.; Fumagalli, M.A.; Bim, E. Evolving fuzzy controller applied in indirect field oriented control of induction motor. In Proceedings of the 2016 IEEE International Conference on Industrial Technology, Taipei, Taiwan, 14–17 March 2016; pp. 1452–1457.
10. Merabet, A.; Ouhrouche, M.; Bui, R.T. Nonlinear predictive control with disturbance observer for induction motor drive. In Proceedings of the 2006 IEEE International Symposium on Industrial Electronics, Montreal, QC, Canada, 9–13 July 2006; Volume 1, pp. 86–91.
11. Tsai, M.-F.; Tseng, C.-S.; Lin, B.-Y. Phase voltage-oriented control of a PMSG wind generator for unity power factor correction. *Energies* **2020**, *13*, 5693. [CrossRef]
12. Raghuwanshi, S.S.; Khare, V.; Gupta, K. Analysis of SPWM VSI fed AC drive using different modulation index. In Proceedings of the International Conference on Information, Communication, Instrumentation and Control (ICICIC), Indore, India, 17–19 August 2017.
13. Hazari, M.R.; Effat Jahan, E.; Mannan, M.A.; Tamura, J. Artificial neural network based speed control of an SPWM-VSI fed induction motor with considering core loss and stray load losses. In Proceedings of the IEEE International Conference on Electrical Machines and Systems (ICEMS), Chiba, Japan, 13–16 November 2016.
14. Szabo, C.; Szoke, E.; Szekely, N.C.; Zacharias, V.; Imecs, M. Analysis of current-feedback PWM procedures based on hysteresis and current-carrier-wave control for VSI-fed induction motor drive. In Proceedings of the International Aegean Conference on Electrical Machines and Power Electronics (ACEMP) & International Conference on Optimization of Electrical and Electronic Equipment (OPTIM), Istanbul, Turkey, 27–29 August 2019.
15. Harikrishnan, R.; Ashni, E.G. Direct torque control of PMSM using hysteresis modulation, PWM and DTC PWM based on PI control for EV—A comparative analysis between the three strategies. In Proceedings of the IEEE International Conference on Intelligent Computing, Instrumentation and Control Technologies (ICICICT), Kerala, India, 5–6 July 2019.
16. Tzou, Y.-Y.; Hsu, H.-J.; Kuo, T.-S. FPGA-based SVPWM control IC for 3-phase PWM inverters. In Proceedings of the 22nd International Conference on Industrial Electronics, Control, and Instrumentation (IECON 1996), Taipei, Taiwan, 9 August 1996; pp. 138–143.
17. Tzou, Y.-Y.; Hsu, H.-J. FPGA realization of space-vector PWM control IC for three-phase PWM inverters. *IEEE Trans. Power Electron.* **1997**, *12*, 953–963. [CrossRef]
18. Bowes, S.R.; Lai, Y.-S. The relationship between space-vector modulation and regular-sampled PWM. *IEEE Trans. Ind. Electron.* **1997**, *44*, 670–679. [CrossRef]
19. Tzou, Y.-Y.; Lin, S.-Y. Fuzzy-tuning current-vector control of a three-phase PWM inverter for high-performance AC drives. *IEEE Trans. Ind. Electron.* **1998**, *45*, 782–791. [CrossRef]
20. Hava, A.M.; Kerkman, R.J.; Lipo, T.A. Simple analytical and graphical methods for carrier-based PWM-VSI drives. *IEEE Trans. Power Electron.* **1999**, *14*, 49–61. [CrossRef]
21. Trzynadlowski, A.M.; Bech, M.M.; Blaabjerg, F.; Pedersen, J.K. An integral space-vector PWM technique for DSP-controlled voltage-source inverters. *IEEE Trans. Ind. Appl.* **1999**, *35*, 1091–1097. [CrossRef]
22. Henriksen, S.; Betz, R.; Cook, B. Digital hardware implementation of a current controller for IM variable-speed drives. *IEEE Trans. Ind. Appl.* **1999**, *35*, 1021–1029. [CrossRef]
23. Marwali, M.N.; Keyhani, A.; Tjanaka, W. Implementation of indirect vector control on an integrated digital signal processor-based system. *IEEE Trans. Energy Convers.* **1999**, *14*, 139–146. [CrossRef]

24. Ma, J.D.; Bin, W.; Zargari, N.R.; Rizzo, S.C. A space vector modulated CSI-based AC drive for multi motor applications. *IEEE Trans. Power Electron.* **2001**, *16*, 535–544. [CrossRef]
25. Tsai, M.-F.; Chen, H.-C. Design and implementation of a CPLD-based SVPWM ASIC for variable-speed control of AC motor drive. In Proceedings of the IEEE International Conference on Power Electronics and Drive Systems (PEDS), Denpasar, Indonesia, 21–25 October 2001.
26. Zhou, K.; Wang, D. Relationship between space-vector modulation and three-phase carrier-based PWM: A comprehensive analysis [three-phase inverters]. *IEEE Trans. Ind. Electron.* **2002**, *49*, 186–196. [CrossRef]
27. Attaianese, C.I.R.O.; Nardi, V.I.T.O.; Tomasso, G. A Novel SVM Strategy for VSI Dead-Time-Effect Reduction. *IEEE Trans. Ind. Appl.* **2005**, *41*, 1667–1674. [CrossRef]
28. Zhang, W.-F.; Yu, Y.-H. Comparison of three SVPWM strategies. *J. Electron. Sci. Technol. China* **2007**, *5*, 283–287.
29. Li, Y.W.; Wu, B.; Xu, D.; Zargari, N. Space Vector Sequence Investigation and Synchronization Methods for Active Front-End Rectifiers in High-Power Current-Source Drives. *IEEE Trans. Ind. Electron.* **2008**, *55*, 1022–1034. [CrossRef]
30. Tonelli, M.; Battaiotto, P.; Valla, M.I. FPGA implementation of a universal space vector modulator. In Proceedings of the 27th Annual conference of the IEEE Industrial Electronics Society, Denver, CO, USA, 29 Novenber–2 December 2001; pp. 1172–1177.
31. Gaballah, M.M. Design and Implementation of Space Vector PWM Inverter Based on a Low Cost Microcontroller. *Arab. J. Sci. Eng.* **2012**, *38*, 3059–3070. [CrossRef]
32. Ahmed, W.; Ali, S.M.U. Comparative study of SVPWM (space vector pulse width modulation) & SPWM (sinusoidal pulse width modulation) based three phase voltage source inverters for variable speed drive. In *IOP Conference Series: Materials Science and Engineering*; IOP Publishing: Bristol, UK, 2013; Volume 51, p. 012027.
33. Avinash Mishra, A.; Save, S.; Sen, R. Space vector pulse width modulation. *Int. J. Sci. Eng. Res.* **2014**, *5*, 2.
34. Fan, Y.; Zhang, L.; Cheng, M.; Chau, K.T. Sensorless SVPWM-FADTC of a New Flux-Modulated Permanent-Magnet Wheel Motor Based on a Wide-Speed Sliding Mode Observer. *IEEE Trans. Ind. Electron.* **2015**, *62*, 3143–3151. [CrossRef]
35. Liu, G.; Qu, L.; Zhao, W.; Chen, Q.; Xie, Y. Comparison of two SVPWM control strategies of five-phase fault-tolerant permanent-magnet motor. *IEEE Trans. Power Electron.* **2016**, *31*, 6621–6630. [CrossRef]
36. Attique, Q.; Li, Y.; Wang, K. A survey on space-vector pulse width modulation for multilevel inverters. *Cpss Trans. Power Electron. Appl.* **2017**, *2*, 226–236. [CrossRef]
37. Zhang, H.; Wang, L.; Chang, W. Speed sensorless vector control of cascaded H-bridge inverter drive PMSM based on MRAS and two-level SVPWM. In Proceedings of the 21st International Conference on Electrical Machines and Systems (ICEMS), Jeju, Korea, 7–10 October 2018.
38. De, S.K.; Baishya, P.; Chatterjee, S. Speed sensor-less rotor flux oriented control of a 3-phase induction motor drive using SVPWM. In Proceedings of the IEEE International Conference on Intelligent Computing, Information and Control Systems (ICICCS 2019), Secunderabad, India, 27–28 June 2019.
39. Inan, R.; Demir, R. Improved speed-sensorless input/output linearisation-based SVPWM-DTC of IM. In Proceedings of the IEEE 1st Global Power, Energy and Communication Conference (GPECOM2019), Cappadocia, Turkey, 12–15 June 2019.
40. Ketenci, G.; Karabacak, M. Comparative performance assessment of hysteresis and constant switching frequency DTC over AC machines. In Proceedings of the 2nd Global Power, Energy and Communication Conference (IEEE GPECOM 2020), Ephesus Izmir, Turkey, 20–23 October 2020.
41. Tzou, Y.-Y.; Tsai, M.-F.; Lin, Y.-F.; Wu, H. Dual DSP based fully digital control of an AC induction motor. In Proceedings of the IEEE International Symposium on Industrial Electronics, Warsaw, Poland, 17–20 June 1996; pp. 673–678.
42. Tzou, Y.-Y.; Lee, W.-A.; Lin, S.-Y. Dual-DSP sensorless speed control of an induction motor with adaptive voltage compensation. In Proceedings of the 27th Annual IEEE Power Electronics Specialists Conference, Baveno, Italy, 23–27 June 1996; pp. 351–357.
43. Chen, C.; Yu, H.; Gong, F.; Wu, H. Induction motor adaptive back stepping control and efficiency optimization based on load observer. *Energies* **2020**, *13*, 3712. [CrossRef]

energies

MDPI

Article

Development of Three-Phase Permanent-Magnet Synchronous Motor Drive with Strategy to Suppress Harmonic Current

Wei-Tse Kao , Jonq-Chin Hwang * and Jia-En Liu

Department of Electrical Engineering, National Taiwan University of Science and Technology, Taipei 106, Taiwan; D10107202@ntust.edu.tw (W.-T.K.); M10307204@ntust.edu.tw (J.-E.L.)
* Correspondence: jchwang@ntust.edu.tw

Abstract: This study aimed to develop a three-phase permanent-magnet synchronous motor drive system with improvement in current harmonics. Considering the harmonic components in the induced electromotive force of a permanent-magnet synchronous motor, the offline response of the induced electromotive force (EMF) was measured for fast Fourier analysis, the main harmonic components were obtained, and the voltage required to reduce the current harmonic components in the corresponding direct (d-axis) and quadrature (q-axis) axes was calculated. In the closed-loop control of the direct axis and quadrature axis current in the rotor reference frame, the compensation amount of the induced EMF with harmonic components was added. Compared with the online adjustment of current harmonic injection, this simplifies the control strategy. The drive system used a 32-bit digital signal processor (DSP) TMS320F28069 as the control core, the control strategies were implemented in software, and a resolver with a resolver-to-digital converter (RDC) was used for the feedback of angular position and speed. The actual measurement results of the current harmonic improvement control show that the total harmonic distortion of the three-phase current was reduced from 5.30% to 2.31%, and the electromagnetic torque ripple was reduced from 15.28% to 5.98%. The actual measurement results verify the feasibility of this method.

Keywords: motor drive; current harmonic reduction; torque ripple reduction

Citation: Kao, W.-T.; Hwang, J.-C.; Liu, J.-E. Development of Three-Phase Permanent-Magnet Synchronous Motor Drive with Strategy to Suppress Harmonic Current. *Energies* **2021**, *14*, 1583. https://doi.org/10.3390/en14061583

Academic Editor: Mario Marchesoni

Received: 30 January 2021
Accepted: 9 March 2021
Published: 12 March 2021

1. Introduction

The induced electromotive force of a permanent-magnet motor may contain harmonic components because of the design, which produces electromagnetic torque ripples after multiplying with the phase current of the motor (which also contains harmonic components) [1,2]. Therefore, reducing the harmonic components of the motor current provides a reduction in torque ripples. By performing the harmonic analysis of the current to obtain the harmonic components contained and then adding them to the current command, the influence of the current on the torque can be reduced [3,4]. To reduce the current ripple due to the induced EMF, the harmonic components of the induced EMF can be analyzed and injected into the voltage command, as shown in [5]. Most methods add the harmonic components of the induced EMF to the three-phase voltage command. The compensation parameters of current or induced EMF are based on the analysis of its harmonic components. It is known that the three-phase signals containing harmonics can be transformed into the rotating rotor reference frame and projected on the 0-axis, direct axis (d-axis), and quadrature axis (q-axis) [6]. With open-end windings and driven each phase current independently, the zero-sequence harmonics of the induced EMF are added to the 0-axis voltage command directly, in conjunction with field-oriented control (FOC) can reduce current harmonics [7]. Torque ripple can be estimated from the calculated energy and co-energy through the voltage and current feedback, which is used as the feed-forward compensation amount to reduce the torque ripple and current ripple component [8]. Recently, different control methods have been applied to reduce current harmonics or torque ripple. By using the Kalman filter, the stator current and permanent-magnet (PM) rotor flux are used to track the flux linkage and compensate for the torque ripple caused by the demagnetization [9]. Artificial neural networks are used to reduce the torque ripple of the permanent-magnet synchronous motor (PMSM) with non-sinusoidal induced EMF and cogging torque [10]. Predictive torque control is also used to reduce torque ripple and improve its control accuracy by compensating for the current prediction errors [11]. Selective current harmonic suppression method is proposed to reduce current harmonics in case of high-speed operation [12].

In this study, the methods presented in [13,14] were used to conduct fast Fourier transform (FFT) analysis of induced EMF. The induced EMF of the fifth and seventh harmonic components was obtained, the contents and offset angles of which were further analyzed. Moreover, the rotor's rotating coordinate system was used to obtain the sixth harmonic component in direct axis and quadrature axis. Compensation was applied to the q-axis and the d-axis to reduce current harmonic components and torque ripple. Compared with the techniques mentioned above, the method proposed in this paper requires less computational burden and does not need to change the control scheme; however, it is necessary to measure the induced EMF of the motor and analyze its harmonic components as the reference of compensation.

In this paper, Section 2 introduces the mathematical model of the permanent-magnet synchronous motor considering the fifth and seventh harmonics; Section 3 proposes a control strategy to reduce torque ripple with two methods of compensation for abc or qd coordinate systems; Section 4 first measures the induced EMF of the permanent-magnet synchronous motor and determines the compensation amount by spectrum analysis, and compares the current spectrum and torque ripple components with different control strategies.

2. Current and Torque Ripples of Permanent-Magnet Synchronous Motors

If the salience effect is neglected and the d-axis and q-axis inductances are the same, the voltages of a three-phase permanent-magnet synchronous motor can be expressed as follows:

$$v_a = R_s i_a + L_s \frac{d}{dt} i_a + e_{af}, \tag{1}$$

$$v_b = R_s i_b + L_s \frac{d}{dt} i_b + e_{bf}, \tag{2}$$

$$v_c = R_s i_c + L_s \frac{d}{dt} i_c + e_{cf}, \tag{3}$$

where v_a, v_b, and v_c are the input phase voltages of the motor; i_a, i_b, and i_c are the three-phase stator currents; R_s is the equivalent resistance and L_s is the equivalent inductance of the stator; e_{af}, e_{bf}, and e_{cf} are the three-phase induced EMF of the motor. Due to the symmetry of the rotor, the induced EMF contains no even harmonic components. For the motor with the Y connection, the zero-sequence harmonics can be neglected as they are coupled to both ends of each phase [15]. Therefore, only the odd harmonics of the positive and negative sequences need to be considered.

The matrix for transforming the three-phase system to the rotor reference frame and its inverse transformation is as follows:

$$T_{qd}^{r} = \frac{2}{3} \begin{bmatrix} \cos\hat{\theta}_r & \cos(\hat{\theta}_r - 120^\circ) & \cos(\hat{\theta}_r + 120^\circ) \\ \sin\hat{\theta}_r & \sin(\hat{\theta}_r - 120^\circ) & \sin(\hat{\theta}_r + 120^\circ) \end{bmatrix}, \tag{4}$$

$$T_{qd}^{r^{-1}} = \begin{bmatrix} \cos\theta_r & \sin\theta_r \\ \cos(\theta_r - 120^\circ) & \sin(\theta_r - 120^\circ) \\ \cos(\theta_r + 120^\circ) & \sin(\theta_r + 120^\circ) \end{bmatrix}. \tag{5}$$

Under balanced conditions, the three-phase motor uses the transformation matrix T^r_{qd} of the rotor reference frame, and the voltages of the q-axis and the d-axis are:

$$v^r_q = R_s i^r_q + L_q \frac{d}{dt} i^r_q + \omega_r(\lambda'_m + L_d i^r_d), \tag{6}$$

$$v^r_d = R_s i^r_d + L_d \frac{d}{dt} i^r_d - \omega_r L_q i^r_q, \tag{7}$$

where v^r_q and v^r_d are the input voltages; i^r_q and i^r_d are the currents of the q-axis and d-axis, respectively.

The three-phase permanent-magnet synchronous motor used in this study mostly contains the fifth and seventh harmonic components, and the other harmonics barely exist. Therefore, the fifth and seventh harmonic components were added to the estimated values in this section. The resultant induced EMF estimates $\hat{e}_a$, $\hat{e}_b$, and $\hat{e}_c$ are:

$$\hat{e}_a = \hat{\omega}_r \lambda'_m \cos\hat{\theta}_r + h_5 \hat{\omega}_r \lambda'_m \cos[5(\hat{\theta}_r + \frac{\delta_5}{5})] + h_7 \hat{\omega}_r \lambda'_m \cos[7(\hat{\theta}_r + \frac{\delta_7}{7})], \tag{8}$$

$$\hat{e}_b = \hat{\omega}_r \lambda'_m \cos(\hat{\theta}_r - 120^\circ) + h_5 \hat{\omega}_r \lambda'_m \cos[5(\hat{\theta}_r + \frac{\delta_5}{5} - 120^\circ)] + h_7 \hat{\omega}_r \lambda'_m \cos[7(\hat{\theta}_r + \frac{\delta_7}{7} - 120^\circ)], \tag{9}$$

$$\hat{e}_c = \hat{\omega}_r \lambda'_m \cos(\hat{\theta}_r + 120^\circ) + h_5 \hat{\omega}_r \lambda'_m \cos[5(\hat{\theta}_r + \frac{\delta_5}{5} + 120^\circ)] + h_7 \hat{\omega}_r \lambda'_m \cos[7(\hat{\theta}_r + \frac{\delta_7}{7} + 120^\circ)], \tag{10}$$

where $\hat{E}_m = \hat{\omega}_r \lambda'_m$ is the estimated value of the peak induced EMF; h_5 and h_7 are the percentages of the fifth and seventh harmonic components of the induced EMF; δ_5 and δ_7 are the phase differences between the fifth and seventh harmonics and the fundamental wave of the induced EMF, respectively.

Using the transformation matrix T^r_{qd} of the rotor's synchronous rotating coordinate system, the induced EMF containing the fifth and seventh harmonic components is transformed into components along the q-axis and d-axis, $\hat{e}^r_q$ and $\hat{e}^r_d$, respectively, expressed as:

$$\begin{aligned} \hat{e}^r_q &= \hat{\omega}_r \lambda'_m + h_5 \hat{\omega}_r \lambda'_m \cos(6\hat{\theta}_r + \delta_5) + h_7 \hat{\omega}_r \lambda'_m \cos(6\hat{\theta}_r + \delta_7) \\ &= \hat{\omega}_r \lambda'_m + h_{6q} \hat{\omega}_r \lambda'_m \cos(6\hat{\theta}_r + \delta_{6q}) \end{aligned}, \tag{11}$$

$$\begin{aligned} \hat{e}^r_d &= h_5 \hat{\omega}_r \lambda'_m \sin(6\hat{\theta}_r + \delta_5) - h_7 \hat{\omega}_r \lambda'_m \sin(6\hat{\theta}_r + \delta_7) \\ &= h_{6d} \hat{\omega}_r \lambda'_m \sin(6\hat{\theta}_r + \delta_{6d}) \end{aligned}, \tag{12}$$

where the harmonics of the induced EMF along the q-axis and d-axis, $\hat{e}^r_{q-h}$ and $\hat{e}^r_{d-h}$, respectively, are expressed as:

$$\hat{e}^r_{q-h} = h_{6q} \hat{\omega}_r \lambda'_m \cos(6\hat{\theta}_r + \delta_{6q}), \tag{13}$$

$$\hat{e}^r_{d-h} = h_{6d} \hat{\omega}_r \lambda'_m \sin(6\hat{\theta}_r + \delta_{6d}). \tag{14}$$

It can be seen from Equations (11) and (12) that, when the fifth and seventh harmonic components of the induced EMF are transformed into the rotor reference frame and further simplified by the polar coordinate system, they will contain the sixth harmonic components. Therefore, the sixth harmonic components can be used for compensation in the rotor reference frame. In the equations above: h_{6q} is the coefficient indicating the percentage of the sixth harmonic component along the q-axis; δ_{6q} is the phase difference between the sixth harmonic and the fundamental wave along the q-axis; h_{6d} is the coefficient indicating the percentage of the sixth harmonic component along the d-axis; δ_{6d} is the phase difference between the sixth harmonic component and the fundamental wave along the d-axis.

3. Current Harmonic Improvement Strategy for Closed-Loop Control of Current along q-Axis and d-Axis

3.1. Closed-Loop Control Strategy for q-Axis and d-Axis Currents with Compensation for Fundamental Wave of Induced EMF

The closed-loop control of the speed and the q-axis and d-axis currents of the three-phase permanent-magnet synchronous motor in this study are shown in Figure 1. The speed feedback $\hat{\omega}_m$ and position feedback $\hat{\theta}_r$ in Figure 1 are obtained from the the resolver and RDC. The three-phase feedback currents $\hat{i}_a$, $\hat{i}_b$, and $\hat{i}_c$ of the motor are transformed to the rotor reference frame to obtain the q-axis current $\hat{i}_q^r$ and the d-axis current $\hat{i}_d^r$. Moreover, the q-axis and d-axis current control strategy is used to obtain the q-axis and d-axis voltage commands v_q^{r*} and v_d^{r*}. Finally, the inverse transformation matrix of the rotor reference frame is used to obtain the three-phase voltage commands v_a^*, v_b^*, and v_c^*, and then feed to the space vector pulse width modulation (SVPWM) module to produce the switching signals.

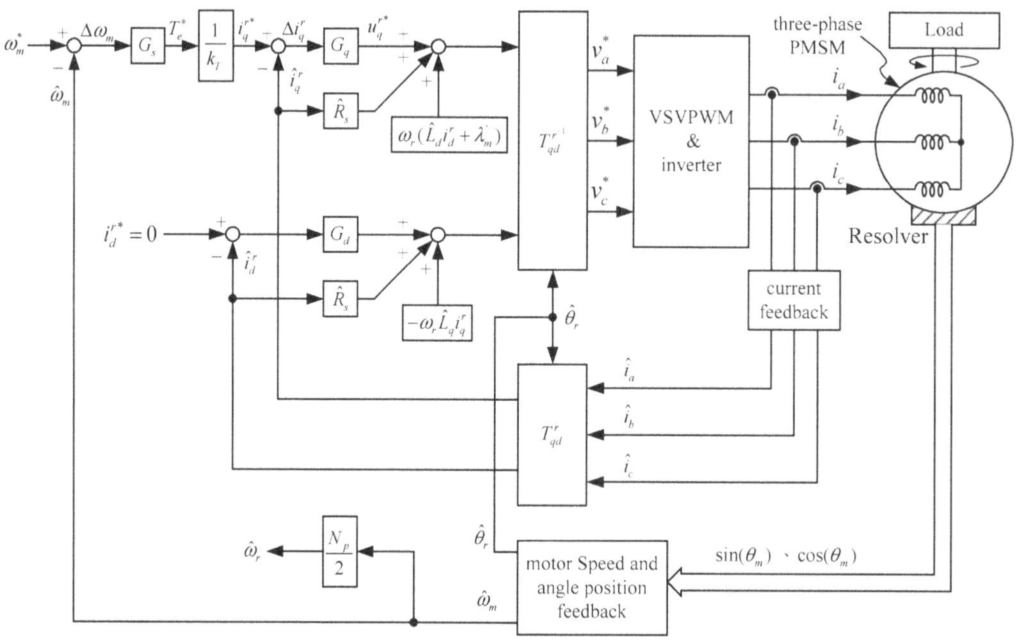

Figure 1. Three-phase permanent-magnet synchronous motor speed and q-axis/d-axis current closed-loop control block.

The closed-loop control block of the q-axis and d-axis currents is shown in Figure 1. The three-phase sinusoidal signal without the harmonic components is transformed into two direct current (DC) signals, direct axis and quadrature axis signals respectively in the rotor reference frame. The transformation of the rotor reference frame can reduce the influence of external disturbance and improve the robustness of the control strategy. The q-axis and d-axis current regulator of the synchronous motor can be obtained according to voltage Equations (6) and (7). By reducing the coupling terms $\omega_r(L_d i_d^r + \lambda'_m)$ and $-\omega_r L_q i_q^r$ in voltage Equations (6) and (7), the linearized voltages u_q^r and u_d^r along the q-axis and d-axis are defined as, respectively:

$$\begin{aligned} u_q^r &= R_s i_q^r + L_q \frac{d}{dt} i_q^r \\ &= v_q^r - \omega_r (L_d i_d^r + \lambda'_m) \end{aligned}, \tag{15}$$

$$\begin{aligned} u_d^r &= R_s i_d^r + L_d \frac{d}{dt} i_d^r \\ &= v_d^r + \omega_r L_q i_q^r \end{aligned}, \tag{16}$$

Decoupling u_q^r and i_q^r in Equation (15), and u_d^r and i_d^r in Equation (16), independent linear systems can be obtained, and the q-axis and d-axis current regulator outputs are, respectively:

$$u_q^{r*} = G_q \circ (i_q^{r*} - \hat{i}_q^r), \tag{17}$$

$$u_d^{r*} = G_d \circ (i_d^{r*} - \hat{i}_d^r), \tag{18}$$

where i_q^{r*} and i_d^{r*} are the q-axis and d-axis current commands. The current regulators G_q and G_d are both proportional-integral controllers with "$\circ$" as the symbol of proportional-integral operation. The transfer function of the current regulator is expressed as

$$G_q = k_{pq} + \frac{k_{iq}}{s}, \tag{19}$$

$$G_d = k_{pd} + \frac{k_{id}}{s}, \tag{20}$$

where k_{pq} and k_{pd} are the proportional gain; k_{iq} and k_{id} are the integral gain. The q-axis and d-axis voltage commands v_q^{r*} and v_d^{r*} can be expressed as

$$v_q^{r*} = u_q^{r*} + \hat{R}_s \hat{i}_q^r + \omega_r (\hat{L}_d \hat{i}_d^r + \hat{\lambda}_m'), \tag{21}$$

$$v_d^{r*} = u_d^{r*} + \hat{R}_s \hat{i}_d^r - \omega_r \hat{L}_q \hat{i}_q^r, \tag{22}$$

The pre-computed feed-forward compensation $R_s i_q^r$ and $R_s i_d^r$ can be added into the q-axis and d-axis voltage commands, respectively, to increase system response speed.

3.2. Compensation Method of Induced EMF Harmonics along Phase-a, Phase-b, and Phase-c

The compensation strategy along the a-axis, b-axis, and c-axis is based on the speed and the q-axis and d-axis current closed-loop control strategy shown in Figure 1, with additional compensation for the fifth and seventh harmonic components of the three-phase induced EMF, as shown in Figure 2. As the induced EMF compensation in Figure 1 contains only that for the fundamental wave, the fifth and seventh harmonic components of the induced EMF are not compensated. After transforming the q-axis and d-axis voltage commands v_q^{r*} and v_d^{r*} with the rotor's synchronous reference system, compensation for the fifth and seventh harmonic components of the induced EMF along the a-axis, b-axis, and c-axis are added, as shown in Figure 2.

3.3. Compensation Method of Induced EMF Harmonics along q-Axis and d-Axis

The fifth and seventh harmonic components of the induced EMF are mapped to the sixth harmonic component after transformed to the rotor reference frame, as shown in Figure 3. By compensating on the qd-axis, the calculation of the control strategy can be simplified. The compensation strategy for harmonics in the induced EMF along the q-axis and d-axis adds the sixth harmonic component in Figure 1, as shown in Equations (13) and (14). By adding Equations (13) and (14) into Figure 1 as the compensation, the resulting control block is as shown in Figure 3. The software was developed based on Figure 3.

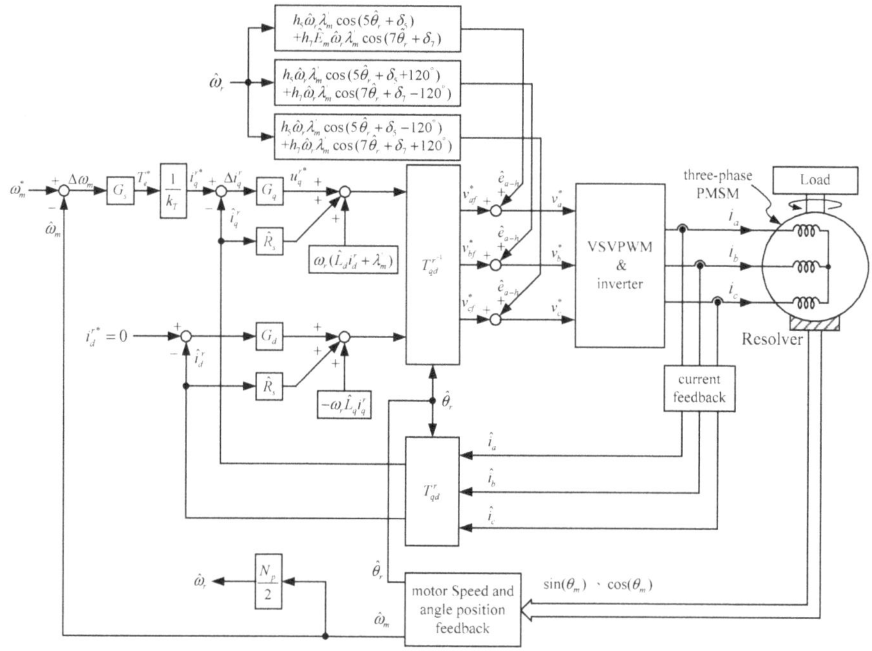

Figure 2. Current harmonics improvement control block of compensation for the induced electromotive force (EMF) harmonics along the a-axis, b-axis, and c-axis in the q-axis/d-axis current closed-loop control.

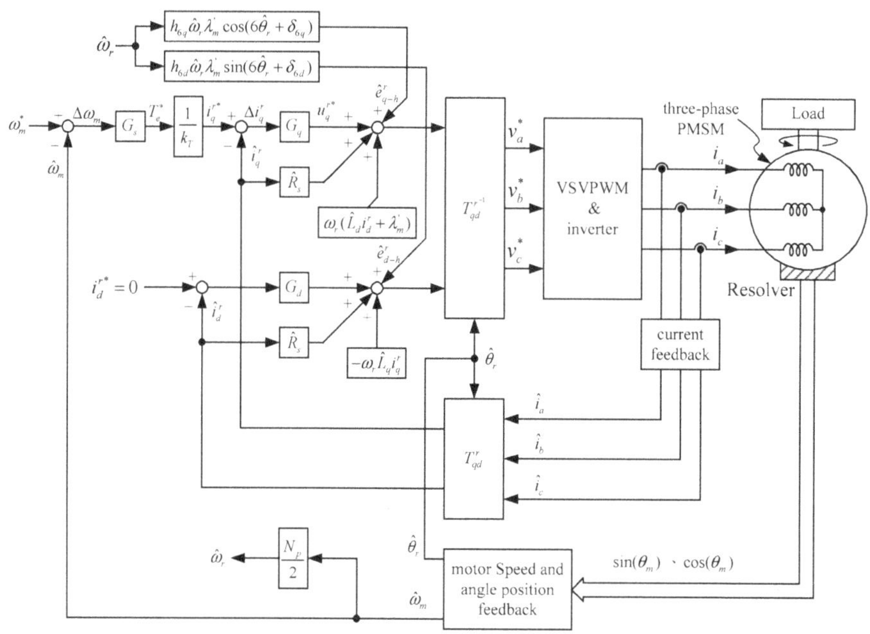

Figure 3. Current harmonics improvement control block of compensation for the induced EMF harmonics along the q-axis and d-axis in the q-axis/d-axis current closed-loop control.

4. Measurement Results

A laboratory prototype based on DSP TMS320F28069 is established to verify the proposed method. The photograph of the motor driver and testbed is shown in Figure 4. The switching frequency of the three-phase inverter is 10 kHz, the sampling time of the closed-loop current control is 100 μs, and the resolution of the A/D converter is 16-bit, corresponding to −5~5 V. The resolution of the D/A converter is 12-bit, corresponding to 0~3.3 V, and the measurement bandwidth of the current probe is 100 kHz. The parameters of the driven motor are shown in Appendix A.

(**a**) (**b**)

Figure 4. Laboratory prototype photos: (**a**) motor driver; (**b**) test bed.

4.1. Measurement of Induced EMF

The prime mover is used to drive the three-phase permanent-magnet synchronous motor at a speed of 1500 rpm. The parameters are $\hat{\omega}_r = 2\pi(\frac{1500}{60})(\frac{N_p}{2})$ (rad/s), $\lambda'_m = 0.0153$ (V/rad/s), $h_5 = 3.30\%$, $h_7 = 1.55\%$, $\delta_5 = 31.51^{\circ}$, and $\delta_7 = 77.35^{\circ}$. After the simplification by the polar coordinate system, $h_{6q} = 4.52\%$, $h_{6d} = 2.48\%$, $\delta_{6q} = 45.76^{\circ}$, $\delta_{6d} = 4.91^{\circ}$ are obtained, and the estimated results are shown in Figure 5. Figure 5a shows the estimated induced EMF $\hat{e}_a$, which is very close to the measured $e_{an'}$ shown in Figure 5e, verifying the correctness of the estimation in this study. In addition, the estimated values of the fundamental wave, fifth harmonic component, and seventh harmonic component are shown in Figure 5b–d, respectively. By using the synchronously rotating reference system to transform from abc to qd reference systems in Equation (4), $\hat{e}_q^r$ and $\hat{e}_d^r$ can be obtained. In addition, $\hat{e}_{q-h}^r$ and

$\hat{e}^r_{d-h}$ can be obtained using Equations (13) and (14) derived in this study, as shown in Figure 6. Figure 6 shows $\hat{e}^r_q$ and $\hat{e}^r_d$. As $\hat{e}^r_q$ contains the fundamental value $\hat{\omega}_r\lambda'_m$ and the sixth harmonic component, whereas $\hat{e}^r_d$ only contains the sixth harmonic component, their respective responses are the same as shown in Equations (11) and (12).

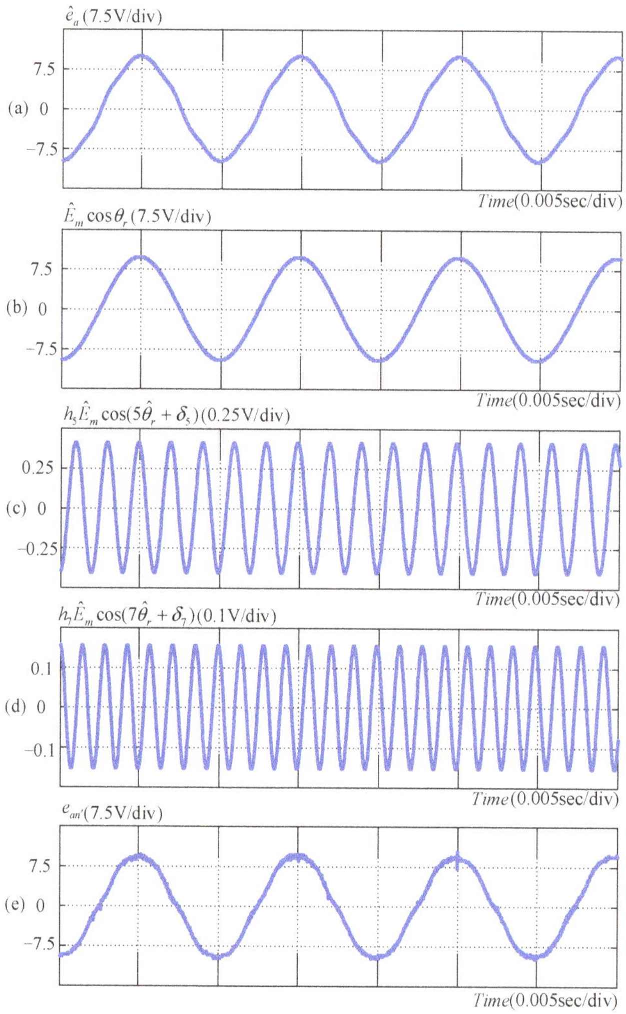

Figure 5. Measured and estimated results of induced EMF of three-phase permanent-magnet synchronous motor driven by the prime mover at a speed of 1500 rpm: (**a**) estimated $\hat{e}_a$ of a-phase induced EMF; (**b**) $\hat{E}_m \cos\hat{\theta}_r$; (**c**) $h_5\hat{E}_m \cos(5\hat{\theta}_r + \delta_5)$; (**d**) $h_7\hat{E}_m \cos(7\hat{\theta}_r + \delta_7)$; (**e**) $e_{an'}$.

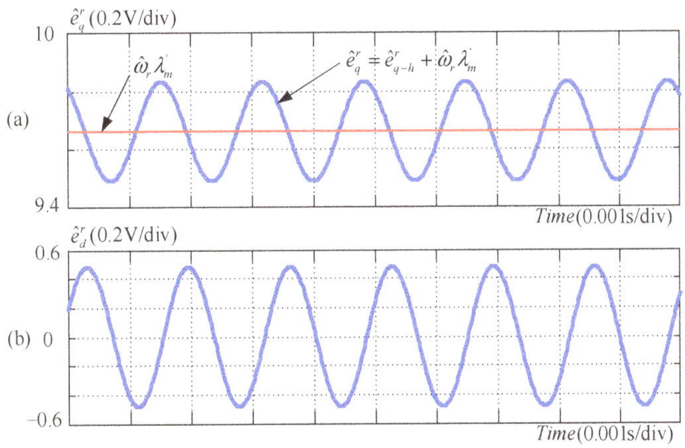

Figure 6. Estimated results of the q-axis and d-axis of the three-phase permanent-magnet synchronous motor driven by the prime mover at a speed of 1500 rpm: (**a**) $\hat{e}_q^r$; (**b**) $\hat{e}_d^r$.

4.2. Measured Results of Induced EMF without Compensation for Harmonic Components

The current harmonic improvement control strategy is not added to the control block of Figure 1. With the effective value of the rated phase current of 23.3 A, the rated rotational speed of 1500 rpm, and the rated electromagnetic torque of 3.5 N·m, the measured results are as shown in Figure 7. It can be seen from Figure 7 that the phase current and the fundamental wave of induced EMF are in phase, and the peak value of the phase current is 33.04 A. The total harmonic distortion rate of the current is 5.30%, with 3.30% for the fifth harmonic component and 2.97% for the seventh harmonic component of the phase current. The current response transformed onto the q-axis and d-axis is shown in Figure 8. It can be seen from Figure 8 that the average value of the q-axis current is 32.75 A, and the average value of the d-axis current is not 0 A. The measurement results for speed response and electromagnetic torque are shown in Figure 9. In this study, electromagnetic torque is taken as equivalent to $k_T\hat{i}_q^r$, which is then measured by the D/A converter. It can be seen from Figure 9 that the average value of the electromagnetic torque is 3.01 N·m, and the peak-to-peak value of the electromagnetic torque is 0.46 N·m. As torque ripple is the peak-to-peak value of electromagnetic torque divided by the average value of electromagnetic torque, the torque ripple is 15.28%.

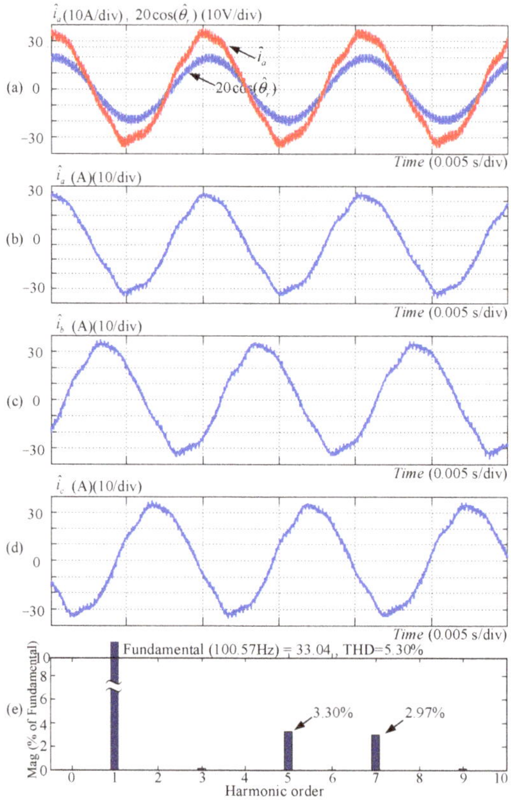

Figure 7. Measured phase currents of a three-phase permanent-magnet synchronous motor driven at 1500 rpm without compensation for harmonic components in induced EMF: (**a**–**d**) measured phase currents $\hat{i}_a$, $\hat{i}_b$ and $\hat{i}_c$; (**e**) spectrum of phase current $\hat{i}_a$.

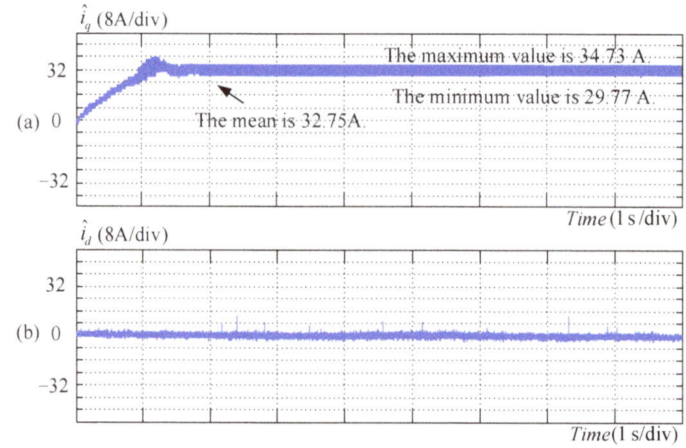

Figure 8. Measured q-axis and d-axis currents of a three-phase permanent-magnet synchronous motor driven at 1500 rpm without compensation for harmonic components in induced EMF: (**a**) measured $\hat{i}^r_q$; (**b**) measured $\hat{i}^r_d$.

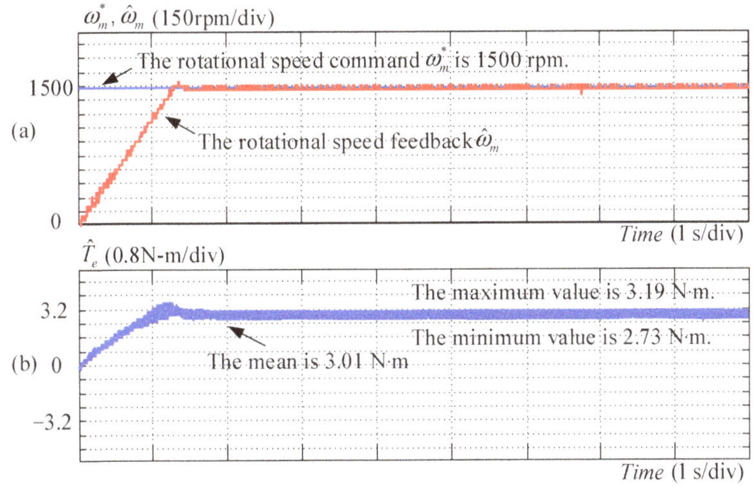

Figure 9. Measured rotational speed and electromagnetic torque of a three-phase permanent-magnet synchronous motor driven at 1500 rpm without compensation for harmonic components in induced EMF: (**a**) measured rotational speed command ω_m^* and rotational speed feedback $\hat{\omega}_m$; (**b**) measured electromagnetic torque $\hat{T}_e$.

4.3. Measured Results of Induced EMF along Phase-a, Phase-b, and Phase-c with Compensation for Harmonic Components

The control block in Figure 2 shows the control strategy for induced EMF along the a-axis, b-axis, and c-axis with compensation for harmonics. The measured results are as shown in Figure 10. It can be seen from Figure 10 that the phase current and the fundamental wave of induced EMF are in phase, and the peak value of the phase current is 33.55 A. The total harmonic distortion rate of the current is 2.57%, with 0.43% for the fifth harmonic component and 0.72% for the seventh harmonic component of the phase current. The current response transformed onto the q-axis and d-axis is as shown in Figure 11. It can be seen from Figure 11 that the average value of the q-axis current is 33.25 A, and the average value of the d-axis current is approximately 0 A. The measurement results of the speed response and electromagnetic torque are shown in Figure 12. It can be seen from Figure 12 that the average value of the electromagnetic torque is 3.05 N·m, the peak-to-peak value of the electromagnetic torque is 0.19 N·m, and the torque ripple is 6.23%.

4.4. Measured Results of Induced EMF along q-Axis and d-Axis with Compensation Strategy for Harmonic Components

The control block in Figure 3 shows the control strategy for induced EMF along the q-axis and d-axis with compensation for harmonics. The measured results are as shown in Figure 13. It can be seen from Figure 13 that the phase current and the fundamental wave of induced EMF are in phase, and the peak value of the phase current is 33.70 A. The total harmonic distortion rate of the current is 2.31%, with 0.61% for the fifth harmonic component and 0.35% for the seventh harmonic component of the phase current. Compared to the result without compensation, the fifth harmonic component is reduced from 3.30% to 0.61%, and the seventh harmonic component is reduced from 2.97% to 0.35%. With reference to the experimental results of the selective current harmonic suppression method [12], the fifth harmonic content is reduced from 19.24% to 5.62%, and the seventh harmonic content is reduced from 12.87% to 4.63%. In both methods, the fifth and seventh harmonic components are reduced.

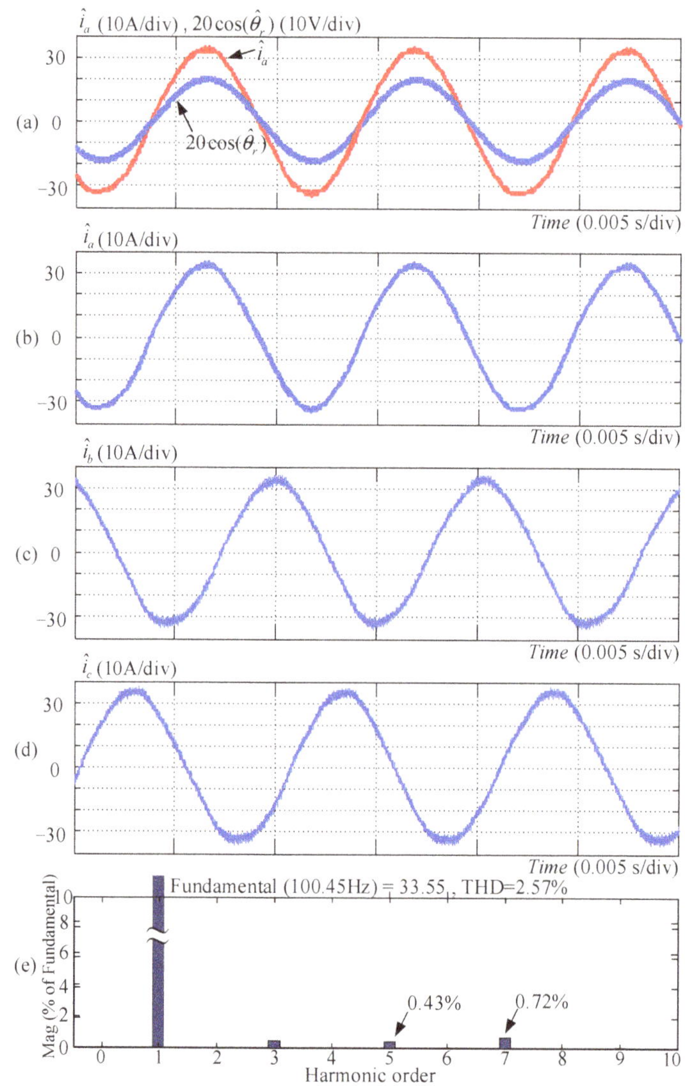

Figure 10. Measured phase currents of a three-phase permanent-magnet synchronous motor driven at 1500 rpm with compensation strategy for harmonic components in induced EMF along the a-axis, b-axis, and c-axis: (**a**–**d**) measured phase current $\hat{i}_a$, $\hat{i}_b$ and $\hat{i}_c$; (**e**) spectrum of phase current $\hat{i}_a$.

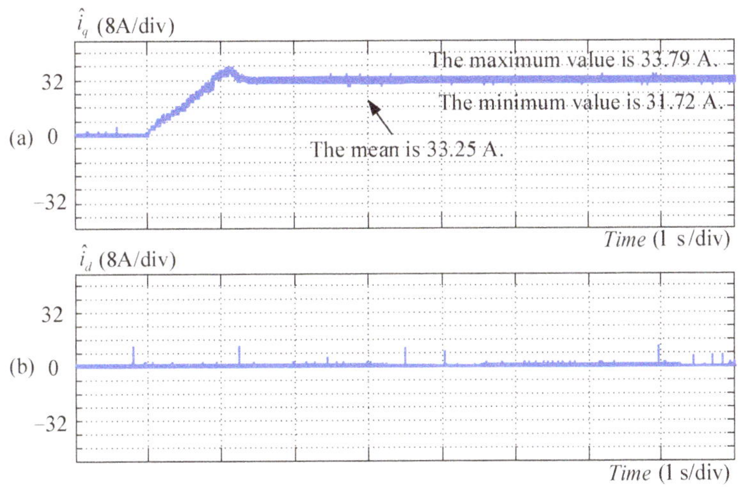

Figure 11. Measured q-axis and d-axis currents of a three-phase permanent-magnet synchronous motor driven at 1500 rpm with compensation strategy for harmonic components in induced EMF along the a-axis, b-axis, and c-axis: (**a**) measured $\hat{i}_q^r$; (**b**) measured $\hat{i}_d^r$.

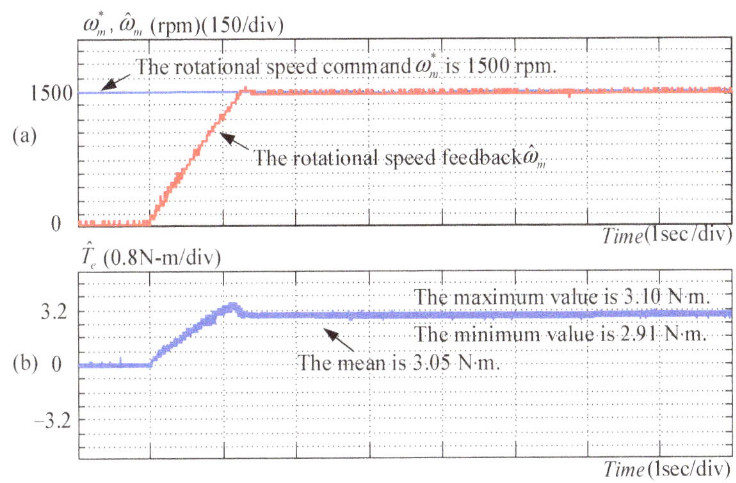

Figure 12. Measured rotational speed and electromagnetic torque of a three-phase permanent-magnet synchronous motor driven at 1500 rpm with a compensation strategy for harmonic components in induced EMF along the a-axis, b-axis, and c-axis: (**a**) measured rotational speed ω_m^* and rotational speed feedback $\hat{\omega}_m$; (**b**) measured electromagnetic torque $\hat{T}_e$.

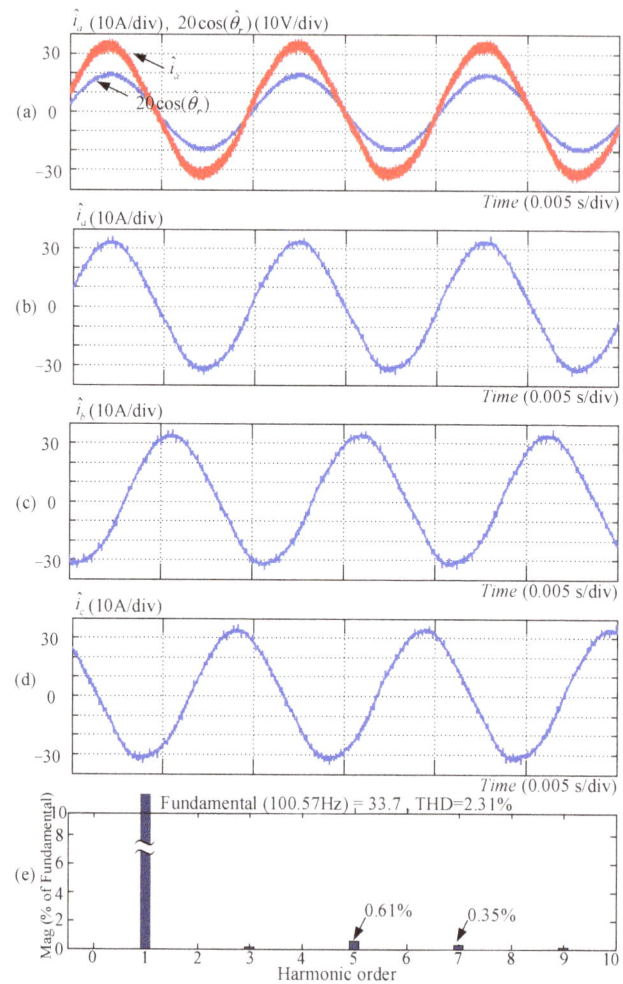

Figure 13. Measured phase currents of a three-phase permanent-magnet synchronous motor driven at 1500 rpm with a compensation strategy for harmonic components in induced EMF along the q-axis and d-axis: (**a**–**d**) measured phase currents $\hat{i}_a$, $\hat{i}_b$ and $\hat{i}_c$; (**e**) spectrum of phase current $\hat{i}_a$.

The current response transformed onto the q-axis and d-axis is as shown in Figure 14. It can be seen from Figure 14 that the average value of the q-axis current is 32.74 A, and the average value of the d-axis current is approximately 0 A. The measurement results of speed response and electromagnetic torque are shown in Figure 15. It can be seen from Figure 15 that the average value of the electromagnetic torque is 3.01 N·m, the peak-to-peak value of the electromagnetic torque is 0.18 N·m, and the torque ripple is 5.98%.

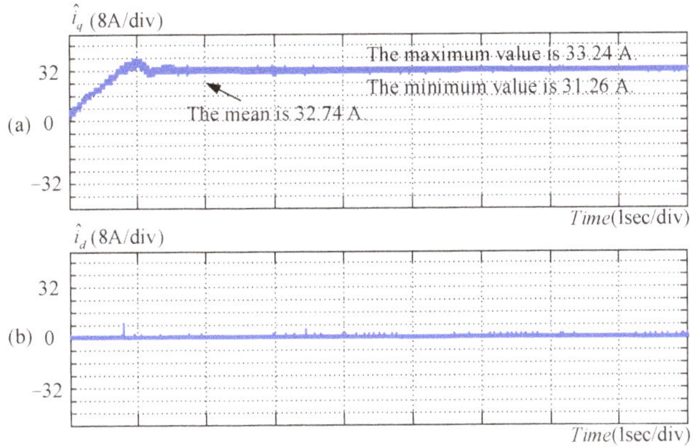

Figure 14. Measured q-axis and d-axis currents of a three-phase permanent-magnet synchronous motor driven at 1500 rpm with a compensation strategy for harmonic components in induced EMF along the q-axis and d-axis: (**a**) measured $\hat{i}_q^r$; (**b**) measured $\hat{i}_d^r$.

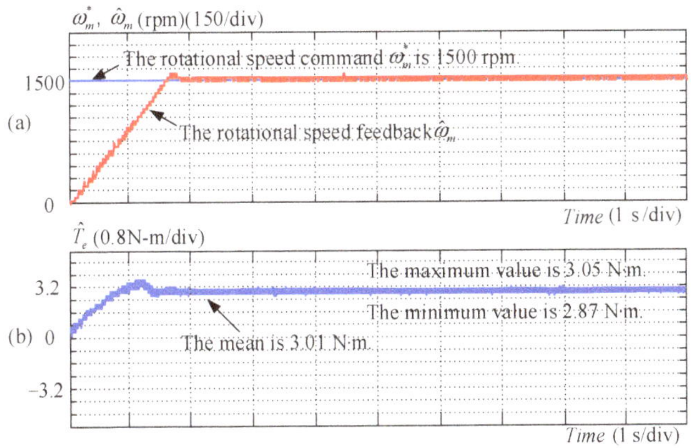

Figure 15. Measured rotational speed and electromagnetic torque of a three-phase permanent-magnet synchronous motor driven at 1500 rpm with a compensation strategy for harmonic components in induced EMF along the q-axis and d-axis: (**a**) measured rotational speed command ω_m^* and rotational speed feedback $\hat{\omega}_m$; (**b**) measured electromagnetic torque $\hat{T}_e$.

5. Conclusions

This study explains the closed-loop control strategy for rotational speed and q-axis and d-axis currents, as well as the improvement strategy for current harmonics, including the compensation strategy for induced EMF along an abc reference frame and a qd reference frame. With the compensation for harmonics in the induced EMF along the q-axis and d-axis, the total harmonic distortion rate is reduced from 5.30% to 2.31%, in which the value is reduced from 3.30% to 0.61% for the fifth harmonic component and from 2.97% to 0.35% for the seventh harmonic component. The peak-to-peak electromagnetic torque is reduced from 0.46 to 0.18 N·m, whereas the electromagnetic torque ripple is reduced from 15.28% to 5.98%. The parameters are summarized in Table 1. It can be seen that the compensation method for harmonics is considerably effective. The method can be used to improve the jitters and ripples of the electromagnetic torque to enhance the performance of the servo drive.

Table 1. Comparison of the closed-loop control strategy for three-phase permanent-magnet synchronous motor speed and q-axis/d-axis currents.

	Without Compensation for Harmonics in Induced EMF	With Compensation for Harmonics in Induced EMF along Phase-a, Phase-b, and Phase-c	With Compensation for Harmonics in Induced EMF along q-Axis and d-Axis	Explanation
Peak-to-peak electromagnetic torque jitter	0.46 N·m	0.19 N·m	0.18 N·m	The current harmonic distortion rate with compensation strategy for the induced EMF along the q-axis and d-axis is smaller, and the torque ripple is lower.
Electromagnetic torque ripple	15.28%	6.23%	5.98%	
Phase-a current total harmonic distortion rate	5.30%	2.57%	2.31%	
Fifth harmonic component of current	3.30%	0.43%	0.61%	
Seventh harmonic component of current	2.97%	0.72%	0.35%	
Additional computational burden	-	$h_5\hat{\omega}_r\lambda'_m\cos(5\hat{\theta}_r+\delta_5)$ $+h_7\hat{E}_m\hat{\omega}_r\lambda'_m\cos(7\hat{\theta}_r+\delta_7)$, $h_5\hat{\omega}_r\lambda'_m\cos(5\hat{\theta}_r+\delta_5+120^\circ)$ $+h_7\hat{\omega}_r\lambda'_m\cos(7\hat{\theta}_r+\delta_7-120^\circ)$, $h_5\hat{\omega}_r\lambda'_m\cos(5\hat{\theta}_r+\delta_5-120^\circ)$ $+h_7\hat{\omega}_r\lambda'_m\cos(7\hat{\theta}_r+\delta_7+120^\circ)$	$h_{6q}\hat{\omega}_r\lambda'_m\cos(6\hat{\theta}_r+\delta_{6q})$, $h_{6d}\hat{\omega}_r\lambda'_m\sin(6\hat{\theta}_r+\delta_{6d})$	
Percentage of execution time in each switching cycle (100 μs = 100%)	33%	60%	45%	

Author Contributions: Conceptualization, J.-C.H. and W.-T.K.; methodology, W.-T.K.; software, W.-T.K. and J.-E.L.; validation, J.-C.H., W.-T.K., and J.-E.L.; formal analysis, W.-T.K.; investigation, J.-C.H. and W.-T.K.; resources, J.-C.H. and W.-T.K.; data curation, J.-E.L.; writing—original draft preparation, W.-T.K.; writing—review and editing, J.-C.H. and W.-T.K.; visualization, J.-E.L.; supervision, J.-C.H.; project administration, J.-C.H.; funding acquisition, J.-C.H. All authors have read and agreed to the published version of the manuscript.

Funding: This research received no external funding.

Institutional Review Board Statement: Not applicable.

Informed Consent Statement: Not applicable.

Acknowledgments: The authors would like to acknowledge the financial support of the Ministry of Science and Technology of Taiwan under grant MOST 107-2221-E-011-109-MY3.

Conflicts of Interest: The authors declare no conflict of interest.

Appendix A

Table A1. Measured results of three-phase permanent-magnet synchronous motor parameters.

Parameter	Value
Equivalent resistance of each phase R_s	45.87 mΩ
Equivalent inductance of each phase L_s	0.338 mH
Rotor flux linkage λ'_m	$15.2 \times 10^{-3}\ \frac{\text{V}}{(\text{rad/s})}$
Fifth harmonic component of induced EMF h_5	3.30%
Seventh harmonic component of induced EMF h_7	1.55%
Rated torque	3.0 N·m
Rated rotational speed	1500 rpm
DC side voltage	24 V
Equivalent viscosity coefficient B_{eq}	$11.6 \times 10^{-3}\ \frac{\text{N·m}}{(\text{rad/s})}$
Equivalent moment of inertia J_{eq}	4.43×10^{-3} kg·m^2

Appendix B

e_{af}, e_{bf}, e_{cf}	Three-phase induced EMFs of the motor.
$\hat{e}_q^r, \hat{e}_d^r$	The voltages observed on the q-axis and d-axis by project the three-phase induced EMFs to the rotor reference frame.
h_5, h_7	The fifth and seventh harmonic components of the induced EMF.
i_a, i_b, i_c	Three-phase currents of the stator.
$\hat{i}_q^r, \hat{i}_d^r$	The currents observed on the q-axis and d-axis by project the three-phase currents to the rotor reference frame.
L_d, L_q	Direct- and quadrature-axis inductance.
R_s	Equivalent resistance of the stator.
T_{qd}^r	The transformation matrix to project the abc to the qd reference frame.
$T_{qd}^{r^{-1}}$	The inverse transformation matrix to project the abc to the qd reference frame.
v_a, v_b, v_c	Three-phase voltages of the stator.
δ_5, δ_7	Phase shifts of each phase from the fifth and seventh harmonic components, respectively.
δ_{6q}, δ_{6d}	Phase shifts of q-axis and d-axis from the sixth harmonic component, respectively.
$\hat{\theta}_r$	Electrical angle of the rotor.

References

1. Hoang, L.H.; Robert, P.; Rene, F. Minimization of Torque Ripple in Brushless DC Motor Drives. *IEEE Trans. Ind. Appl.* **1986**, *4*, 748–755.
2. Liang, W.; Wang, J.; Luk, P.C.; Fang, W.; Fei, W. Analytical Modeling of Current Harmonic Components in PMSM Drive with Voltage-Source Inverter by SVPWM Technique. *IEEE Trans. Energy Convers.* **2014**, *29*, 673–680. [CrossRef]
3. Mattavelli, P.; Tubiana, L.; Zigliotto, M. Torque-Ripple Reduction in PM Synchronous Motor Drives Using Repetitive Current Control. *IEEE Trans. Power Electron.* **2005**, *20*, 1423–1431. [CrossRef]
4. Lee, G.H.; Kim, S.I.; Hong, J.P.; Bahn, J.H. Torque Ripple Reduction of Interior Permanent Magnet Synchronous Motor Using Harmonic Injected Current. *IEEE Trans. Magn.* **2005**, *44*, 1582–1585.
5. Zhu, H.; Xiao, X.; Li, Y.D. Permanent Magnet Synchronous Motor Current Ripple Reduction with Harmonic Back-EMF Compensation. In Proceedings of the International Conference on Electrical Machines and Systems, Incheon, Korea, 10–13 October 2010; pp. 1094–1097.
6. Hwang, J.C.; Wei, H.T. The current harmonics elimination control strategy for six-leg three-phase permanent magnet synchronous motor drives. *IEEE Trans. Power Electron.* **2014**, *29*, 3032–3040. [CrossRef]
7. Lin, J.H. Current Harmonics Improvement of Three-phase Permanent-Magnet Synchronous Motor Drives. Master's Thesis, Department of Electrical Engineering, National Taiwan University of Science and Technology, Taipei, Taiwan, 2015.
8. Nakao, N.; Akatsu, K. A new control method for torque ripple compensation of permanent magnet motors. In Proceedings of the International Power Electronics Conference, Sapporo, Japan, 21–24 June 2010; pp. 1421–1427.
9. Xiao, X.; Chen, C. Reduction of Torque Ripple Due to Demagnetization in PMSM Using Current Compensation. *IEEE Trans. Appl. Supercon.* **2010**, *20*, 1068–1071. [CrossRef]
10. Flieller, D.; Nguyen, N.K.; Wira, P.; Sturtzer, G.; Abdeslam, D.O.; Mercklé, J. A Self-Learning Solution for Torque Ripple Reduction for Nonsinusoidal Permanent-Magnet Motor Drives Based on Artificial Neural Networks. *IEEE Trans. Ind. Electron.* **2014**, *61*, 655–666. [CrossRef]
11. Siami, M.; Khaburi, D.A.; Rodríguez, J. Torque Ripple Reduction of Predictive Torque Control for PMSM Drives with Parameter Mismatch. *IEEE Trans. Power Electron.* **2017**, *32*, 7160–7168. [CrossRef]
12. Liu, G.; Chen, B.; Wang, K.; Song, X. Selective Current Harmonic Suppression for High-Speed PMSM Based on High-Precision Harmonic Detection Method. *IEEE Trans Ind. Inform.* **2019**, *15*, 3457–3468. [CrossRef]
13. Hwang, J.C.; Lim, C.Y.; Wei, H.T. The Current Harmonics Reduction of Six-phase Permanent-magnet Synchronous Motor Drives. *IEEE Trans. Power Electron.* **2013**, *29*, 3032–3040. [CrossRef]
14. Zhang, P.; Sizov, G.Y.; Demerdash, N.A.O. Comparison of Torque Ripple Minimization Control Techniques in Surface-mounted Permanent Magnet Synchronous Machines. In Proceedings of the IEEE International Electronics Machines and Drives Conference, Niagara Falls, ON, Canada, 15–18 May 2011; pp. 188–193.
15. Hanselman, D. *Brushless Permanent Magnet Motor Design*, 2nd ed.; Magna Physics Pub: Lebanon, OH, USA, 2006.

energies

MDPI

Article

Inverse Optimal Control in State Derivative Space System with Applications in Motor Control

Feng-Chi Lee [1], Yuan-Wei Tseng [2,*], Rong-Ching Wu [2], Wen-Chuan Chen [1] and Chin-Sheng Chen [1]

1 Graduate Institute of Automation Technology, National Taipei University of Technology, Taipei City 10608, Taiwan; lifengchi@itri.org.tw (F.-C.L.); JasonChen@itri.org.tw (W.-C.C.); saint@mail.ntut.edu.tw (C.-S.C.)
2 Department of Electrical Engineering, I-Shou University, Kaohsiung City 840, Taiwan; rcwu@isu.edu.tw
* Correspondence: yuanwei@isu.edu.tw; Tel.: +886-7-6577711 (ext. 6637)

Abstract: This paper mathematically explains how state derivative space (SDS) system form with state derivative related feedback can supplement standard state space system with state related feedback in control designs. Practically, inverse optimal control is attractive because it can construct a stable closed-loop system while optimal control may not have exact solution. Unlike the previous algorithms which mainly applied state feedback, in this paper inverse optimal control are carried out utilizing state derivative alone in SDS system. The effectiveness of proposed algorithms are verified by design examples of DC motor tracking control without tachometer and very challenging control problem of singular system with impulse mode. Feedback of direct measurement of state derivatives without integrations can simplify implementation and reduce cost. In addition, the proposed design methods in SDS system with state derivative feedback are analogous to those in state space system with state feedback. Furthermore, with state derivative feedback control in SDS system, wider range of problems such as singular system control can be handled effectively. These are main advantages of carrying out control designs in SDS system.

Keywords: inverse optimal control; state derivative space (SDS) system; state derivative feedback; DC motor control; singular system; nonlinear control

Citation: Lee, F.-C.; Tseng, Y.-W.; Wu, R.-C.; Chen, W.-C.; Chen, C.-S. Inverse Optimal Control in State Derivative Space System with Applications in Motor Control. *Energies* **2021**, *14*, 1775. https://doi.org/10.3390/en14061775

Academic Editor: Frede Blaabjerg

Received: 31 January 2021
Accepted: 18 March 2021
Published: 23 March 2021

1. Introduction

In modern control, state space system is used to carry out state related feedback control design. In state space system, state derivative vector $\dot{x}(t)$ is a dependent function of both control input vector $u(t)$ and state vector $x(t)$ as follows.

$$\dot{x}(t) = f(x(t), u(t)) \tag{1}$$

Previously, in most researches, state related feedback control algorithms $u(t) = \phi(x(t))$ were developed in state space system form so that the following is a stable closed loop system.

$$\dot{x}(t) = f(x(t), \phi(x(t))) \tag{2}$$

However, in reality the control design approach of carrying out state related feedback in state space system has some limitations. For instance, not every system can have its state space system form. Singular systems [1] with pole at infinity are such cases. For example, electrical circuits [2], aerospace vehicles [3], piezoelectric smart structures [4], and chemical systems [5] are actually singular systems. Control design of singular systems were mainly developed in the following generalized state space system or descriptor system form [6,7] where E is a singular matrix.

$$E\dot{x}(t) = F(x(t), u(t)). \tag{3}$$

Control designs for such systems are carried out in large augmented systems and usually require feedbacks of both state and state derivative variables [7–11].

Therefore, comparing with the design processes for standard state space system, control design processes for singular systems are much more complicate. In the analysis, singular systems are further classified into impulse-free mode and impulse mode [7,12]. When a singular system has impulse mode, designers have to further investigate if the system is impulse controllable and if the impulse mode can be eliminated [7]. In the best case, applying state feedback control only can stabilize singular systems with impulse mode. Therefore, state feedback control design for a singular systems with impulse mode is usually considered as very challenging task.

Moreover, in many systems, the direct measurements by sensor are not state signals but state derivative signals. For example, accelerations sensed by accelerometers [13] and voltages or more precisely speaking current derivatives sensed by inductors are directly measured state derivative related signals in many applications. Especially, velocities and accelerations which can be modelled as state derivative vector are easily available from measurements in vehicle dynamic systems [14–18] and piezoelectric smart structure systems [4]. For those applications, we should not insist to apply state related feedback in control designs because additional numerical integrations or integrators are needed in implementation that result in complex and expensive controllers. Instead, state derivative related feedback should be applied. However, it is not convenient to develop state derivative related feedback algorithms under standard state space system form. Another system form for people to conveniently develop state derivative feedback is needed.

Inspired by the above analysis, the correspondence author of this paper proposed the following state derivative space system form, abbreviated as SDS systems and dedicated for state derivative related feedback control designs.

$$x(t) = F(\dot{x}(t), u(t)). \tag{4}$$

In SDS systems, state vector $x(t)$ is an explicit function vector of state derivative vector $\dot{x}(t)$ and control vector $u(t)$.

When state derivative related feedback control law $u(t) = \phi(\dot{x}(t))$ is properly designed and applied, one can obtain a stable closed loop system as

$$x(t) = F(\dot{x}(t), \phi(\dot{x}(t))). \tag{5}$$

The linear time invariant system of SDS system, namely, Reciprocal State Space (RSS) system can be described as

$$x = f\dot{x} + gu \tag{6}$$

where f and g are constant matrices. When $u = -K\dot{x}$ is properly designed and applied, the following closed loop system is stable.

$$x = (f - gK)\dot{x} = f_c\dot{x} \tag{7}$$

It is well known that the eigenvalues of an invertible matrix and the eigenvalues of its inverse matrix are actually reciprocals to each other and that was why the name of Reciprocal State Space system was given. Therefore, closed loop system poles are the reciprocals of the eigenvalues of matrix f_c in (7). To construct a stable closed loop RSS system, all eigenvalues of matrix f_c in (7) must have negative real parts by design of feedback gain K.

Both SDS system and RSS system forms were proposed by the correspondence author of this paper. State derivative related feedback control algorithms such as sliding mode control [19,20], H infinity control [21,22], optimal, and LQR control [23] have been developed in SDS system or RSS system form. Even the complicated singular system with impulse mode were successfully controlled in SDS system with state derivative related feedback control laws [22,23].

When systems' global operations are accurately modeled, they are mostly nonlinear systems. However, control design of nonlinear systems are more difficult than of linear systems. Optimal control is among the handful design approaches that can systematically handle nonlinear systems.

Mathematically speaking, problems of nonlinear optimal control can be solved based on a Hamilton–Jacobi–Bellman (HJB) equation to obtain a Lyapunov function of closed-loop system (or control Lyapunov function) and correlated optimal control law that minimize a given performance functional. However, it is not easy to solve this equation. In general cases, exact solution may not even exist [24,25]. For unstable nonlinear systems, the fundamental requirement is to find control laws to stabilize them but this requirement may not be achieved with optimal control. In 1964, Kalman proposed inverse optimal control (IOC) as the alternative for finding control laws that can stabilize nonlinear systems. In design approach of inverse optimal control, a control Lyapunov function is selected at the beginning. Therefore, solving a HJB equation is circumvented. Followed by design steps according to the Lyapunov stability theorem and the coupling in HJB equation, one can find an optimal controller related to a meaningful performance integrand [24,26]. More precisely speaking, the performance integrand to be constructed is related to the control Lyapunov function, system dynamic and feedback control law because they are coupled in the HJB equation. Therefore, Inverse optimal control has great design flexibility by varying parameters in both the performance integrand and the control Lyapunov function to characterize globally stabilizing controller to meet response constraints of closed loop system [27]. Hence, for unstable nonlinear systems, inverse optimal control is usually considered as the last resort to stabilize them.

Inverse optimal control has been widely applied in robotic control [28,29], biological systems [30,31], aerospace vehicles [24,32–34] and power systems [35–37]. In this paper, inverse optimal control in SDS system with state derivative related feedback is presented. To authors' best knowledge, this type of research have not been reported before.

To verify the proposed algorithm, a non-traditional speed tracking controller and torque tracking controller of a DC motor without tachometer by feeding back the voltage of a small inductor externally connected in series with armature circuit of a DC motor are provided as one application example. The small inductor serves as sensor in the DC motor tracking control. Unlike the traditional DC motor controls which apply state related feedback of speed or current, the inductor's voltage is state derivative related measurement feedback of current which is well suitable to apply the proposed IOC algorithm based on state derivative feedback. The advantages of the proposed controllers with inductor's voltage include 1. Inductor's average power is zero so it does not damage the armature circuit. 2. No tachometer is needed so it can save the implementation cost. Another example of a challenging singular system with impulse mode and bounded disturbance is also provided.

The organization of paper is described as follows. In Section 2, we introduce the inverse optimal control design algorithms for SDS systems with state derivative related feedback. In Section 3, we present illustrative examples and simulation results. Finally, we discuss the results and potential of constructing compact and cheap controller for system with direct state derivative measurement in Section 4 and conclusions in Section 5.

2. Inverse Optimal Control in State Derivative Space (SDS) System with State Derivative Related Feedback

This section first introduces stability analysis of SDS system, followed by the algorithms of carrying out inverse optimal control in SDS system with state derivative related feedback building on the inspirations of inverse optimal control deign in state space system with state feedback in [26,27,31].

2.1. Stability Analysis of SDS Systems

Consider the following SDS system with proper dimensions.

$$x = f(\dot{x}(t)), x(0) = x_0,\ t \geq 0 \tag{8}$$

A Lyapunov function $V(x)$ should be continuously differentiable and meet the following requirements.

$$V(x) > 0, \text{ if and only if } x \neq 0 \text{ and } V(0) = 0. \tag{9}$$

For $x \neq 0$, taking derivative of $V(x)$ with respect to time t and substituting system Equation in (8), if the result is negative, the SDS system is stable.

$$\dot{V}(x) = \frac{dV(x)}{dt} = \frac{dV(x)}{dx}\frac{dx}{dt} = V'(x)\dot{x} = \dot{x}^T V'^T(f(\dot{x}(t))) < 0 \tag{10}$$

For a stable system, when $t \to \infty$, $V(x(\infty)) \to 0$ as $x(\infty) \to 0$.

For simplicity of presentation and for people to better understand that in formula derivation of SDS system control designs, state x should be substituted by its SDS system equation. In this paper, a popular quadratic Lyapunov function $V(x)$ that meets the requirements in (9) is selected for formula derivation as follows.

$$V(x) = \frac{1}{2}x^T Px > 0, \tag{11}$$

where P is a positive definite and symmetric matrix.

Consequently, using SDS system Equation in (8), if

$$\dot{V}(x) = \frac{dV(x)}{dt} = x^T P\dot{x} = \dot{x}^T Px = \dot{x}^T Pf(\dot{x}) < 0 \tag{12}$$

the SDS system is stable.

Hence, for a stable SDS system, let the performance integrand as

$$L(\dot{x}(t)) = -\dot{x}^T Pf(\dot{x}) = -\dot{V}(x(t)) > 0 \tag{13}$$

We have the following positive performance functional

$$J(x_0) = \int_0^\infty L(\dot{x}(t))dt = -V(x(\infty)) + V(x(0)) = V(x_0) = \frac{1}{2}x_0^T Px_0 > 0 \tag{14}$$

The value of performance functional is bounded, greater than zero and related to the initial condition $x(0) = x_0$.

2.2. Inverse Optimal Control for SDS Systems with State Derivative Related Feedback

In this section, we explain the inverse optimal nonlinear control design process for SDS systems with state derivative feedback.

Consider the following nonlinear controlled dynamic SDS system with proper dimensions and initial condition.

$$x = F(\dot{x}(t), u(t)), x(0) = x_0, \ t \geq 0, \tag{15}$$

with performance functional as

$$J(x_0, u(\cdot)) = \int_0^\infty L(\dot{x}(t), u(t))dt \tag{16}$$

where $u(\cdot)$ is an admissible control.

The following process is to construct an inverse optimal globally stabilizing control law.

$$u(t) = \phi(\dot{x}(t)) \tag{17}$$

First, a symmetric and positive definite matrix P serving as the design parameter should be selected for Lyapunov function in (11).

Therefore, when the control law $u(t)$ is properly designed and substitute SDS system Equation in (15), we should have

$$\dot{V}(x) = \frac{dV(x)}{dt} = x^T P\dot{x} = \dot{x}^T Px = \dot{x}^T PF(\dot{x}(t), u(t)) < 0 \tag{18}$$

For simplicity of presentation, we omit (*t*) in the following formula derivation.

Second, select another design parameter, namely the performance integrand $L(\dot{x}, u)$ and applying (15) so that we have the following Hamiltonian for the SDS system in (15) with the performance functional in (16).

$$H(\dot{x}, u) = L(\dot{x}, u) + \dot{V}(x) = L(\dot{x}, u) + \dot{x}^T Px = L(\dot{x}, u) + \dot{x}^T PF(\dot{x}, u) \geq 0 \tag{19}$$

Third, one can have the inverse optimal feedback control law in (17) by setting

$$\frac{\partial H(\dot{x}, u)}{\partial u} = 0 \tag{20}$$

Fourth, applying the obtained inverse optimal feedback control law in the third step, if

$$\dot{V}(x) = \frac{dV(x)}{dt} = x^T P\dot{x} = \dot{x}^T Px = \dot{x}^T PF(\dot{x}, \phi(\dot{x})) < 0 \tag{21}$$

and the steady-state Hamilton–Jacobi–Bellman Equation is zero as follows.

$$H(\dot{x}, \phi(\dot{x})) = 0 \tag{22}$$

Then, the following closed-loop *SDS* system is stable.

$$x = F(\dot{x}(t), \phi(\dot{x}(t))) \tag{23}$$

Therefore, the selection of design parameter $L(\dot{x}, u)$ should meet the requirement of Hamiltonian in (19). Consequently, the inverse optimal feedback control law in (17) obtained from solving (20) should satisfy both (21) and (22) to guarantee the global asymptotic stability of the closed-loop SDS system in (23).

Furthermore, from (19), we have the following performance integrand.

$$L(\dot{x}, u(t)) = -\dot{V}(x) + H(\dot{x}, u(t)). \tag{24}$$

Taking integrals of both sides of (24) and using (19) and (22), it follows that

$$\begin{aligned} J(x_0, u(\cdot)) &= \int_0^\infty L(\dot{x}(t), u(t))dt = \int_0^\infty \left[-\dot{V}(x) + H(\dot{x}(t), u(t))\right]dt \\ &= -\lim_{t\to\infty} V(x(t)) + V(x_0) + \int_0^\infty H(\dot{x}(t), u(t))dt \\ &= V(x_0) + \int_0^\infty H(\dot{x}(t), u(t))dt \geq V(x_0) + \int_0^\infty H(\dot{x}(t), \phi(\dot{x}(t)))dt \\ &\geq V(x_0) = J(x_0, \phi(\dot{x}(\cdot))), \end{aligned} \tag{25}$$

Hence, when we apply inverse optimal control law in (17), we have (22). Consequently, performance functional is the minimum as follows.

$$J(x_0, \phi(\dot{x})) = \min J(x_0, u(\cdot)) = V(x_0) = \frac{1}{2}x_0^T Px_0 > 0 \tag{26}$$

2.3. *Inverse Optimal Control for Affine SDS Systems with State Derivative Related Feedback*

Affine systems are nonlinear systems that are linear in the input. Consider the nonlinear affine *SDS* system with dimension notations given by

$$x_{n\times1} = f_{n\times1}(\dot{x}(t)) + g_{n\times m}(\dot{x}(t))u_{m\times1}(t), x(0) = x_0, t \geq 0, \tag{27}$$

with performance functional as

$$J(x_0, u(\cdot)) = \int_0^{\infty} L(\dot{x}(t), u(t))dt. \tag{28}$$

The following process is to construct an inverse optimal globally stabilizing control law.

First, a symmetric and positive definite matrix P serving as the first design parameter should be selected for Lyapunov function $V(x)$ in (11). Therefore, when the control law $u(t)$ is properly designed and substitute SDS system Equation in (27), we should have

$$\dot{V}(x) = \frac{dV(x)}{dt} = x^T P\dot{x} = \dot{x}^T Px = \dot{x}^T P[f(\dot{x}(t)) + g(\dot{x}(t))u(t)] < 0 \tag{29}$$

Second, we consider the performance integrand $L(\dot{x}(t), u(t))$ which is also a design parameter of the form

$$L_{1\times 1}(\dot{x}(t), u(t)) = L_{1_{1\times 1}}(\dot{x}(t)) + L_{2_{1\times m}}(\dot{x}(t))u_{m\times 1} + u_{1\times m}{}^T(t)R_{m\times m}(\dot{x})u_{m\times 1}(t). \tag{30}$$

$$\begin{gathered}\text{Therefore, } L(\dot{x}(t), u(t)) \text{ is decomposed into three design parameters,} \\ \text{namely, } L_{1_{1\times 1}}(\dot{x}(t)), L_{2_{1\times m}}(\dot{x}(t)) \text{ and } R_{m\times m}(\dot{x}).\end{gathered} \tag{31}$$

For simplicity of presentation, we omit (t) and dimension notations in the following formula derivation.

Third, use (29) and define following Hamiltonian for the SDS system in (27) with the performance functional specified in (28).

$$H(\dot{x}, u) = L(\dot{x}, u) + \dot{V}(x) = L_1(\dot{x}) + L_2(\dot{x})u + u^T R(\dot{x})u + \dot{x}^T P[f(\dot{x}) + g(\dot{x})u] \geq 0 \tag{32}$$

We should first select a positive definite $R(\dot{x})$ so that $u^T R(\dot{x})u > 0$ in (32).

Setting the partial derivative of the Hamiltonian with respect to u to zero,

$$\frac{\partial H(\dot{x}, u)}{\partial u} = L_2^T(\dot{x}) + 2R(\dot{x})u + g^T(\dot{x})P\dot{x} = 0, \tag{33}$$

the inverse optimal state derivative related feedback control law is obtained as follows.

$$u = \phi(\dot{x}) = \frac{-1}{2}R^{-1}(\dot{x})\left[L_2^T(\dot{x}) + g^T(\dot{x})P\dot{x}\right]. \tag{34}$$

From (34), we have

$$\left[L_2(\dot{x}) + \dot{x}^T Pg(\dot{x})\right] = -2\phi^T(\dot{x})R(\dot{x}). \tag{35}$$

Fourth, substituting (34) into (29), we should have

$$\begin{aligned}\dot{V}(x) &= \tfrac{dV(x)}{dt} = x^T P\dot{x} = \dot{x}^T Px = \dot{x}^T P[f(\dot{x}) + g(\dot{x})\phi(\dot{x})] \\ &= \dot{x}^T P\left[f(\dot{x}) - \tfrac{1}{2}g(\dot{x})R^{-1}(\dot{x})L_2^T(\dot{x}) - \tfrac{1}{2}g(\dot{x})R^{-1}(\dot{x})g^T(\dot{x})P\dot{x}\right].\end{aligned} \tag{36}$$

Therefore, to ensure (34) is a stabilizing control law, $L_2(\dot{x})$ should be selected such that

$$\dot{x}^T P\left[f(\dot{x}) - \frac{1}{2}g(\dot{x})R^{-1}(\dot{x})L_2^T(\dot{x}) - \frac{1}{2}g(\dot{x})R^{-1}(\dot{x})g^T(\dot{x})P\dot{x}\right] < 0. \tag{37}$$

Fifth, using (32) and (35), $L_1(\dot{x})$ should be selected as

$$L_1(\dot{x}) = \phi^T(\dot{x})R(\dot{x})\phi(\dot{x}) - \dot{x}^T Pf(\dot{x}) \tag{38}$$

The following is the proof.

Substituting (35) and (38) into (32), it can be shown that

$$\begin{aligned} H(\dot{x},u) &= L_1(\dot{x}) + L_2(\dot{x})u + u^T R(\dot{x})u + \dot{x}^T P[f(\dot{x}) + g(\dot{x})u] \\ &= \phi^T(\dot{x})R(\dot{x})\phi(\dot{x}) - \dot{x}^T P f(\dot{x}) + L_2(\dot{x})u + u^T R(\dot{x})u + \dot{x}^T P[f(\dot{x}) + g(\dot{x})u] \\ &= \phi^T(\dot{x})R(\dot{x})\phi(\dot{x}) + \left[L_2(\dot{x}) + \dot{x}^T P g(\dot{x})\right]u + u^T R(\dot{x})u \\ &= \phi^T(\dot{x})R(\dot{x})\phi(\dot{x}) - 2\phi^T(\dot{x})R(\dot{x}) + u^T R(\dot{x})u \\ &= [u - \phi(\dot{x})]^T R(\dot{x})[u - \phi(\dot{x})] \geq 0 \end{aligned} \tag{39}$$

Based on (39), applying the inverse optimal control law in (34), the steady-state Hamilton–Jacobi–Bellman equation is zero as follows.

$$H(\dot{x}, \phi(\dot{x})) = 0 \tag{40}$$

Consequently, the performance integrand in (30) is obtained as

$$\begin{aligned} L(\dot{x},u) &= \phi^T(\dot{x})R(\dot{x})\phi(\dot{x}) - \dot{x}^T P f(\dot{x}) + L_2(\dot{x})u + u^T R(\dot{x})u \\ &= \phi^T(\dot{x})R(\dot{x})\phi(\dot{x}) + L_2(\dot{x})u + u^T R(\dot{x})u - \dot{x}^T P[f(\dot{x}) + g(\dot{x})\phi(\dot{x})] + \dot{x}^T P g(\dot{x})\phi(\dot{x}) \end{aligned} \tag{41}$$

From (35), we have

$$L_2(\dot{x}) = -2\phi^T(\dot{x})R(\dot{x}) - \dot{x}^T P g(\dot{x}). \tag{42}$$

Substituting (34) and (42) into (41) and using (29) yields

$$\begin{aligned} L(\dot{x},\phi(\dot{x})) &= \phi^T(\dot{x})R(\dot{x})\phi(\dot{x}) + \left(-2\phi^T(\dot{x})R(\dot{x}) - \dot{x}^T P g(\dot{x})\right)\phi(\dot{x}) + \phi^T(\dot{x})R(\dot{x})\phi(\dot{x}) \\ &\quad -\dot{x}^T P[f(\dot{x}) + g(\dot{x})\phi(\dot{x})] + \dot{x}^T P g(\dot{x})\phi(\dot{x}) \\ &= -\dot{x}^T P[f(\dot{x}) + g(\dot{x})\phi(\dot{x})] = -\dot{V}(x) \end{aligned} \tag{43}$$

Therefore, based on (43) when inverse optimal law in (34) is applied, the closed loop SDS system is stable, performance functional in (28) becomes

$$J(x_0, \phi(\dot{x})) = \int_0^\infty L(\dot{x}, \phi(\dot{x}))dt = -\int_0^\infty \dot{V}(x)dt = -\lim_{t\to\infty} V(x(t)) + V(x_0) = \frac{1}{2}x_0^T P x_0 > 0 \tag{44}$$

Hence, to have a small value of performance functional, one may consider to select a diagonal P matrix with positive but small diagonal elements.

2.4. Inverse Optimal Control for Affine SDS Systems with L_2 Disturbance

Consider the nonlinear affine SDS system with bounded L_2 input disturbance $\omega(t)$ [27] in the following form.

$$x(t) = f(\dot{x}(t)) + g(\dot{x}(t))u + J_1(\dot{x}(t))\omega(t), x(0) = x_0, \omega(\cdot) \in L_2, t \geq 0 \tag{45}$$

with the following performance variables.

$$z = h(\dot{x}(t)) + J(\dot{x}(t))u(t), \tag{46}$$

We consider the non-expansivity case [27] so that the supply rate is given by

$$r(z, w) = \gamma^2 \omega^T \omega - z^T z \tag{47}$$

where $\gamma > 0$.

An inverse optimal globally stabilizing control law should be designed so that the closed loop system satisfies the non-expansivity constraint [27].

$$\int_0^T z^T z dt \leq \int_0^T \gamma^2 \omega^T \omega dt + V(x_0),\ T \geq 0,\ \omega(\cdot) \in L_2 \tag{48}$$

The performance integrand is considered as

$$L(\dot{x},u) = L_1(\dot{x}) + L_2(\dot{x})u + u^T R(\dot{x})u \tag{49}$$

Therefore, the performance functional becomes

$$J(x_0,u(\cdot)) = \int_0^{\infty}\left[L_1(\dot{x}) + L_2(\dot{x})u + u^T R(\dot{x})u\right]dt \tag{50}$$

The following process is to construct an inverse optimal globally stabilizing control law $\phi(\dot{x})$ with state derivative feedback.

First, a symmetric and positive definite matrix P serving as the first design parameter should be selected for Lyapunov function in (11). Consequently, substituting SDS system Equation in (45), we have

$$\dot{V}(x) = \frac{dV(x)}{dt} = x^T P\dot{x} = \dot{x}^T Px = \dot{x}^T P[f(\dot{x}(t)) + g(\dot{x}(t))u + J_1(\dot{x}(t))\omega(t)] \tag{51}$$

In [22], H_∞ control has been carried out for the same SDS system in (45), and the $\omega^T\omega$ is maximum when disturbance is

$$\omega = \omega^* = \frac{1}{2\gamma^2}J_1^T(\dot{x})P\dot{x} \text{ and } \omega^{*T}\omega^* = \frac{1}{4\gamma^2}\dot{x}^T PJ_1(\dot{x})J_1^T(\dot{x})P\dot{x} \tag{52}$$

Considering (46) and (52), when an inverse optimal globally stabilizing control law $\phi(\dot{x})$ is obtained, we should have the following conditions in (53)–(55).

$$\Gamma(\dot{x},\phi(\dot{x})) \geq 0 \text{with } \Gamma(\dot{x},u) = \frac{1}{4\gamma^2}\dot{x}^T PJ_1(\dot{x})J_1^T(\dot{x})P\dot{x} + [h(\dot{x}) + J(\dot{x})u]^T[h(\dot{x}) + J(\dot{x})u] \tag{53}$$

$$\dot{x}^T P[f(\dot{x}(t)) + g(\dot{x}(t))\phi(\dot{x})] < 0 \tag{54}$$

$$\dot{x}^T PJ_1(\dot{x})\omega \leq \gamma^2\omega^T\omega - z^T z + L(\dot{x},\phi(\dot{x})) + \Gamma(\dot{x},\phi(\dot{x})) \tag{55}$$

Therefore, applying (55) yields

$$\begin{gathered}\dot{V}(x) = \tfrac{dV(x)}{dt} = \dot{x}^T P[f(\dot{x}) + g(\dot{x})\phi(\dot{x}) + J_1(\dot{x})\omega(t)] \\ \leq \dot{x}^T P[f(\dot{x}) + g(\dot{x})\phi(\dot{x})] + \gamma^2\omega^T\omega - z^T z + L(\dot{x},\phi(\dot{x})) + \Gamma(\dot{x},\phi(\dot{x}))\end{gathered} \tag{56}$$

Second, an auxiliary cost functional is specified as

$$\Im(x_0,u(\cdot)) = \int_0^{\infty}[L(\dot{x},u) + \Gamma(\dot{x},u)]dt \tag{57}$$

From (50), (53), and (57) yields

$$J(x_0,\phi(\dot{x})) \leq \Im(x_0,\phi(\dot{x})) = \int_0^{\infty}[L(\dot{x},\phi(\dot{x})) + \Gamma(\dot{x},\phi(\dot{x}))]dt \tag{58}$$

Third, with (45), (49), and (53), and the Hamiltonian has the form

$$\begin{gathered}H(\dot{x},u) = L_1(\dot{x}) + L_2(\dot{x})u + u^T R(\dot{x})u + \dot{x}^T P[f(\dot{x}) + g(\dot{x})u] \\ +\tfrac{1}{4\gamma^2}\dot{x}^T PJ_1(\dot{x})J_1^T(\dot{x})P\dot{x} + [h(\dot{x}) + J(\dot{x})u]^T[h(\dot{x}) + J(\dot{x})u] \geq 0\end{gathered} \tag{59}$$

Then, with the feedback control law $\phi(\dot{x})$, there exists a neighborhood of the origin such that if x_0 within this neighborhood and when SDS system in (45) is undisturbed $(\omega(t) \equiv 0)$, the zero solution $x(t) \equiv 0$ of the closed loop system is locally asymptotically stable.

We should select a positive definite $R(\dot{x})$ so that $u^T R(\dot{x})u > 0$ in (49), followed by setting

$$\frac{\partial H(\dot{x},u)}{\partial u} = L_2^T(\dot{x}) + 2R(\dot{x})u + g^T(\dot{x})P\dot{x} + 2J^T(\dot{x})J(\dot{x})u + 2J(\dot{x})h(\dot{x}) = 0, \quad (60)$$

and define

$$R_a(\dot{x}) = R(\dot{x}) + J^T(\dot{x})J(\dot{x}) \quad (61)$$

the inverse optimal state derivative related feedback control law is obtained as follows.

$$u = \phi(\dot{x}) = \frac{-1}{2}R_a^{-1}(\dot{x})\left[L_2^T(\dot{x}) + g^T(\dot{x})P\dot{x} + 2J^T(\dot{x})h(\dot{x})\right] \quad (62)$$

Consequently, from (62) yields

$$\left[L_2(\dot{x}) + \dot{x}^T Pg(\dot{x}) + 2h^T(\dot{x})J(\dot{x})\right] = -2\phi(\dot{x})R_a(\dot{x}) \quad (63)$$

$L_2(\dot{x})$ should be selected such that

$$\dot{x}^T P\left[f(\dot{x}) - \frac{1}{2}g(\dot{x})R_a^{-1}\left(L_2^T(\dot{x}) + g^T(\dot{x})P\dot{x} + 2J^T(\dot{x})h(\dot{x})\right)\right] + \Gamma(\dot{x},\phi(\dot{x})) < 0 \quad (64)$$

According to (53), (64) implies (54).
In addition, the auxiliary cost functional in (57), with

$$L_1(\dot{x}) = \phi^T(\dot{x})R_a(\dot{x})\phi(\dot{x}) - \dot{x}^T Pf(\dot{x}) - h^T(\dot{x})h(\dot{x}) - \frac{1}{4\gamma^2}\dot{x}^T PJ_1(\dot{x})J_1^T(\dot{x})P\dot{x} \quad (65)$$

in the sense that

$$\Im(x_0,\phi(\dot{x})) = \min\Im(x_0,u(\cdot)) \quad (66)$$

Applying (53), (59), (63), and (65), we have

$$\begin{aligned} H(\dot{x},u) &= L_1(\dot{x}) + L_2(\dot{x})u + u^T R(\dot{x})u + \Gamma(\dot{x},u) + \dot{x}^T P[f(\dot{x}) + g(\dot{x})u] \\ &= \phi^T(\dot{x})R_a(\dot{x})\phi(\dot{x}) + \left[L_2(\dot{x}) + \dot{x}^T Pg(\dot{x}) + 2h^T(\dot{x})J(\dot{x})\right]u + u^T\left[R(\dot{x}) + J_1^T(\dot{x})J_1(\dot{x})\right]u \\ &= \phi^T(\dot{x})R_a(\dot{x})\phi(\dot{x}) - 2\phi(\dot{x})R_a(\dot{x})u + u^T R_a(\dot{x})u \\ &= [u-\phi(\dot{x})]^T R_a(\dot{x})[u-\phi(\dot{x})] \geq 0 \end{aligned} \quad (67)$$

Hence,

$$H(\dot{x},\phi(\dot{x})) = 0 \quad (68)$$

Furthermore, (67) and (68) imply that

$$\begin{aligned} L(\dot{x},\phi(\dot{x})) + \Gamma(\dot{x},\phi(\dot{x})) &= L_1(\dot{x}) + L_2(\dot{x})\phi(\dot{x}) + \phi(\dot{x})R(\dot{x})\phi(\dot{x}) + \Gamma(\dot{x},\phi(\dot{x})) \\ &= -\dot{x}^T P[f(\dot{x}) + g(\dot{x})\phi(\dot{x})] > 0 \end{aligned} \quad (69)$$

Substituting (69) into (56) yields

$$\dot{V}(x) = \frac{dV(x)}{dt} = \dot{x}^T P[f(\dot{x}) + g(\dot{x})\phi(\dot{x}) + J_1(\dot{x})\omega(t)] \leq \gamma^2\omega^T\omega - z^T z \quad (70)$$

Integrating over $[\ 0,\ \ T\ ]$

$$V(x(T)) - V(x_0) \leq \int_0^T \left(\gamma^2\omega^T\omega - z^T z\right)dt \quad (71)$$

$$V(x(T)) + \int_0^T z^T z dt \leq \int_0^T \gamma^2\omega^T\omega dt + V(x_0) \quad (72)$$

$$\because V(x(T)) \geq 0 \quad (73)$$

$$\therefore \int_0^T z^T z dt \leq \int_0^T \gamma^2 \omega^T \omega dt + V(x_0) \tag{74}$$

Therefore, applying the inverse optimal control law in (62), the closed loop system satisfies the non-expansivity constraint in (74).

2.5. Brief Mathematical Review of Singular System with Impulse Mode

As mentioned in the introduction, singular systems with impulse mode are difficult in control designs with state related feedback alone. However, some of singular systems with impulse mode can be expressed in RSS system form and can be fully controlled with state derivative feedback. An example of such system will be provided in next section to verify the proposed design process, but before that, the limitation of applying state feedback alone to control singular system with impulse mode is reviewed in this subsection.

The researches of linear singular system control mainly focus on the impulse-free mode. For the following linear and time invariant singular system

$$E\dot{x} = Fx + Nu \tag{75}$$

When matrix E in (75) has zero eigenvalues, it cannot be expressed in state space system form. For people to better understand the nature of this kind of system, singular value decomposition (SVD) can be applied to convert the system to new coordinates as follows.

$$\begin{bmatrix} I & 0 \\ 0 & 0 \end{bmatrix} \begin{bmatrix} \dot{q}_1 \\ \dot{q}_2 \end{bmatrix} = \begin{bmatrix} F_{11} & F_{12} \\ F_{21} & F_{22} \end{bmatrix} \begin{bmatrix} q_1 \\ q_2 \end{bmatrix} + \begin{bmatrix} N_1 \\ N_2 \end{bmatrix} u \tag{76}$$

The singular system is *impulse-free* when matrix F_{22} is invertible. Consequently, q_2 and q_1 are coupled by the following equation.

$$q_2 = -F_{22}^{-1} F_{21} q_1 - F_{22}^{-1} N_2 u \tag{77}$$

Substituting (77) into the first equation in (76) gives the following subsystem in state space system form with state vector of q_1.

$$\dot{q}_1 = (F_{11} - F_{12} F_{22}^{-1} F_{21}) q_1 + (N_1 - F_{12} F_{22}^{-1} N_2) u \tag{78}$$

Therefore, the state vector q_1 can be fully controlled with state feedback design if the subsystem in (78) is controllable. However, through the coupling in (77), state vector q_2 is only stabilized but not fully controlled.

When matrix F_{22} is noninvertible, the singular system has impulse mode. This kind of system is usually very difficult to be controlled with state feedback alone. If it is impulse controllable, applying proper state feedback control may obtain a stabilizing closed loop systems as follows.

$$\begin{bmatrix} I & 0 \\ 0 & 0 \end{bmatrix} \begin{bmatrix} \dot{q}_1 \\ \dot{q}_2 \end{bmatrix} = \begin{bmatrix} F_{c11} & F_{c12} \\ F_{c21} & F_{c22} \end{bmatrix} \begin{bmatrix} q_1 \\ q_2 \end{bmatrix} \tag{79}$$

In such case, matrix F_{c22}^{-1} must exist. Similarly, state vector q_2 still can only be stabilized through its coupling with q_1. If the system is impossible to apply state feedback to obtain an invertible matrix F_{c22}, it is called impulse uncontrollable in the research literature. Consequently, the system is not stabilizable by applying state feedback alone. In short, applying state feedback alone cannot control the entire singular system. Singular systems can be stabilized with state feedback control only if it is impulse free or impulse controllable.

However, if matrix F of the singular system with impulse mode in (75) is invertible, we may express it in the following SDS form to fully control it with state derivative feedback alone.

$$x = F^{-1}E\dot{x} - F^{-1}Nu = f\dot{x} + gu. \tag{80}$$

In next section, we will provide an example that verify the proposed design process in this section.

3. Examples and Results

In this section, we provide four examples to verify the effectiveness of the proposed design method. In addition, from both the implementation and mathematical point of views, the advantages of using direct state derivative measurement in control design of SDS system are explained.

Example 1. *There are part (a) and part (b) in this example for different purposes.*

(a) *To illustrate the utility of the proposed design process for SDS systems and to emphasize that some SDS systems have no equivalent state space form, we consider*

$$x(t) = \begin{bmatrix} x_1(t) \\ x_2(t) \end{bmatrix} = \begin{bmatrix} -\dot{x}_1^5(t) + \dot{x}_2^2(t) \\ \dot{x}_1^2(t) \end{bmatrix} + \begin{bmatrix} 0 \\ 1 \end{bmatrix} u(t) = f(\dot{x}(t)) + g(\dot{x}(t))u(t) \quad (81)$$

Please note that we cannot convert the SDS form in (81) to state space form because the characteristics of original SDS system will be lost after using square root or power of even number order operation.

First, we select $R(\dot{x}) = 1$ *and* $V(x) = \frac{1}{2}x^T Px = \frac{1}{2}\begin{bmatrix} x_1 & x_2 \end{bmatrix}\begin{bmatrix} 2 & 0 \\ 0 & 2 \end{bmatrix}\begin{bmatrix} x_1 \\ x_2 \end{bmatrix} = x_1^2 + x_2^2$.

Based on (36) and (37), $L_2(\dot{x})$ *should be selected such that*

$$\dot{V}(x) = \begin{bmatrix} 2\dot{x}_1 & 2\dot{x}_2 \end{bmatrix}\begin{bmatrix} -\dot{x}_1^5 + \dot{x}_2^2 \\ \dot{x}_1^2 - \frac{1}{2}L_2^T(\dot{x}) - \frac{1}{2}(2\dot{x}_2) \end{bmatrix} = -2\dot{x}_1^6 + 2\dot{x}_1\dot{x}_2^2 + 2\dot{x}_1^2\dot{x}_2 - L_2^T(\dot{x})\dot{x}_2 - 2\dot{x}_2^2 < 0$$

Select $L_2(\dot{x}) = 2\left(\dot{x}_1\dot{x}_2 + \dot{x}_1^2\right)$ *, we have*

$$\dot{V}(x) = \begin{bmatrix} 2\dot{x}_1 & 2\dot{x}_2 \end{bmatrix}\begin{bmatrix} -\dot{x}_1^5 + \dot{x}_2^2 \\ \dot{x}_1^2 - \frac{1}{2}L_2^T(\dot{x}) - \frac{1}{2}(2\dot{x}_2) \end{bmatrix} = -2\dot{x}_1^6 + 2\dot{x}_1\dot{x}_2^2 + 2\dot{x}_1^2\dot{x}_2 - L_2^T(\dot{x})\dot{x}_2 - 2\dot{x}_2^2 < 0\dot{V}(x) = -2\dot{x}_1^6 - 2\dot{x}_2^2 < 0$$

Therefore, the closed loop system is stable and the corresponding inverse optimal control law $\phi(\dot{x})$ *is obtained using (34) as*

$$\phi(\dot{x}) = \frac{-1}{2}R^{-1}(\dot{x})\left[L_2^T(\dot{x}) + g^T(\dot{x})P\dot{x}\right] = -\dot{x}_1\dot{x}_2 - \dot{x}_1^2 - \dot{x}_2.$$

Next, using (38) obtains

$$L_1(\dot{x}) = \phi^T(\dot{x})R(\dot{x})\phi(\dot{x}) - \dot{x}^T Pf(\dot{x}) = \left(\dot{x}_1\dot{x}_2 + \dot{x}_1^2 + \dot{x}_2\right)^2 - 2\dot{x}_1\left(-\dot{x}_1^5 + \dot{x}_2^2\right) - 2\dot{x}_2\dot{x}_1^2.$$

Consequently, using (30) obtains the performance integrand in (28) as

$$\begin{aligned} L(\dot{x}, u) &= L_1(\dot{x}) + L_2(\dot{x})u + u^T R(\dot{x})u \\ &= \left(\dot{x}_1\dot{x}_2 + \dot{x}_1^2 + \dot{x}_2\right)^2 + 2\dot{x}_1^6 - 2\dot{x}_1\dot{x}_2^2 - 2\dot{x}_2\dot{x}_1^2 + 2\left(\dot{x}_1\dot{x}_2 + \dot{x}_1^2\right)u + u^T u. \end{aligned}$$

Furthermore, it can be shown that

$$H(\dot{x}, \phi(\dot{x})) = L(\dot{x}, \phi(\dot{x})) + \dot{V}(x) = 0.$$

(b) *To emphasize the possibility of constructing a stable closed loop system in SDS form with state feedback control, we consider*

$$x = \begin{bmatrix} x_1 \\ x_2 \end{bmatrix} = \begin{bmatrix} -\dot{x}_1 + \dot{x}_2 \\ \dot{x}_1 + \dot{x}_2^3 \end{bmatrix} + \begin{bmatrix} 0 \\ 1 \end{bmatrix} u = f(\dot{x}) + g(\dot{x})u \tag{82}$$

Please note that the spirit of handling nonlinear control in SDS system is to construct a stable closed loop SDS system with feedback control regardless of using state derivative feedback or state feedback. In this example, if we use state feedback control $u = 2x_2$, *the closed loop SDS system becomes*

$$x = \begin{bmatrix} x_1 \\ x_2 \end{bmatrix} = \begin{bmatrix} -\dot{x}_1 + \dot{x}_2 \\ -\dot{x}_1 - \dot{x}_2^3 \end{bmatrix}$$

If

$$P = \begin{bmatrix} 2 & 0 \\ 0 & 2 \end{bmatrix},$$

we have

$$\dot{V}(x) = \dot{x}^T P x = \begin{bmatrix} 2\dot{x}_1 & 2\dot{x}_2 \end{bmatrix} \begin{bmatrix} -\dot{x}_1 + \dot{x}_2 \\ -\dot{x}_1 - \dot{x}_2^3 \end{bmatrix} = -2\dot{x}_1^2 - 2\dot{x}_2^4 < 0.$$

Therefore, closed loop SDS system is stable.

Similarly, it is also possible to apply state derivative feedback to stabilize a nonlinear system in state space form. For example, for the following nonlinear state space system

$$\dot{x} = \begin{bmatrix} \dot{x}_1 \\ \dot{x}_2 \end{bmatrix} = \begin{bmatrix} -x_1 + x_2 \\ x_1 + x_2^3 \end{bmatrix} + \begin{bmatrix} 0 \\ 1 \end{bmatrix} u$$

If we apply the following state derivative feedback control $u = 2\dot{x}_2$, *the closed loop SDS system becomes*

$$\dot{x} = \begin{bmatrix} \dot{x}_1 \\ \dot{x}_2 \end{bmatrix} = \begin{bmatrix} -x_1 + x_2 \\ -x_1 - x_2^3 \end{bmatrix}$$

and also select the same $V(x)$, *it follows*

$$\dot{V}(x) = x^T P \dot{x} = \begin{bmatrix} 2x_1 & 2x_2 \end{bmatrix} \begin{bmatrix} -x_1 + x_2 \\ -x_1 - x_2^3 \end{bmatrix} = -2x_1^2 - 2x_2^4 < 0$$

Therefore, closed loop state space system is stable.

Example 2. *A popular actuator widely used in control systems is DC motor. The DC motor model in [38] is modified and adapted for this example. Figure 1 shows the free-body diagram of the rotor and the equivalent circuit of the armature. As seen in the figure, a small inductor L1 externally connected in series with armature circuit of a DC motor serves as the only sensor of the control system and we assume that tachometer is not installed to measure the rotational speed* $\dot{\theta}$ *of the shaft. The voltage of L1 is measured and used in feedback controller design. Unlike the traditional DC motor controls which apply state related feedback of angular velocity or current, the inductor's voltage* ($\mathrm{L1}\,\frac{di}{dt}$) *is state derivative related measurement feedback of current derivative which is well suitable to apply the proposed IOC algorithm based on state derivative feedback. Furthermore, inductor's average power is zero so it does not damage the armature circuit. Therefore, the controller can save implementation cost and avoid power lose.*

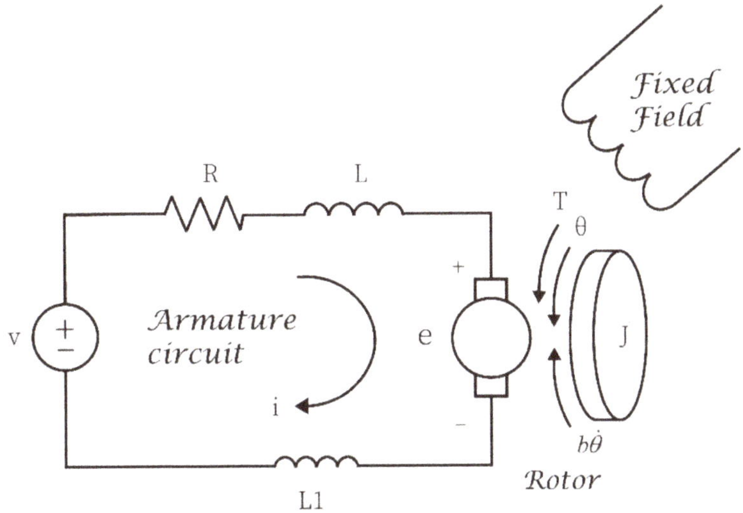

Figure 1. The rotor free-body and the electric equivalent armature circuit diagram.

In this example, torque tracking controller and rotational speed tracking controller are constructed based on the design algorithm of inverse optimal control for affine SDS Systems with state derivative related feedback. This example is used to illustrate the design method in Section 2.3 for linear time invariant SDS systems, namely RSS systems.

From the above figure, according to Newton's second law and Kirchhoff's voltage law, we can get the following governing Equations.

$$\mathrm{J}\ddot{\theta} + b\dot{\theta} = K\mathrm{i}$$

$$(\mathrm{L} + \mathrm{L1})\frac{di}{dt} + \mathrm{i}R = \mathrm{v} - K\dot{\theta}$$

where i is armature circuit's current, θ is rotor's rotational angle, R is electric resistance, L is electric inductance, J is rotor's moment of inertia, b is motor viscous friction constant, L1 is the external inductor sensor connected in series with armature circuit, and K represents both electromotive force constant and motor torque constant in SI unit.

For simulation purpose, the DC motor's physical parameters in this example are given as R:1 Ω, L: 0.49 H, J: 0.01 $kg.m^2$, L1: 0.01 H and K : 0.01 $V/rad/sec$ for electromotive force constant and 0.01 $N.m/Amp$ for motor torque constant.

Defining state vector as $x = \begin{bmatrix} \dot{\theta} \\ \mathrm{i} \end{bmatrix}$, using the governing equations, one can obtain the following SDS system.

$$x = \begin{bmatrix} \dot{\theta} \\ \mathrm{i} \end{bmatrix} = \frac{1}{\mathrm{R}b + K^2}\begin{bmatrix} -\mathrm{RJ} & -K(\mathrm{L}+\mathrm{L1}) \\ K\mathrm{J} & -(\mathrm{L}+\mathrm{L1}) \end{bmatrix}\begin{bmatrix} \ddot{\theta} \\ \frac{di}{dt} \end{bmatrix} + \frac{1}{\mathrm{R}b + K^2}\begin{bmatrix} K \\ b \end{bmatrix}\mathrm{v} \tag{83}$$

Substituting the physical parameters of the DC motor into above SDS system, one obtains

$$x = \begin{bmatrix} \dot{\theta} \\ \mathrm{i} \end{bmatrix} = \begin{bmatrix} -0.0999 & -0.05 \\ 0.001 & -4.995 \end{bmatrix}\begin{bmatrix} \ddot{\theta} \\ \frac{di}{dt} \end{bmatrix} + \begin{bmatrix} 0.0999 \\ 0.9990 \end{bmatrix}\mathrm{v} = f\dot{x} + gu \tag{84}$$

Since the L1 inductor's voltage is state derivative related measurement, the measurement of the system is given as

$$y = \begin{bmatrix} 0 & L1 \end{bmatrix} \begin{bmatrix} \ddot{\theta} \\ \frac{di}{dt} \end{bmatrix} = \begin{bmatrix} 0 & 0.01 \end{bmatrix} \begin{bmatrix} \ddot{\theta} \\ \frac{di}{dt} \end{bmatrix} = C\dot{x} \tag{85}$$

The measurement feedback control law is given as

$$u = -ky = -kC\dot{x}$$

where k is the measurement feedback gain.

If the rotational speed of the shaft $\dot{\theta}$ needs to track a reference command r_0, the performance output equation is given as follows.

$$z = \begin{bmatrix} 1 & 0 \end{bmatrix} \begin{bmatrix} \dot{\theta} \\ \mathrm{i} \end{bmatrix} = H_s x \tag{86}$$

If the torque $T = K\mathrm{i}$ needs to track a reference command r_0, the performance output equation is given as follows.

$$z = \begin{bmatrix} 0 & K \end{bmatrix} \begin{bmatrix} \dot{\theta} \\ \mathrm{i} \end{bmatrix} = \begin{bmatrix} 0 & 0.01 \end{bmatrix} \begin{bmatrix} \dot{\theta} \\ \mathrm{i} \end{bmatrix} = H_t x \tag{87}$$

From (83) to (84), the DC motor model is linear and time invariant. Therefore, the SDS system is also a RSS system as mentioned in introduction section. Therefore, the open loop system poles: -9.9975 and -2.0025 are the reciprocals of the eigenvalues of matrix f in the system. Although the open loop RSS system is stable, its tracking performance can be further improved.

To carry out tracking control, the closed loop system should be stable.

First, a symmetric and positive definite matrix P serving as the design parameter for Lyapunov function $V(x)$ in (11) is selected as follows.

$$P = \begin{bmatrix} 1 & 0 \\ 0 & 0.5 \end{bmatrix}.$$

Followed by selecting another design parameter $R(\dot{x})$ as

$$R(\dot{x}) = 0.0004.$$

To ensure the closed loop system is stable, based on (37), the $L_2(\dot{x})$ is selected as

$$L_2(\dot{x}) = \begin{bmatrix} \ddot{\theta} & \frac{di}{dt} \end{bmatrix} \begin{bmatrix} -0.0999 \\ -0.4999 \end{bmatrix}.$$

Using (34), the inverse optimal law $\phi(\dot{x})$ is obtained as

$$\phi(\dot{x}) = 0.497976\frac{di}{dt} = -kC\dot{x} = -k\begin{bmatrix} 0 & 0.01 \end{bmatrix} \begin{bmatrix} \ddot{\theta} \\ \frac{di}{dt} \end{bmatrix} = -0.01k\frac{di}{dt}.$$

Therefore, we obtain the measurement feedback gain k in (85) as

$$k = -49.7976.$$

Consequently, use (38) to obtain $L_1(\dot{x})$ as

$$L_1(\dot{x}) = \begin{bmatrix} \ddot{\theta} & \frac{di}{dt} \end{bmatrix} \begin{bmatrix} 0.0999 & 0.0500 \\ -0.0005 & 0.2498 \end{bmatrix} \begin{bmatrix} \ddot{\theta} \\ \frac{di}{dt} \end{bmatrix}$$

and use (41) to obtain the performance integrand $L(\dot{x}, u)$.

Applying the obtained control law with feedback gain, the close loop system poles have been moved to better locations at -10.0102 and -493.9479. We use the following tracking controller to illustrate the improvement of closed loop system.

$$u = -ky + Nr_0 = -kC\dot{x} + Nr_0$$

where r_0 is given reference command vector to be tracked by performance output and N is a feedforward gain to be designed.

The steady state derivative is zero $(\dot{\overline{x}}(\infty) = 0)$ when the RSS closed loop system is stable. In that case, the steady state is

$$\overline{x} = (f - gkC)\dot{\overline{x}} + gNr_0 = gNr_0.$$

If the performance output is $z = Hx$, to have zero tracking error of steady state, let

$$\overline{\varepsilon}(\infty) = r_0 - H\overline{x} = r_0 - HgNr_0 = (I - HgN)r_0 = 0.$$

Consequently, the feedforward gain N is obtained as

$$N = (Hg)_{right}^{-1} \tag{88}$$

where $(Hg)_{right}^{-1}$ is the right inverse of matrix Hg.

If Hg is a full rank matrix with the size of $m \times n$ and $m \leq n$, we have

$$(Hg)_{right}^{-1} = (Hg)^T (Hg(Hg)^T)^{-1}$$

The feedforward gain N and feedback gain k can be designed separately because they are independent of each other.

Therefore, according to (86), for rotational speed tracking, we have feedforward gain N_s as

$$N_s = (H_s g)_{right}^{-1} = 10.01$$

Similarly, for motor torque tracking, from (87) we have feedforward gain N_t as

$$N_t = (H_t g)_{right}^{-1} = 100.1$$

Since the open loop system is stable, we can give the reference command for system to track. As seen in Figure 2, there are obvious attenuation and phase lag for tracking a $\sin t$ reference command of rotational speed $\dot{\theta}$.

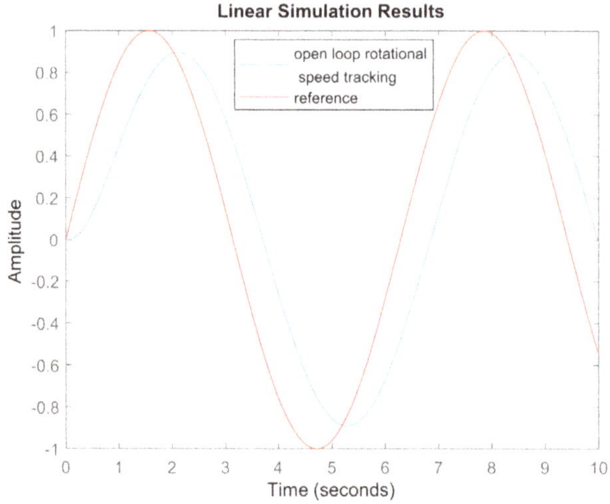

Figure 2. Open loop tracking result of rotational speed $\dot{\theta}$ with reference command $\sin t$.

Furthermore, as seen in Figure 3, both attenuation and phase lag are large for tracking a $\sin 50t$ waveform of reference command of torque Ki.

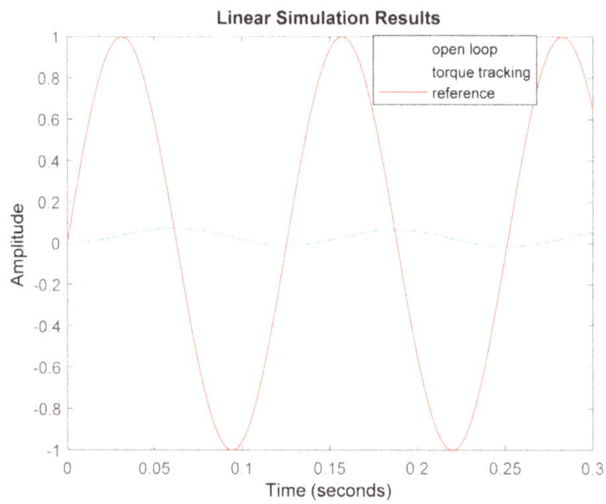

Figure 3. Open loop tracking result of torque Ki with reference command $\sin 50t$.

Therefore, tracking performance should be improved. Applying the obtained inverse optimal law, as seen in Figure 4, rotational speed $\dot{\theta}$ can better track $\sin t$ reference command. There is no attenuation and phase lag is only 5.72°.

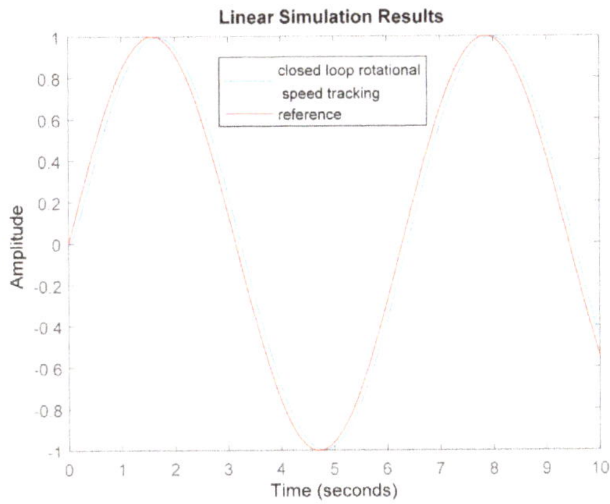

Figure 4. Closed loop tracking result of rotational speed $\dot{\theta}$ with reference command $\sin t$.

Applying the obtained inverse optimal law, as seen in Figure 5, torque Ki can track $\sin 50t$ reference command much better. There is no attenuation and phase lag is only 5.72°.

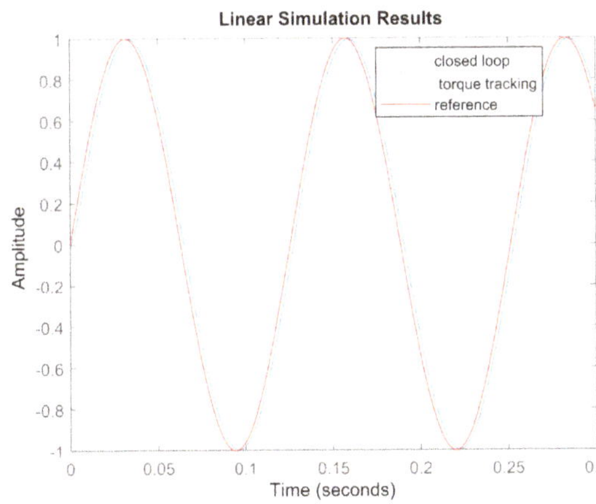

Figure 5. Closed loop tracking result of torque Ki with reference command $\sin 50t$.

Therefore, the control law works well to improve the tracking performance.

Example 3. *The following circuit in Figure 6 is a typical singular system with impulse mode from [39] with C = 1. It is unstable and has pole at infinity. This example is used to illustrate the design approach of Inverse Optimal Control for Affine SDS Systems with L_2 Disturbance in Section 2.4.*

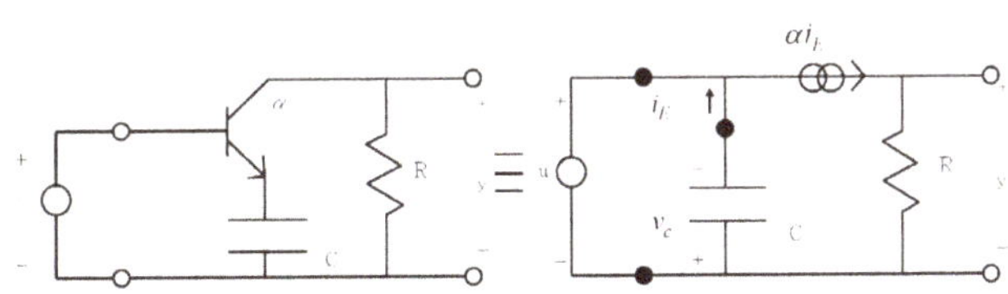

Figure 6. Singular circuit with impulse mode.

$$E\dot{x} = \begin{bmatrix} 1 & 0 \\ 0 & 0 \end{bmatrix}\dot{x} = \begin{bmatrix} 0 & 1 \\ 1 & 0 \end{bmatrix}x + \begin{bmatrix} 0 \\ 1 \end{bmatrix}u = Fx + Nu,\ x = \begin{bmatrix} v_C \\ i_E \end{bmatrix} = \begin{bmatrix} x_1 \\ x_2 \end{bmatrix}$$

Since matrix F is invertible, the singular system with impulse mode can be expressed in the following SDS system form.

$$x = F^{-1}E\dot{x} - F^{-1}Nu = f\dot{x} + gu \tag{89}$$

For verifying the proposed algorithm, external disturbance ω is added to the system as follows.

$$\begin{gathered} x = \begin{bmatrix} 0 & 0 \\ 2 & 0 \end{bmatrix}\dot{x} + \begin{bmatrix} -1 \\ 0 \end{bmatrix}u + \begin{bmatrix} 0.1 \\ -0.1 \end{bmatrix}\omega = f\dot{x} + gu + J_1\omega \\ z = \begin{bmatrix} 2 & 1 \end{bmatrix}\dot{x} + 0.8u = h\dot{x} + Ju \\ u = -K_\infty\dot{x},\ \gamma = 0.6 \end{gathered}$$

First, a symmetric and positive definite matrix P serving as the design parameter for Lyapunov function $V(x)$ in (11) is selected as follows.

$$P = \begin{bmatrix} 1 & 0 \\ 0 & 1 \end{bmatrix}.$$

Followed by selecting another design parameter $R(\dot{x})$ as

$$R(\dot{x}) = 1.$$

Consequently,

$$\Gamma(\dot{x}, u) = \dot{x}^T\begin{bmatrix} 4.0069 & 1.9931 \\ 1.9931 & 1.0069 \end{bmatrix}\dot{x} + \dot{x}^T\begin{bmatrix} 3.2 \\ 1.6 \end{bmatrix}u + 0.64u^Tu$$

To ensure the closed loop system is stable, based on (64), the $L_2(\dot{x})$ is selected as

$$L_2(\dot{x}) = \begin{bmatrix} \dot{x}_1 & \dot{x}_2 \end{bmatrix}\begin{bmatrix} -9.9759 \\ -4.9702 \end{bmatrix} \tag{90}$$

Then using (62) yields

$$\phi(\dot{x}) = \begin{bmatrix} 2.3707 & 1.0275 \end{bmatrix}\dot{x} = -K_\infty\dot{x} \tag{91}$$

Therefore, the obtained full state derivative feedback gain is

$$K_\infty = \begin{bmatrix} -2.3707 & -1.0275 \end{bmatrix}.$$

Applying the control law, the closed loop system poles locate at $-0.5768 \pm 0.3923i$. Since their real parts are all negative, the closed loop system are stable.

Applying (65), we have

$$L_1(\dot{x}) = \dot{x}^T\begin{bmatrix} 2.2102 & 2.0018 \\ 0.0018 & 0.7245 \end{bmatrix}\dot{x}$$

Furthermore, applying (91) obtains

$$\Gamma(\dot{x}, \phi(\dot{x})) = \dot{x}^T\begin{bmatrix} 15.1901 & 7.0926 \\ 7.0926 & 3.3266 \end{bmatrix}\dot{x} \tag{92}$$

Substituting (90) and (92) into (64) yields

$$-12.5019x^T x < 0$$

It proved that $L_2(\dot{x})$ is properly selected.

In addition, this SDS system is actually controllable with state derivative feedback control. For example, if we want to assign the closed loop poles at -2 and -4 using full state derivative feedback control law $u = -K\dot{x}$, the closed loop SDS system becomes

$$x = f\dot{x} + gu = (f - gK)\dot{x}.$$

The gain K should be designed such that matrix $(f - gK)$ has eigenvalues at -0.5 and -0.25 because they are the reciprocals of -2 and -4, respectively. In this case, using Matlab command *place*, one can easily find $K = \begin{bmatrix} -0.7500 & -0.0625 \end{bmatrix}$. Therefore, using state derivative feedback control, it is possible to assign all closed loop poles for some singular systems with impulse mode if they can be expressed in a controllable SDS system form. However, using state feedback control, it is impossible to assign all closed loop poles for any singular systems. It is only possible to stabilize some of the singular systems with state feedback control.

Example 4. *It is interesting to compare the proposed inverse optimal control (IOC) in Section 2.4 with sliding mode control (SMC) design for SDS system with matched disturbance because they are both developed based on Lyapunov stability theorem. Consider the following unstable SDS system with matched disturbance given in [20].*

$$x(t) = f\dot{x}(t) + gu(t) + J_1\omega(t) = f\dot{x} + g(u + \omega(t)),\ t \geq 0 \tag{93}$$

where $f = \begin{bmatrix} 1 & -0.5 & 0.25 \\ 0 & 0.5 & -0.25 \\ 0 & 0 & 0.5 \end{bmatrix}$, $g = J_1 = \begin{bmatrix} -0.25 \\ 0.25 \\ -0.5 \end{bmatrix}$ (for matched disturbance), $x(t) = \begin{bmatrix} x1 \\ x2 \\ x3 \end{bmatrix}$, and $\omega(t) = 0.2\sin 0.3333t$, with the following full state derivative performance variables.

$$z = h\dot{x}(t) + Ju(t) = \dot{x}(t)$$

where $h = I_{3\times3}$ (identity matrix) and $J = [0\ 0\ 0]^{\mathrm{T}}$.

(a) For sliding mode control (SMC) [20], the sliding surface is selected as

$$s = \begin{bmatrix} -84 & -180 & -50 \end{bmatrix} x = Cx$$

Consequently, we have $Cg = 1$.

The ideal controller is given as

$$u(t) := -(Cg)^{-1}Cf\dot{x}(t) - (Cg)^{-1}(\gamma + \alpha)sign(\dot{s}(t)) \tag{94}$$

where $\|\gamma\| > \|Cg\omega(t)\| = 0.2$ in this example for countermeasure the matched disturbance and $\alpha > 0$ should be selected to ensure that the approaching condition can happen.

In [20], it has been proven that applying the ideal controller in (94), the following approaching condition happens.

$$s^T(t) \cdot \dot{s}(t) < -\alpha \cdot \|\dot{s}(t)\| < 0$$

To avoid or reduce "chattering phenomenon" due to $sign(\dot{s}(t))$ switching function in (94), the following modified controller is used.

$$u(t) := -(Cg)^{-1}Cf\dot{x}(t) - (Cg)^{-1}(\gamma + \alpha)sat(\dot{s}(t), \varepsilon) \quad (95)$$

where "*sat*" is a saturation function to smoothly handle the switching as follows.

$$sat(\dot{s}, \varepsilon) = \begin{cases} 1 & \dot{s} > \varepsilon \\ \frac{\dot{s}}{\varepsilon} & |\dot{s}| \le \varepsilon \\ -1 & \dot{s} < -\varepsilon \end{cases} = \begin{cases} sign(\dot{s}) & |\dot{s}| > \varepsilon \\ \frac{\dot{s}}{\varepsilon} & |\dot{s}| \le \varepsilon \end{cases}$$

Here ε is a small positive value as the bound of the differential sliding surface $\dot{s}$ such that

$$|\dot{s}| \le \varepsilon$$

In this example $\varepsilon = 0.5, \gamma = 0.4$ and $\alpha = 0.05$ are used in the simulation. When sliding surface is selected, we need to tune design parameters of ε, γ and α to have a SMC controller with good enough performance.

(b) For applying the inverse optimal control (IOC) design method in Section 2.4, we can follow the same design steps and use the same notations in example 3. In this example, we have

$$h = I_{3\times3} \text{ (identity matrix) and } J = [0\ 0\ 0]^{\mathrm{T}}$$

First, we select the following design parameters as

$$\gamma = 0.6,\ P = I_{3\times3} \text{ (identity matrix) and} R(\dot{x}) = 1$$

To ensure the closed loop system is stable, based on (64), the $L_2(\dot{x})$ is then selected as

$$L_2(\dot{x}) = \begin{bmatrix} \dot{x}1 & \dot{x}2 & \dot{x}3 \end{bmatrix} \begin{bmatrix} -79.75 \\ -76.25 \\ -17.50 \end{bmatrix}$$

Consequently, using (62) yields

$$\phi(\dot{x}) = \begin{bmatrix} 40.00 & 38.00 & 9.00 \end{bmatrix} \dot{x} = -K_\infty \dot{x}$$

Therefore, the obtained full state derivative feedback gain is

$$K_\infty = \begin{bmatrix} -40.00 & -38.00 & -9.00 \end{bmatrix}.$$

Applying this control law, the closed loop system poles locate at -0.5 ± 0.5i and -1. Since their real parts are all negative, the closed loop system are stable.

The state responses and control effort of both SMC and IOC are plotted in the following figures for comparisons.

The unstable system in this example can be properly controlled by both sliding mode control (SMC) and inverse optimal control (IOC) to have bounded closed loop state responses. As shown in Figures 7–10, for matched disturbance case in this example, when SMC is properly designed, its performance could be better than that of IOC because SMC can have smaller state responses in Figures 7–9 by applying smaller control effort in Figure 10.

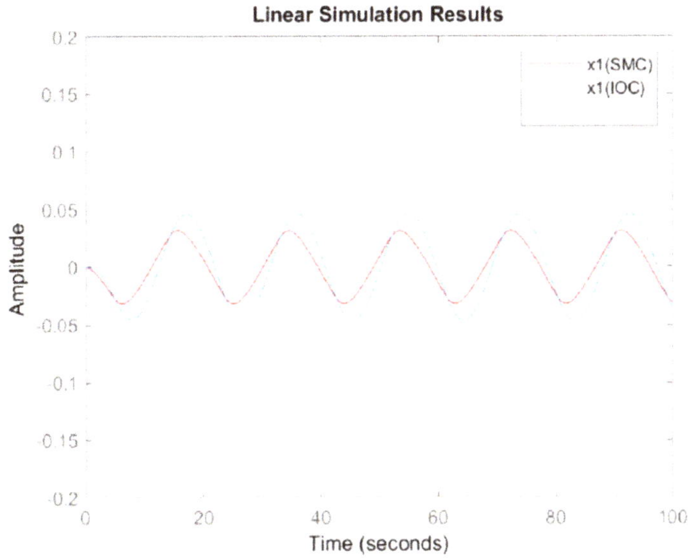

Figure 7. State ×1 responses.

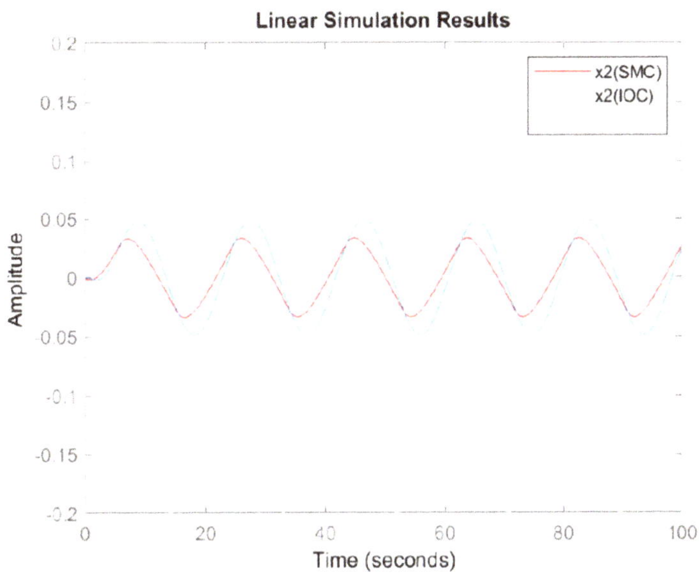

Figure 8. State ×2 responses.

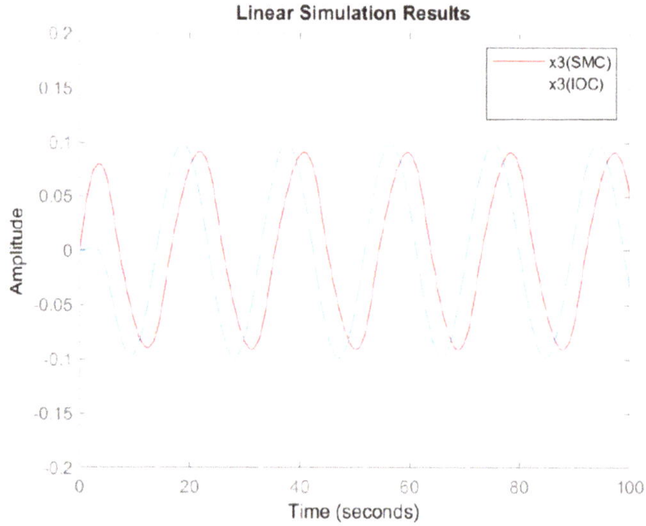

Figure 9. State ×3 responses.

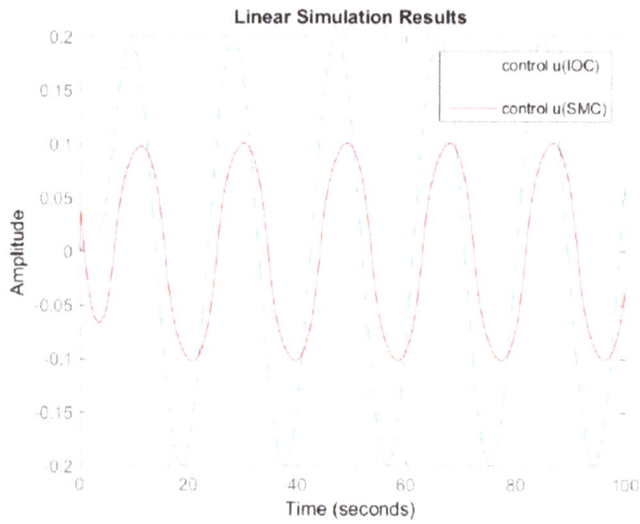

Figure 10. Control u responses.

4. Discussion

The proposed design methods are inspired by other previous work [26,27,31] in inverse optimal control in state space system with state feedback design approach. The results show that the use of state derivative feedback for control design in an SDS system is as simple as the use of state feedback for control design in a state space system.

Therefore, with the understanding of SDS systems, many design tools developed in state space system with state feedback can be modified and adapted for control designs in SDS system with state derivative feedback. Since the Hamilton–Jacobi–Bellman equation is not always solvable to obtain the control Lyapunov function, the existence of optimal control solution is not always guaranteed. On the other hand, if we can solve for a control Lyapunov function from HJB equation, we can find from it a control law that achieves the minimum of performance functional and the resulting closed loop system has a unique solution forward in time. Please refer to Chapter 6 in [27] for details, the descriptions and formula derivations are analogous to those for SDS system case. Since the control Lyapunov function is predefined in inverse optimal control design process and no need to solve HJB equation, it is very suitable to find stabilizing control laws for unstable nonlinear systems.

Regarding the examples to verify the proposed methods, Example 1 demonstrates the design steps for the design approach of inverse optimal control for affine SDS systems with state derivative related feedback. In the same example, it also suggests that people should free their mindset in control designs. No matter the system in state space form or SDS form, the possibilities of applying state feedback or state derivative feedback should be both checked so that the controller can be simple while perform well. The DC motor tracking control without tachometer in Example 2 discusses the possibility of using alternative measurement of inductor voltage in DC motor control based on the obtained state derivative feedback algorithms. This idea seems promising, because the method we propose can construct a cheap and compact controller without the need for an expensive tachometer. Furthermore, unlike resistor sensor, the average power of inductor sensor is zero, it will not damage the armature circuit or cause power loss. Example 3 is a singular system with impulse mode. This is a very challenging design problem in previous researches using state feedback in control design of the generalized state space system. Since it can be expressed in SDS system, the design approach is straightforward. Therefore, some systems that are difficult to control through state feedback can be controlled through state derivative feedback in SDS system form.

Since both sliding mode control (SMC) [19,20] and the inverse optimal control in SDS system form are developed based on Lyapunov stability theorem and their performances are dependent on tuning their design parameters, we compare them in terms of conditions to use and limitations. SMC methods in [19,20] have low sensitivity to parameter uncertainties, can work with matched uncertainties as well as matched disturbance that enter into control inputs and apply discontinuous switching control law to ensure the finite time convergence. Those are their advantages. However, SMC methods in [19,20] may suffer from chattering phenomenon and when uncertainties and disturbance are not matched ones, the performance could be downgraded. On the contrary, the inverse optimal control method in Section 2.4 can handle bounded disturbance which are not from control inputs. As shown in Example 4, for system with matched disturbance, the SMC controller could use smaller control effort to obtain smaller state responses than IOC controller. However, from the implementation point of view, the structure of IOC controller is simpler than that of SMC controller and consequently the implementation cost of IOC controller could be cheaper. The inverse optimal control methods in this paper are suitable for controlling the system with precise parameters, such as the DC motor used for tracking control in Example 2. Therefore, designers can make tradeoff between IOC and SMC in terms of performance and cost.

Regarding the future works, for implementation of the DC motor application in example 2, as seen in Figure 1, there is large electric inductance L in armature circuit in the model of DC motor. No matter what kind of sensor we use, one potential problem that is common in DC motor control is inductive kick (kickback) phenomenon or so called Ldi/dt voltages [39]. Since the windings of the DC motor will produce current conversion during commutation, the current conversion will cause the inductive kickback that disturbs both the voltage and current of armature circuit as shown in Figure 11. Therefore the measured voltage of L1 sensor in Figure 1 will also be disturbed.

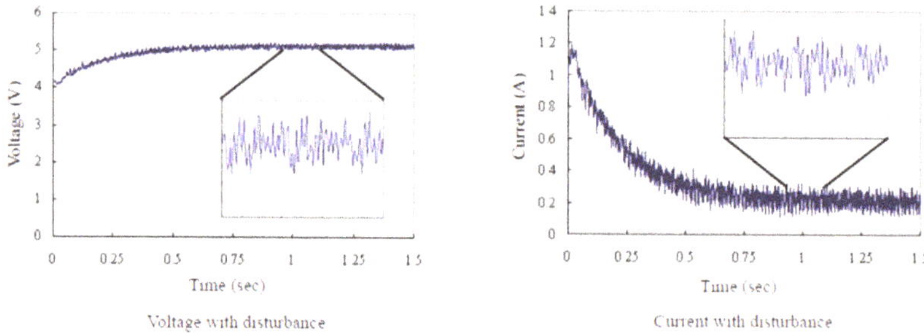

Figure 11. Disturbances due to inductive kickback.

Conventionally, in implementation, various absorption circuits of inductive load kickback [40] can be used as the countermeasure for disturbance caused by inductive kick. Other than applying absorption circuits, we are considering another solution of eliminating disturbance of L1 sensor voltage due to inductive kick in our future research. The basic idea is as follows. Since the inductive kick is formed by periodic commutation, it has an average of 0 characteristics, if the T is the time of every commutator segment passes through a brush, selecting the signal window of L1 sensor voltage with a period as a multiple of T and calculating the average voltage value of window, the L1 voltage disturbance caused by the inductive kick of commutators could be considerably decreased. The control voltage is then generated by feedback of L1 voltage with reduced disturbance. In addition, since this disturbance of L1 sensor voltage is through control input channel, it can be considered as matched disturbance. Hence, we will also consider to apply sliding mode control with state derivative measurement feedback to control it in future.

In this paper, we have proven that the inverse optimal control can be carried out in SDS system form with state derivative feedback. In future, more challenging problems such as stochastic systems can be explored. If a system has randomness associated with it, it is called a stochastic system and does not always produce the same output for a given input. Stochastic systems exist in many applications such as communication systems, markets, social systems, and epidemiology. Optimal control [41,42] and inverse optimal control [42,43] for stochastic systems in state space system form have been solved with stochastic Hamilton–Jacobi–Bellman equation to obtain state related feedback control laws. In future, for people who want to develop inverse optimal control for stochastic systems with state derivative feedback, it is highly recommended to first study [42] because the design approaches in [42] and this paper are both built on [27].

5. Conclusions

In this paper we have discussed about how SDS system with state derivative feedback can be supplement of state space system with state feedback in control designs. Followed by developing inverse optimal control methods in SDS systems with solely state derivative feedback. As far as the authors know, no similar results have been reported. Inverse optimal control can construct a stable closed-loop system while nonlinear optimal control may not have exact solution. Hence, inverse optimal control should be collaboratively used together with optimal control for designs. The proposed methods are very suitable to find stabilizing control laws for unstable nonlinear systems. The correctness of proposed methods has been properly verified by numerical examples and simulations. Especially, in the third example, a classic difficult problem in control, namely singular system with impulse mode is fully controllable by state derivative feedback in SDS system form and satisfy the non-expansivity constraint when the system is subjected to disturbance. On the contrary, the same system can only be stabilized by state feedback control. The above is the summary of academic contributions of this paper.

From application points of view, in vibration systems of vehicle dynamics and smart structure, accelerations and velocities are available measurements of state derivative vector. In addition, the inductor voltages in electrical systems are also state derivative related measurement. For those systems, using state derivative feedback design in SDS system form are very likely to have more simple, cheap and compact controllers because integrators or numerical integrations are not needed. Therefore, the idea of connecting a small inductor in series with an armature circuit as the only sensor of a DC motor control system in Example 3 is very promising because average power loss of inductor is zero and no tachometer is needed. DC motors are widely used in many industries and facilities for daily life. For example, in automotive body electronics, the estimated demand for automotive DC motors in body domain was 2 billion units in 2020 [44]. So the proposed design approach of this paper can have a wide range of practical applications.

With understandings and awareness of SDS system form, state derivative feedback and inverse optimal control, designers can solve more control problems and develop more new applications based on their previous knowledge and experience in state feedback designs in state space system without applying too much of advanced mathematics.

6. Patents

Authors, Yuan-Wei Tseng and Rong-Ching Wu hold the following patents related to state derivative feedback control in SDS systems.

1. US Patent NO. US 10,598,688 B2, "Oscillation Control System and Oscillation Control Method", 2020.03.24~2038.09.20 (https://patents.google.com/patent/US10598688B2/en?oq=US+10%2c598%2c688+B2) (accessed on 16 March 2021).
2. Taiwan Patent NO. I670924, "Vibrational Control System", 2019.09.01~2038.07.01

Author Contributions: Conceptualization, Y.-W.T. and F.-C.L.; methodology, F.-C.L.; software, F.-C.L.; validation, R.-C.W., F.-C.L., and W.-C.C.; formal analysis, Y.-W.T.; investigation, F.-C.L. and C.-S.C.; resources, W.-C.C.; data curation, F.-C.L.; writing—original draft preparation, Y.-W.T.; writing—review and editing, C.-S.C. and R.-C.W.; visualization, F.-C.L.; supervision, Y.-W.T.; project administration, W.-C.C. All authors have read and agreed to the published version of the manuscript.

Funding: This research received no external funding.

Conflicts of Interest: The authors declare no conflict of interest.

References

1. Campbell, S.L. *Singular Systems of Differential Equations II*; Pitman: Marshfield, MA, USA, 1982.
2. Newcomb, R.W. The semistate description of nonlinear time variable circuits. *IEEE Trans. Circuits Syst.* **1981**, *28*, 62–71. [CrossRef]
3. Brenan, K.E. Numerical simulation of trajectory prescribed path control problems by the backward differentiation formulas. *IEEE Trans. Autom. Control* **1986**, *31*, 266–269. [CrossRef]
4. Tseng, Y.-W. Vibration control of piezoelectric smart plate using estimated state derivatives feedback in reciprocal state space form. *Int. J. Control Theory Appl.* **2009**, *2*, 61–71.
5. Pantelides, C.C. The consistent initialization of differential algebraic systems. *Siam J. Sci. Stat. Comput.* **1988**, *9*, 213–231. [CrossRef]
6. Verghese, G.C.; L'evy, B.C.; Kailath, T. A generalized state space for singular systems. *IEEE Trans. Autom. Control* **1981**, *26*, 811–831. [CrossRef]
7. Liu, D.; Zhang, G.; Xie, Y. Guaranteed cost control for a class of descriptor systems with uncertainties. *Int. J. Inf. Syst. Sci.* **2009**, *5*, 430–435.
8. Cobb, D. State feedback impulse elimination for singular systems over a Hermite Do-main. *SIAM J. Control Optim.* **2006**, *44*, 2189–2209. [CrossRef]
9. Saadni, M.S.; Chaabane, M.; Mehdi, D. Robust stability and stabilization of a class of singular systems with multiple time varying delays. *Asian J. Control* **2006**, *8*, 1–11. [CrossRef]
10. Varga, A. Robust pole assignment for descriptor systems. In Proceedings of the Mathematical Theory of Networks and Systems, Perpignan, France, 23–27 August 2000.
11. Cobb, D. Eigenvalue conditions for convergence of singularly perturbed matrix exponential functions. *Siam J. Control Optim.* **2010**, *48*, 4327–4351. [CrossRef]
12. Yang, C.; Sun, J.; Zhang, Q.; Ma, X. Lyapunov Stability and Strong Passivity Analysis for Nonlinear Descriptor Systems. *IEEE Trans. Circuits Syst. I* **2013**, *60*, 1003–1012. [CrossRef]

13. Kulah, H.; Chae, J.; Yazdi, N.; Najafi, K. Noise analysis and characterization of a sigma-delta capacitive microaccelerometer. *IEEE J. Solid-State Circuits* **2006**, *41*, 352–361. [CrossRef]
14. Martini, A.; Bonelli, G.P.; Rivola, A. Virtual Testing of Counterbalance Forklift Trucks: Implementation and Experimental Validation of a Numerical Multibody Model. *Machines* **2020**, *8*, 26. [CrossRef]
15. Rebelle, J.; Mistrot, P.; Poirot, R. Development and validation of a numerical model for predicting forklift truck tip-over. *Veh. Syst. Dyn.* **2009**, *47*, 771–804. [CrossRef]
16. Cattabriga, S.; De Felice, A.; Sorrentino, S. Patter instability of racing motorcycles in straight braking manoeuver. *Veh. Syst. Dyn.* **2021**, *59*, 33–55. [CrossRef]
17. Martini, A.; Bellani, G.; Fragassa, C. Numerical assessment of a new hydro-pneumatic suspension system for motorcycles. *Int. J. Automot. Mech. Eng.* **2018**, *15*, 5308–5325. [CrossRef]
18. Romualdi, L.; Mancinelli, N.; De Felice, A.; Sorrentino, S. A new application of the extended kalman filter to the estimation of roll angles of a motorcycle with inertial measurement unit. *FME Trans.* **2020**, *48*, 255–265. [CrossRef]
19. Tseng, Y.-W.; Wang, Y.-N. Sliding Mode Control with State Derivative Output Feedback in Reciprocal State Space Form. *Abstr. Appl. Anal.* **2013**, *2013*, 1–12. Available online: https://www.hindawi.com/journals/aaa/2013/590524/ (accessed on 23 March 2021). [CrossRef]
20. Tseng, Y.-W. *Sliding Mode Control with State Derivative in Novel Reciprocal State Space Form, Advances and Applications in Nonlinear Control Systems*; Springer International Publishing: Cham, Switzerland, 2016.
21. Tseng, Y.-W. Finding the Minimum of H Infinity Norm in Novel State Derivative Space Form. *Adv. Mater. Res.* **2013**, *740*, 45–50. [CrossRef]
22. Tseng, Y.-W. H Infinity Control of Descriptor System with Impulse Mode in Novel State Derivative Space Form. *Adv. Mater. Res.* **2013**, *740*, 39–44. [CrossRef]
23. Tseng, Y.-W.; Hsieh, J.-G. Optimal Control for a Family of Systems in Novel State Derivative Space Form with Experiment in a Double Inverted Pendulum System. *Abstr. Appl. Anal.* **2013**, *2013*, 1–8. Available online: https://www.hindawi.com/journals/aaa/2013/715026/ (accessed on 23 March 2021). [CrossRef]
24. Krstic, M.; Tsiotras, P. Inverse optimal stabilization of a rigid spacecraft. *IEEE Trans. Autom. Control* **1999**, *44*, 1042–1049. [CrossRef]
25. Almobaied, M.; Eksin, I.; Guzelkaya, M. Inverse Optimal Controller Design Based on Multi-Objective Optimization Criteria for Discrete-Time Nonlinear Systems. In Proceedings of the IEEE 7th Palestinian International Conference on Electrical and Computer Engineering (PICECE), Gaza, Palestine, 26–27 March 2019.
26. Prasanna, P.; Jacob, J.; Nandakumar, M.P. Inverse optimal control of a class of affine nonlinear systems. *Trans. Inst. Meas. Control* **2019**, *41*, 2637–2650. [CrossRef]
27. Haddad, W.M.; Chellabonia, V. *Nonlinear Dynamical Systems and Control: A Lyapunov-Based Approach*; Princeton University Press: Princeton, NJ, USA, 2008.
28. Mainprice, J.; Hayne, R.; Berenson, D. Goal Set Inverse Optimal Control and Iterative Replanning for Predicting Human Reaching Motions in Shared Workspaces. *IEEE Trans. Robot.* **2016**, *32*, 897–908. [CrossRef]
29. Quintal, G.; Sanchez, E.N.; Alanis, A.Y.; Arana-Daniel, N.G. Real-time FPGA decentralized inverse optimal neural control for a Shrimp robot. In Proceedings of the 10th System of Systems Engineering Conference (SoSE), San Antonio, TX, USA, 17–20 May 2015; pp. 250–255.
30. Clever, D.; Schemschat, R.M.; Felis, M.L.; Mombaur, K. Inverse optimal control based identification of optimality criteria in whole-body human walking on level ground. In Proceedings of the 6th IEEE International Conference on Biomedical Robotics and Biomechatronics (BioRob), Singapore, 26–29 June 2016; pp. 1192–1199.
31. Priess, M.C.; Conway, R.; Choi, J.; Popovich, J.M.; Radcliffe, C. Solutions to the Inverse LQR Problem With Application to Biological Systems Analysis. *IEEE Trans. Control Syst. Technol.* **2015**, *23*, 770–777. [CrossRef] [PubMed]
32. Guerrero-Castellanos, J.F.; González-Díaz, V.; Vega-Alonzo, A.; Mino-Aguilar, G.; López-López, M.; Guerrero-Sánchez, W.F.; Maya-Rueda, S.E. CLF based design for attitude control of VTOL-UAVs: An inverse optimal control approach. In Proceedings of the 2015 Workshop on Research, Education and Development of Unmanned Aerial Systems (RED-UAS), Cancun, Mexico, 23–25 November 2015; pp. 162–171.
33. Long, H.H.; Zhao, J.K. Anti-disturbance inverse optimal attitude control design for flexible spacecraft with input saturation. In Proceeding of the 11th World Congress on Intelligent Control and Automation, Shenyang, China, 29 June–4 July 2014; pp. 2961–2966.
34. Luo, W.C.; Chu, Y.C.; Ling, K.V. Inverse optimal adaptive control for attitude tracking of spacecraft. *IEEE Trans. Autom. Control* **2005**, *50*, 1639–1654.
35. Ruiz-Cruz, R.; Sanchez, E.N.; Ornelas-Tellez, F.; Loukianov, A.G.; Harley, R.G. Particle Swarm Optimization for Discrete-Time Inverse Optimal Control of a Doubly Fed Induction Generator. *IEEE Trans. Cybern.* **2013**, *43*, 1698–1709. [CrossRef]
36. Vega, C.; Alzate, R. Inverse optimal control on electric power conversion. In Proceedings of the 2014 IEEE International Autumn Meeting on Power, Electronics and Computing (ROPEC), Ixtapa, Mexico, 5–7 November 2014; pp. 1–5.
37. Ruiz-Cruz, R.; Sanchez, E.N.; Loukianov, A.G.; Ruz-Hernandez, J.A. Real-Time Neural Inverse Optimal Control for a Wind Generator. *IEEE Trans. Sustain. Energy* **2019**, *10*, 1172–1183. [CrossRef]
38. Control Tutorials. University of Michigan. Available online: https://ctms.engin.umich.edu/CTMS/index.php?example=MotorSpeed&section=SystemModeling (accessed on 6 March 2021).

39. Wu, R.-C.; Pan, C.-L.; Chen, C.-Y. Parameter Estimation for Permanent Magnet DC Machine by Least Square Method. *ICIC Express Lett. Part B Appl.* **2017**, *11*, 1571–1578.
40. Sun, J.; Zhang, B.; Wang, H.; Ming, X.; Xiao, K.; Ye, P.; Cao, L. An Inductive Kickback Absorption Scheme without Power Zener and Large Capacitor. *IEEE Trans. Ind. Electron.* **2011**, *58*, 709–716. [CrossRef]
41. Moon, J. Generalized Risk-Sensitive Optimal Control and Hamilton-Jacobi-Bellman Equation. *IEEE Trans. Autom. Control* **2020**. [CrossRef]
42. Rajpurohit, T.; Haddad, W.M. Nonlinear–nonquadratic optimal and inverse optimal control for stochastic dynamical systems. *Int. J. Robust Nonlinear Control* **2017**, *27*, 4723–4751. Available online: https://haddad.gatech.edu/journal/Nonlinear_Stocastic_Optimal_RNC.PDF (accessed on 7 March 2021).
43. Haddad, W.M.; Jin, X. Universal Feedback Controllers and Inverse Optimality for Nonlinear Stochastic Systems. *J. Dyn. Syst. Meas. Control* **2020**, *142*, 10. Available online: http://haddad.gatech.edu/journal/Universal_Stochastic_Control_ASME.pdf (accessed on 7 March 2021).
44. DC Motors Trends in Automotive Body Electronics. Available online: https://www.edomtech.com/en/article/ins.php?index_id=42 (accessed on 6 March 2021).

MDPI
St. Alban-Anlage 66
4052 Basel
Switzerland
Tel. +41 61 683 77 34
Fax +41 61 302 89 18
www.mdpi.com

Energies Editorial Office
E-mail: energies@mdpi.com
www.mdpi.com/journal/energies

www.ingramcontent.com/pod-product-compliance
Lightning Source LLC
Chambersburg PA
CBHW040757250926
51850CB00007BA/1019
* 9 7 8 3 0 3 6 5 9 8 2 1 5 *